Experimental
Hematology
Today 1978

Experimental Hematology Today 1978

Edited by
Siegmund J. Baum
G. David Ledney

With 90 illustrations

 Springer-Verlag   New York   Heidelberg   Berlin

Siegmund J. Baum
Chairman, Experimental Hematology Department
Armed Forces Radiobiology Research Institute
Defense Nuclear Agency
Bethesda, Maryland 20014

G. David Ledney
Head, Immunology Division
Experimental Hematology Department
Armed Forces Radiobiology Research Institute
Defense Nuclear Agency
Bethesda, Maryland 20014

Library of Congress Cataloging in Publication Data

Main entry under title:

Experimental hematology today 1978.

"Presented at the sixth annual meeting of the International Society for Experimental Hematology, held in Basel, Switzerland, 1977."
Includes bibliographical references and index.
1. Hematology, Experimental—Congresses. I. Baum, Siegmund J. II. Ledney, G. David. III. International Society for Experimental Hematology. [DNLM: 1. Hematology—Congresses. WH20 I6le 1977]
RB145.E98   616.1'5   78-8054
ISBN-13: 978-1-4612-6306-7

Softcover reprint of the hardcover 1st edition 1978
9 8 7 6 5 4 3 2 1

ISBN-13: 978-1-4612-6306-7   e-ISBN-13: 978-1-4612-6304-3
DOI: 10.1007/978-1-4612-6304-3

Preface

Experimental Hematology Today 1978 represents a group of papers demonstrating significant advances in the field for the past year, which were presented at the Sixth Annual Meeting of the International Society for Experimental Hematology held in Basel, Switzerland, in 1977, and selected by the local committee. The papers are divided into six parts, three of which primarily represent the basic sciences while the other three show advances in clinical application. The greater emphasis on clinical application in this book may indicate the final fruition and utilization of previously completed basic research.

The first part, chaired by Dr. L. G. Lajtha, Manchester, England, is entitled "Regulation of Stem Cell Proliferation" and deals with further characterization of the still rather elusive multipotential stem cell. This is followed by Part II entitled "Physiology of Committed Stem Cells" under the chairmanship of Dr. D. Metcalf, Victoria, Australia. In this divison of papers, experimenters attempt to create order and better understanding of the various stages of committed stem cells, utilizing the most recent laboratory data. Dr. S. Thierfelder, Munich, Germany, chaired the important part III on "Transplantation Immunology." In this part, authors discuss recent findings in clinical observations and animal experimentation for possible improvement of bone marrow transplantation in man. Chairman Dr. B. Speck, Basel, Switzerland, led Part IV, "Therapeutic Approaches to Aplastic Anemia." Authors of papers in this group attempt to give a better understanding of the pathogenesis of this disease and of ways of improvement for better therapy. Part V, entitled "Treatment of Leukemia" and chaired by Dr. G. S. Santos, Baltimore, Maryland, deals with recent advances in the treatment of this disease. Finally, Part VI entitled "Advances in Experimental Hematology" is comprised of a set of papers in which the authors point toward possibly new horizons in hematological research. This section was chaired by Dr. D. H. Pluznik, Ramat-Gan, Israel.

Descriptions of the most recent advances in basic as well as in clinical hematology are contained in this yearbook that will be valuable to experimental and clinical scientists.

Acknowledgment

The International Society for Experimental Hematology wishes to acknowledge the generous assistance provided by the following agencies of the

United States government. Defense Nuclear Agency, Armed Forces Radiobiology Research Institute.

Furthermore, we greatly appreciate the generous financial support by the Swiss pharmaceutical firms, Hoffman–LaRoche and Ciba–Geigy, and by the American firm, Travenol.

Siegmund J. Baum
G. David Ledney

Contents

Contents

List
of
Contributors

N. Abuaf, Institut de Cancérologie et d'Immunogénétique, Groupe Hospitalier Paul Brousse, 94800 Villejuif, France

Solomon S. Adler, Section of Hematology, Department of Medicine, Rush–Presbyterian–St. Luke's Medical Center, Chicago, Illinois, 60612

A. Ahmed, Naval Medical Research Institute, Bethesda, Maryland, 20014

Ron N. Apte, Department of Life Sciences, Bar-Ilan University, Ramat-Gan, Israel

J. J. Ballet, Groupe de recherches d'Immunologie et de Rhumatologie Pédiatriques, Unité IN-SERM U 132, Pr. P. Mozziconacci, Hôpital des Enfants Malades, 75015, 149 Rue de Sèvres, Paris, France

Siegmund J. Baum, Ph.D., Chairman, Experimental Hematology Department, Defense Nuclear Agency, Armed Forces Radiobiology Institute, Bethesda, Maryland, 20014

B. M. Berardet, Institute de Cancérologie et d'Immunogénétique, Groupe Hospitalier Paul Brousse, 94800 Villejuif, France

L. Berumen, Institut de Cancérologie et d'Immunogénétique, Groupe Hospitalier Paul Brousse, 94800 Villejuif, France

A. B. Bijnen, Laboratory for Experimental Surgery, Erasmus University, Rotterdam, The Netherlands

B. A. Bradley, Department of Immunohaematology, University Hospital Leiden, The Netherlands

Hal E. Broxmeyer, Memorial Sloan-Kettering Institute for Cancer Research, New York, New York, 10021

S. Bruno, M.D., Departments of Medicine A and Grace Cancer Drug Center, Roswell Park Memorial Institute, Buffalo, New York, 14263

C. D. Buckner, Fred Hutchinson Cancer Research Center, Seattle, Washington 98104, and the Departments of Medicine and Pediatrics, Division of Oncology, University of Washington School of Medicine, Seattle, Washington, 98195

M. Castes, Institute de Cancérologie et d'Immunogénétique, Groupe Hospitalier Paul Brousse, 94800 Villejuif, France

G. Christoff, Departments of Medicine A and Grace Cancer Drug Center, Roswell Park Memorial Institute, Buffalo, New York, 14263

Diane de Cicco, Imperial Cancer Research Fund, London, United Kingdom

R. A. Clift, Fred Hutchinson Cancer Research Center, Seattle, Washington, 98104, and the Departments of Medicine and Pediatrics, Division of Oncology, University of Washington School of Medicine, Seattle, Washington, 98195

T. J. Contreras, Physiology Division Experimental Hematology Department, Armed Forces Radiobiology Research Institute, Bethesda, Maryland, 20014

P. Cornu, Division of Hematology, Department of Internal Medicine, Kantonsspital Basel and Transplantation Immunology Unit, Hôpital Cantonal, Geneva, Switzerland

A. Cruchaud, Division of Immunology and Allergology, Department of Medicine, Geneva University Medical School, Hôpital Cantonal, Geneva 4, Switzerland

F. Daguillard, Groupe de recherches d'Immunologie et de Rhumatologie Pédiatriques, Unité IN-SERM U 132, Pr. P. Mozziconacci, Hôpital des Enfants Malades, 75015, Paris, France

P. Dörmer, Abt. Experim. Hämatologie des Institutes für Hämatologie, Landwehrstr. 61, 8000 München, West Germany

A. Durandy, Groupe de recherches d'Immunologie et de Rhumatologie Pédiatriques, Unité IN-SERM U 132, Pr. P. Mozziconacci, Hôpital des Enfants Malades, 75015, Paris, France

Ger van den Engh, Radiobiological Institute TNO, 151 Lange Kleiweg, Rijswijk, The Netherlands

A. Fefer, Fred Hutchinson Cancer Research Center, Seattle, Washington, 98104, and the Departments of Medicine and Pediatrics, Division of Oncology, University of Washington School of Medicine, Seattle, Washington, 98195

V. Von Fliedner, M.D., Departments of Medicine A and Grace Cancer Drug Center, Roswell Park Memorial Institute, Buffalo, New York, 14263

G. L. Floersheim, Kantonsspital Zurich, Dermatologische Universitäts-Klinik, Zurich, Switzerland

W. J. Flor, Physiology Division, Experimental Hematology Department, Armed Forces Radiobiology Research Institute, Bethesda, Maryland, 20014

J. E. French, Physiology Division, Experimental Hematology Department, Armed Forces Radiobiology Research Institute, Bethesda, Maryland, 20014

Ralph van Furth, Department of Infectious Diseases, University Hospital, Leiden, The Netherlands

Robert Peter Gale, UCLA Bone Marrow Transplant Unit, UCLA School of Medicine, The Center for the Health Sciences, Los Angeles, California, 90024

A. L. Goldstein, University of Texas Medical Branch, Galveston, Texas, 77550

Ch. Bender Götze, Kinderpoliklinik der Universität München, Munich, West Germany

B. W. Goodell, Fred Hutchinson Cancer Research Center, Seattle, Washington, 98104, and the Departments of Medicine and Pediatrics, Division of Oncology, University of Washington School of Medicine, Seattle, Washington, 98195

Theo J. L. M. Goud, Department of Infectious Diseases, University Hospital, Leiden, The Netherlands

E. Goulmy, Department of Immunohaematology, University Hospital Leiden, The Netherlands

C. Griscelli, Groupe de recherches d'Immunologie et de Rhumatologie Pédiatriques, Unité IN-SERM U 132, Pr. P. Mozziconacci, Hôpital des Enfants Malades, 75015, Paris, France

M. P. Grissom, Experimental Hematology Department, Armed Forces Radiobiology Research Institute, Bethesda, Maryland, 20014

P. Groff, Division of Hematology, Department of Internal Medicine, Kantonsspital Basel and Transplantation Immunology Unit, Hôpital Cantonal, Geneva, Switzerland

H. Gross-Wilde, Institut für Hämatologie der Gesellschaft für Strahlen-und Umweltforschung, München, 8 München 2, Landwehrstr. 61, West Germany

R. J. Haas, Universitätskinderklinik im Dr. von Haunerschen Kinderspital, Munich, West Germany

O. Halle-Pannenko, Institut de Cancérologie et d'Immunogénétique, Groupe Hospitalier Paul Brousse, 94800 Villejuif, France

George Hausner, Departments of Medicine A and Grace Cancer Drug Center, Roswell Park Memorial Institute, Buffalo, New York, 14263

Eliahu Heller, Department of Life Sciences, Bar-Ilan University, Ramat-Gan, Israel

Chanita F. Hertogs, Department of Life Sciences, Bar-Ilan University, Ramat-Gan, Israel

G. Hoffmann-Fezer, Abt. Immunologie des Institutes für Hämatologie, Landwehrstr. 61, 8000 München, West Germany

D. Huhn, Abt. Klin. Hämatologie des Institutes für Hämatologie, GSF, München, West Germany

Niels Jacobsen, Memorial Sloan-Kettering Institute for Cancer Research, New York, New York, 10021

G. Jäger, Abt. Experim. Hämatologie des Institutes für Hämatologie, Landwehrstr. 61, 8000 München, West Germany

M. Jeannet, Transplantation Immunology Unit, Hôpital Cantonal, Geneva, and Division of Haematology, Kantonsspital Basel, Switzerland

W. W. Jedrzejczak, Naval Medical Research Institute, Bethesda, Maryland, 20014

J. F. Jemionek, Physiology Division, Experimental Hematology Department, Armed Forces Radiobiology Research Institute, Bethesda, Maryland, 20014

F. L. Johnson, Fred Hutchinson Cancer Research Center, Seattle, Washington, 98104, and the Departments of Medicine and Pediatrics, Division of Oncology, University of Washington School of Medicine, Seattle, Washington, 98195

G. R. Johnson, Cancer Research Unit, Walter and Eliza Hall Institute, Royal Melbourne Hospital, P.O. Box 3050, Victoria, Australia

A. C. M. Van Kessel, Radiobiological Institute GO-TNA, Rijswijk, The Netherlands

H. Kolb, Institut für Biologie der Gesellschaft für Strahlen- und Umweltforschung, München, Abteilung für allgemeine und experimentelle Pathologie, 8042 Neuherberg, Ingolstädter Landstrasse 1

H. J. Kolb, Institut für Hämatologie der Gesellschaft für Strahlen- und Umweltforschung, München, 8 München 2, Landwehrstr. 61, West Germany

Jeffrey I. Kurland, Department of Developmental Hematopoiesis, Gar Reichman Laboratory for Advanced Cancer Research, Sloan-Kettering Institute for Cancer Research, New York, New York, 10021 and Cornell University Medical College, New York, New York, 10021

Richard D. Kuznetsky, Section of Hematology, Department of Medicine, Rush–Presbyterian–St. Luke's Medical Center, Chicago, Illinois, 60612

L. G. Lajtha, Paterson Laboratories, Christie Hospital and Holt Radium Institute, Withington, Manchester, England M20 9BX

B. Lau, Abt. Experim. Hämatologie des Institutes für Hämatologie, Landwehrstr. 61, 8000 München, West Germany

G. D. Ledney, Division of Immunology, Experimental Hematology Department, Armed Forces Radiobiology Research Institute, National Naval Medical Center, Bethesda, Maryland, 20014

A. van Leeuwen, Department of Immunohaematology, University Hospital Leiden, The Netherlands

T. J. MacVittie, Division of Hematology, Experimental Hematology Department, Armed Forces Radiobiology Research Institute, National Naval Medical Center, Bethesda, Maryland, 20014

G. Mathe, Institut de Cancérologie et d'Immunogénétique, Groupe Hospitalier Paul Brousse, 94800 Villejuif, France

D. Metcalf, Cancer Research Unit, Walter and Eliza Hall Institute, Royal Melbourne Hospital, P.O. Box 3050, Victoria, Australia

M. A. S. Moore, Memorial Sloan-Kettering Institute for Cancer Research, New York, New York, 10021

A. Munro, Department of Immunohaematology, University Hospital Leiden, The Netherlands

P. E. Neiman, Scholar of the Leukemia Society of America, Fred Hutchinson Cancer Research Center, Seattle, Washington, 98104, and the Departments of Medicine and Pediatrics, Division of Oncology, University of Washington School of Medicine, Seattle, Washington, 98195

B. Netzel, Abt. Immunologie des Institutes für Hämatologie, Landwehrstr. 61, 8000 München, West Germany

C. Nissen, Division of Hematology, Department of Internal Medicine, Kantonsspital Basel and Transplantation Immunology Unit, Hôpital Cantonal, Geneva, Switzerland

H. Obertop, Laboratory for Experimental Surgery, Erasmus University, Rotterdam, The Netherlands

S. Orbach-Arbouys, Institute de Cancérologie et d'Immunogénétique, Groupe Hospitalier Paul Brousse, 94800 Villejuif, France

G. A. Parker, Division of Comparative Pathology, Armed Forces Radiobiology Research Institute, National Naval Medical Center, Bethesda, Maryland, 20014

Dov H. Pluznik, Department of Life Sciences, Bar-Ilan University, Ramat-Gan, Israel

H. D. Preisler, M.D., Departments of Medicine A and Grace Cancer Drug Center, Roswell Park Memorial Institute, Buffalo, New York, 14263

J. Renick, M.D., Departments of Medicine A and Grace Center Drug Center, Roswell Park Memorial Institute, Buffalo, New York, 14263

I. Rieder, Institut für Hämatologie der Gesellschaft für Strahlen- und Umweltforschung, München, 8 München 2, Landwehrstr. 61, West Germany

H. Rodt, Abt. Immunologie des Institutes für Hämatologie, Landwehrstr. 61, 8000 München, West Germany

J. J. van Rood, Department of Immunohaematology, University Hospital Leiden, The Netherlands

Jim Russell, Department of Biological Sciences, Purdue University, West Lafayette, Indiana

Y. M. Rustum, M.D., Departments of Medicine A and Grace Cancer Drug Center, Roswell Park Memorial Institute, Buffalo, New York, 14263

J. E. Sanders, Fred Hutchinson Cancer Research Center, Seattle, Washington, 98104, and the Department of Medicine, Division of Oncology, University of Washington School of Medicine, Seattle, Washington, 98195

George W. Santos, M.D., Oncology Center, The Johns Hopkins University School of Medicine, Baltimore, Maryland, 21205

E. Schäffer, Institut für Biologie der Gesellschaft für Strahlen- und Umweltforschung München, Abteilung für allgemeine und experimentelle Pathologie, 8042 Neuherberg, Ingolstädter Landstrasse 1, West Germany

S. Scholz, Kinderpoliklinik der Universität München, 8 München 2, Pettenkoferstr., West Germany

K. W. Sell, Naval Medical Research Institute, Bethesda, Maryland, 20014

L. L. Sensenbrenner, Oncology Center, The Johns Hopkins University School of Medicine, Baltimore, Maryland, 21205

S. J. Sharkis, Oncology Center, The Johns Hopkins University School of Medicine, Baltimore, Maryland, 21205

J. Singer, Fred Hutchinson Cancer Research Center, Seattle, Washington, 98104, and the Departments of Medicine and Pediatrics, Division of Oncology, University of Washington School of Medicine, Seattle, Washington, 98195

B. Speck, Division of Hematology, Department of Internal Medicine, Kantonsspital Basel and Transplantation Immunology Unit, Hôpital Cantonal, Geneva, Switzerland

D. A. Stewart, Division of Immunology, Experimental Hematology Department, Armed Forces Radiobiology Research Institute, National Naval Medical Center, Bethesda, Maryland, 20014

Gisela Stinner, Anecamp-Street 6 D-300 Hannover-72 West Germany

R. Storb, Fred Hutchinson Cancer Research Center, Seattle, Washington, 98104, and the Departments of Medicine and Pediatrics, Division of Oncology, University of Washington School of Medicine, Seattle, Washington, 98195

A. Termijtelen, Department of Immunohaematology, University Hospital Leiden, The Netherlands

E. Thiel, Abt. Immunologie des Institutes für Hämatologie, Landwehrstr. 61, 8000 München, West Germany

S. Thierfelder, Abt. Immunologie des Institutes für Hämatologie, Landwehrstr. 61, 8000 München, West Germany

E. D. Thomas, Fred Hutchinson Cancer Research Center, Seattle, Washington, 98104, and the Departments of Medicine and Pediatrics, Division of Oncology, University of Washington School of Medicine, Seattle, Washington, 98195

J. Torhorst, Research Laboratories, Dept. of Surgery and Dept. of Pathology, University of Basel, Switzerland

Frank E. Trobaugh, Jr., Section of Hematology, Department of Medicine, Rush–Presbyterian–St. Luke's Medical Center, Chicago, Illinois, 60612

Robert L. Truitt, Ph.D., May and Sigmund Winter Research Laboratory, Mount Sinai Medical Center, Milwaukee, Wisconsin, 53233

J. L. Virelizier, Groupe de recherches d'Immunologie et de Rhumatologie Pédiatriques, Unité INSERM U 132, Pr. P. Mozziconacci, Hôpital des Enfants Malades, 75015, Paris, France

H. M. Vriesendorp, Radiobiological Institute GO-TNO, Rijswijk, The Netherlands

Dick van Waarde, Department of Infectious Diseases, University Hospital, Leiden, The Netherlands

I. Walczak, Departments of Medicine A and Grace Cancer Drug Center, Roswell Park Memorial Institute, Buffalo, New York, 14263

W. Weber, Division of Hematology, Department of Internal Medicine, Kantonsspital Basel and Transplantation Immunology Unit, Hôpital Cantonal, Geneva, Switzerland

P. L. Weiden, Fred Hutchinson Cancer Research Center, Seattle, Washington, 98104, and the Departments of Medicine and Pediatrics, Division of Oncology, University of Washington School of Medicine, Seattle, Washington, 98195

D. L. Westbroek, Laboratory for Experimental Surgery, Erasmus University, Rotterdam, The Netherlands

E. G. Wright, Paterson Laboratories, Christie Hospital and Holt Radium Institute, Withington, Manchester, England

G. F. Wündisch, Kinderklinik Schwabing der Techn. Universität München, West Germany

PART I

Regulation of Stem Cell Proliferation

L. G. Lajtha

1

Regulation of Stem Cell Proliferation

L.G. Lajtha and E.G. Wright

INTRODUCTION

Despite vast demands for the daily replacement of blood cells, the proliferative rate of the pluripotent hemopoietic stem cells (CFU-s) (20), the ultimate source of these cells, is very low—less than 10 percent in DNA synthesis (2). The high cell output is effected by two intercalated amplifying transit populations, the committed precursors and the "recognizable" precursors (10). The degree of amplification—number of cell cycles during transit—is such that, in the steady state, one stem cell induced to differentiate may give rise to 10^3 to 10^4 mature blood cells (19). On demand, this amplification can increase easily by another order of magnitude. Under conditions of hemopoietic regeneration or in cases of sudden greatly increased (or sustained) differentiation demand, the proliferation rate of the stem cells is increased such that 30 to 40 percent of the cells may be in DNA synthesis (2,4,9,11).

It is only during the last 10 to 15 years that techniques for making quantitative assessments of the numbers and proliferative states of the various hemopoietic cell populations have been developed. Consequently, the problem of how the proliferation of the stem cell is controlled has only recently become approachable experimentally.

EXPERIMENTAL EVIDENCE

Evidence from partial body irradiation experiments (4,7) and from studies of CFU-s proliferation in phenylhydrazine (PHZ)-treated mice (17,21) clearly indicates that stem cell proliferation is, at least in part, locally controlled: in the same animal, proliferation of CFU-s in one hemopoietic site may be minimal at a time when CFU-s in other areas are proliferating actively. In general, it appears that CFU-s are stimulated into a high proliferative state when their population decreases in size and, in particular, when the committed precursor population also decreases. We have recently considered the possibility that stem cell proliferation control might involve some cell concentration feedback factor(s) and have conducted two series of experiments: (a) the investigation of cell extracts for evidence of specific inhibitors and/or stimulators of CFU-s proliferation and (b) a study of the interactions of the cell populations containing proliferating and nonproliferating CFU-s found, respectively, in the bone marrow and spleens of PHZ-treated mice.

Crude bone marrow extracts (BME), i.e., the cell-free supernatants obtained from bone marrow cell suspensions after incubation for 2 hr at 37°C, have been partially purified by passing them

3

through Amicon Diaflo ultra-filtration membranes to give a series of fractions of different molecular weights. Extracts from normal bone marrow cells (NBME) and from regenerating bone marrow cells (RBME) obtained from mice given 450 rad x-rays 7 days earlier have been prepared. The fractions are assayed for CFU-s proliferation-modifying ability by means of the tritiated thymidine ($[^3H]TdR$) suicide technique (2,13). Normal bone marrow cell suspensions containing resting CFU-s or regenerating hemopoietic tissue cell suspensions containing rapidly proliferating CFU-s have been incubated with NBME or RBME fractions for a period of 5 hr and $[^3H]TdR$ suicide measurements have been made over the final 30 min of incubation. The results are summarized in Table 1-1. As previously shown (15), the 50,000 to 100,000 molecular weight fraction from normal bone marrow (NBME-IV) either prevents CFU-s from entering DNA synthesis or stops CFU-s from synthesizing DNA, i.e., it switches them into a nonproliferative state. It does not, however, have any effect on the overall number of CFU-s in the population and must therefore be considered noncytotoxic. No other NBME fraction has this inhibitory effect, and the material is not detectable in any fraction from regenerating bone marrow. Furthermore, the proliferative rate of CFU-c (committed granulocyte precursors and direct descendants of CFU-s) remains steady (45 ± 5 percent in DNA synthesis) during the incubation period—as measured by the $[^3H]TdR$ suicide technique described by Iscove et al. (8)—and no inhibitory effect is demonstrable on other bone marrow cells. It can be concluded, therefore, that normal bone marrow contains a factor that is capable of producing specific, noncytotoxic proliferation inhibition of stem cells.

Frindel and Guigon (6) have also reported CFU-s proliferation inhibitory material in bone marrow. However, their material is a dialysate from a fetal calf bone marrow extract. At present, it is difficult to correlate these results; the dialysate is, of course, much lower in molecular weight than the NBME-IV, and we have found no inhibitory activity in low molecular weight NBME fractions from mouse, rat, and porcine tissues. Furthermore, fetal hemopoietic tissue might be expected to contain proliferating stem cells, which in the case of mouse bone marrow, does not yield significant inhibitory material. The calf bone marrow extract has not been investigated for strict cell specificity.

With regard to the proliferation modifying ability of material taken from bone marrow containing proliferating stem cells, the data presented in Table 1-1 indicate that regenerating bone marrow yields material (RBME-III) capable of stimulating resting CFU-s into DNA synthesis (14). Since the committed precursor populations are all proliferating cells (11), it is difficult to test this material for cell-line specificity. However, it is not detectable in NBME-III and the equivalent fractions of extracts prepared from red blood cells or granulocytes do not stimulate resting CFU-s. Furthermore, experiments utilizing the measurement of structuredness of the cytoplasmic matrix (3)—a technique that gives a very early indication of changes in the proliferative state of cell populations—have shown that RBME-III does not stimulate lymphocytes nor does it block the stimulation of lymphocytes by phytohemagglutinin as does the same molecular weight fraction from lymph node cell extracts. The stimulator is, therefore, at least source specific. Bone marrow damaged by hydroxyurea has similarly been shown to stimulate CFU-s proliferation

TABLE 1-1 Effect of Cell Extracts on the Proliferative Activity of Hemopoietic Stem Cells

TREATMENT[a,b]	ASSAY CELLS	
	Resting CFU-s (percent killed by $[^3H]TdR$)	Proliferating CFU-s (percent killed by $[^3H]TdR$)
—	< 10	32 ± 2
Fraction I NBME	—	37 ± 2
RBME	< 10	—
Fraction II NBME	—	27 ± 2
RBME	< 10	—
Fraction III NBME	<10	25 ± 3
RBME	37 ± 7	33 ± 3
Fraction IV NBME	—	< 10
RBME	12 ± 8	35 ± 3
Fraction V NBME[c]	—	31 ± 5
RBME	—	29 ± 1

[a]NBME, normal bone marrow extract; RBME regenerating bone marrow extract.
[b]Molecular weights of fraction I, 500-10,000; II, 10,000-30,000; III, 30,000-50,000; IV, 50,000-100,000; and V, > 100,000.
[c]NBME IV has no detectable effect on the plating efficiency or $[^3H]TdR$ kill of CFU-c in the assay cell population.

by means of a diffusible factor (5), although no attempt was made to isolate or to characterize the responsible material. It would seem possible that this is another example of regenerating bone marrow producing the material we recognize as RBME-III. However, until further information is available, the possibility of nonspecific proliferation stimulation cannot be excluded.

It is probable that the inhibitory and stimulatory factors may be produced *in vivo* by normal physiological control mechanisms. On days 5 to 10 following PHZ injections in normal mice, CFU-s are proliferating rapidly in the bone marrow but not in the spleen (17). Using such treated mice, we have investigated the presence and role of factors responsible for the control of CFU-s proliferation. This has been done by measuring changes in proliferative activity of CFU-s in spleen and bone marrow cells following the addition of radiation-sterilized bone marrow or spleen cells (21). Table 1-2 shows that, for these animals, about 40 percent of the femoral CFU-s are in DNA synthesis seven days after the first injection of PHZ. This proportion is significantly reduced (to less than 10 percent) by a 2-hr incubation with irradiated spleen cells. The proportion of CFU-c in DNA synthesis is unaffected. In the converse experiments, the percentage of splenic CFU-s in DNA synthesis is increased from less than 10 percent to about 40 percent by the addition of irradiated bone marrow cells. Both these effects are dose (i.e., number of irradiated cells added) dependent (21), and the results quoted in Table 1-2 represent plateau values. Similar changes in CFU-s proliferation have also been demonstrated in other situations in which cell populations containing proliferating and nonproliferating CFU-s were mixed (Table 1-2). This is

not, therefore, a phenomenon specifically related to the PHZ-treated mouse. It is probable that these results represent the same findings as the experiments utilizing cell extracts, since recent unpublished experiments have shown that fraction III from PHZ bone marrow and fraction IV from PHZ spleen show, respectively, stimulatory and inhibitory activity.

The changes in CFU-s proliferation observed in these cell interaction experiments are, in fact, exactly analogous to those that result from the use of bone marrow containing rapidly or minimally proliferating CFU-s. When used in competition with each other, one of the extractable inhibitory and stimulatory factors (NBME-IV and RBME-III) is capable of blocking or reversing the changes in the proliferative status of CFU-s produced by the other factor (14,16). These results are summarized in Table 1-3. Thus, the interactions of cell populations or cell extracts can change the proliferative state of CFU-s, which suggests that stem cells may be maintained at an appropriate level of proliferation by a physiologically determined balance of stimulatory and inhibitory factors.

Experiments to study this possibility are in progress, and using PHZ-treated mice, we have investigated the production of factors responsible for the control of CFU-s proliferation. The stimulatory and inhibitory effects exerted by modifier cells are lost by washing the irradiated cells prior to incubation (Table 1-4). However, they are retained in the supernatant media obtained from the first washes. Media from non-irradiated modifier cells are equally as effective as irradiated cells or their supernatants, in altering the proportion of assay CFU-s killed by [^{3}H]TdR. Thus, this phenomenon is not related to the irradiation of the modifier cells.

TABLE 1-2 Effect of Cell Interactions on the Proliferative Activity of Hemopoietic Stem Cells

ASSAY POPULATION	"MODIFIER"[a]	ASSAY CFU-s (PERCENT KILLED BY [^{3}H]TdR)
	—	30 – 45
PHZ[b] bone marrow		
Regenerating bone marrow	PHZ bone marrow	40 ± 5
Regenerating spleen		
Fetal liver	PHZ spleen	
	Normal bone marrow	< 10[c]
	—	< 10
	PHZ spleen	< 10
PHZ spleen	Normal bone marrow	
Normal bone marrow		
	PHZ bone marrow	
	Regenerating bone marrow	
	Regenerating spleen	30–40
	Fetal liver	

[a]All "modifiers" are irradiated (900 rad γ-rays) cell populations.
[b]PHZ, cells taken from mice 7 days after the first injection of phenylhydrazine.
[c]No change in plating efficiency or [^{3}H]TdR kill of CFU-c in the assay cell population.

TABLE 1-3 Effect of Interactions of Bone Marrow Cell Extracts on the Proliferative Activity of Hemopoietic Stem Cells

ASSAY POPULATION	"MODIFIER"[a]	CFU-s KILLED BY [³H]TdR (PERCENT)
Normal bone marrow	—	< 10
	RBME III	39 ± 6
	RBME III + NBME IV[b]	< 10
Regenerating bone marrow	—	27 ± 7
	NBME IV	< 10
	NBME IV + RBME III[b]	38 ± 9

[a]RBME III, 30,000 to 50,000 Ds molecular weight fraction of regenerating bone marrow extract; NBME IV, 50,000 to 100,000 Ds molecular weight fraction of normal bone marrow extract.
[b]Extracts added together or separated by 1 to 2 hr.

The observations that washed cells are not effective in modifying CFU-s proliferation suggests that the simple washing procedure removes the factors responsible, and that, therefore, these factors may be lightly bound to cell surfaces. It is of interest that lymphocyte "chalone" is reported (1) to be lightly bound to the cell surface. Also, granulocyte chalone has been repeatedly collected by washing granulocytes every hour (18). Lord et al. (12) have demonstrated reversibility of the effects of erythrocyte, granulocyte, and lymphocyte extracts simply by washing the extract out. These data suggest that most, if not all, of the reported endogenous proliferation inhibitors may be lightly bound to cell surfaces.

When washed cells (which now lack proliferation-modifying ability) are resuspended in fresh medium and incubated at 37°C, both stimulatory

TABLE 1-4 Effect of Cell Interactions and Conditioned Media from Modifier Cell Suspensions on the Proliferative Activity of Hemopoietic Stem Cells

ASSAY POPULATION[a]	"MODIFIER"	ASSAY CFU-s KILLED BY [³H]TdR (PERCENT)
PHZ bone marrow	—	45 ± 7
	PHZ spleen	
	1. Irradiated cells	< 10
	2. Washed irradiated cells	46 ± 9
	3. Supernatant medium from irradiated cells	< 10
	4. Supernatant medium from non-irradiated cells	< 10
	5. Supernatant medium from washed, resuspended incubated cells[b]	< 10
PHZ spleen	—	< 10
	PHZ bone marrow	
	1. Irradiated cells	41 ± 5
	2. Washed irradiated cells	< 10
	3. Supernatant medium from irradiated cells	42 ± 2
	4. Supernatant medium from non-irradiated cells	45 ± 9
	5. Supernatant medium from washed, resuspended incubated cells[c]	40 ± 7

[a]All cells are taken from mice 7 days after the first injection of phenylhydrazine.
[b]Supernatant medium obtained from washed, resuspended cells incubated at 37°C for 4 to 5 hr prior to centrifugation
[c]Supernatant medium obtained from washed, resuspended cells incubated at 37°C for 2 to 3 hr prior to centrifugation

and inhibitory activities reappear, but after different incubation periods. It is thus possible that the effective material is resynthesized. However, the exact cellular source of these factors is, as yet, unknown. Experiments to investigate the production of the regulatory factors have shown that both stimulatory and inhibitory factors acting on CFU-s can be produced by the same hemopoietic cell suspension and that it is likely that the relative levels of such materials determine the proportion of stem cells in DNA synthesis (Wright and Lord, to be published).

Several questions are as yet unanswered, e.g., Is the source of the controlling factors the stem cell or another cell population? What is the nature of the controlling factors? Are they cell-specific molecules or only cell-specific carriers of nonspecific material? What is the mode of their production and their mechanism of action? With the techniques available, it should be possible to obtain at least partial answers to these questions and to begin to understand cell-specific proliferation regulation mechanisms.

SUMMARY

In a healthy normal mouse, the majority of the pluripotent hemopoietic stem cells are not in a state of proliferation. The "local" nature of stem cell proliferation control indicates the possibility of some short range inhibitory action. Cell-mixing experiments, i.e., co-incubation of cycling and noncycling stem cells indicated that two effects are involved: inhibition and initiation of cycling. The experiments also indicate that both effects are produced by substances eluted from (or leaking out of) living hemopoietic cells. Subsequently, large-scale incubations of normal or regenerating bone marrow in serum-free Fischer's medium (the normal bone marrow containing predominantly non-cycling, the regenerating, cycling stem cells) produced extracts that, when fractionated by molecular sieving, gave a 50 to 100,000 molecular weight fraction from normal marrow that inhibited stem cell cycling without inhibiting committed precursor cell cycling and a 30 to 50,000 molecular weight fraction from regenerating marrow, that induced stem cell cycling. The source and the cell line specificity of these effects suggest that physiological control substances are involved.

REFERENCES

1. Attalah, A.M., Sunshine, G., Hung, C.V., and Houck, J.C. The specific and endogenous mitotic inhibitor of lymphocytes (chalone). *Exp. Cell Res.*, 93:283, 1975.
2. Becker, A.J., McCulloch, E.A., Siminovitch, L., and Till, J.E. The effect for differing demands for blood cell production on DNA synthesis by hemopoietic colony-forming cells of mice. *Blood*, 26:296, 1965.
3. Cercek, L., Cercek, B., and Ockey, C.H. Structuredness of the cytoplasmic matrix and Michaelis-Menten constants for the hydrolysis of FDA during the cell cycle in Chinese hamster ovary cells. *Biophysik*, 10:187, 1973.
4. Croizat, H., Frindel, E., and Tubiana, M. Proliferative activity of the stem cells in the bone marrow of mice after single and multiple irradiations (total or partial body exposure). *Int. J. Radiat. Biol.*, 18:347, 1970.
5. Frindel, E., Croizat, H., and Vassort, F. Stimulating factors liberated by treated bone marrow: *In vitro* effect on CFU kinetics. *Exp. Hemat.*, 4:56, 1976.
6. Frindel, E., and Guigon, M. Inhibition of CFU entry into cycle by a bone marrow extract. *Exp. Hematol.*, 5:74, 1977.
7. Gidali, J., and Lajtha, L.G. Regulation of haemopoietic stem cell turnover in partially irradiated mice. *Cell Tissue Kinet.*, 5:147, 1972.
8. Iscove, N.N., Till, J.E., and McCulloch, E.A. The proliferative states of mouse granulopoietic progenitor cells. *Proc. Soc. Exp. Biol. Med.*, 134:33, 1970.
9. Lahiri, S.K., and Van Putten, L.M. Location of the G_0 phase in the cell cycle of the mouse haemopoietic spleen colony forming cells. *Cell Tissue Kinet.*, 5:365, 1972.
10. Lajtha, L.G., and Schofield, R. On the problem of differentiation in hemopoiesis. *Differentiation*, 2:313, 1974.
11. Lajtha, L.G., Pozzi, L.V., Schofield, R., and Fox, M. Kinetic properties of haemopoietic stem cells. *Cell Tissue Kinet.*, 2:39, 1969.
12. Lord, B.I., Cercek, L., Cercek, B., Shah, G.P., and Lajtha, L.G. Inhibitors of haemopoietic cell proliferation: Reversibility of action. *Brit. J. Cancer*, 29:407, 1974.
13. Lord, B.I., Lajtha, L.G., and Gidali, J. Measurement of the kinetic status of bone marrow precursor cells: Three cautionary tales. *Cell Tissue Kinet.*, 7:507, 1974.
14. Lord, B.I., Mori, K.J., and Wright, E.G. A stimulator of stem cell proliferation in regenerating bone marrow. *Biomedicine*, 27:223, 1977.
15. Lord, B., Mori, K.J., Wright, E.G., and Lajtha, L.G. An inhibitor of stem cell proliferation in normal bone marrow. *Brit. J. Haemat.*, 34:441, 1976.
16. Lord, B.I., Wright, E.G., and Mori, K.J. The role of proliferation inhibitors in the regulation of haemopoiesis. The second symposium of the British Society for Cell Biology on "Stem cells and tissue homeostasis." Ed. Lord, B.I., Potten, C.S. and Cole, R. Cambridge University Press. In press, 1977.
17. Rencricca, N.G., Rizzoli, V., Howard, D., Duffy, P., and Stohlman, F. Jr. Stem cell migration and proliferation during severe anemia. *Blood*, 36:764, 1970.
18. Rytömaa, T., and Kiviniemi, T. Control of granulo-

cyte production. II Mode of action of chalone and anti-chalone. *Cell Tissue Kinet.*, *1*:341, 1965.

19. Schofield, R. Hemopoietic Cell Kinetics. *Proceedings of the XI International Cancer Congress, Florence Vol. 1*, p. 18 Excerpta Medica, Amsterdam, 1974.

20. Till, J.E., and McCulloch, E.A. A direct measurement of the radiation sensitivity of normal mouse bone marrow cells. *Radiation Research*, *14*:212, 1961.

21. Wright, E.G., and Lord, B.I. Regulation of CFU-s proliferation by locally produced endogenous factors. *Biomedicine*, *27*:215, 1977.

2

Surface Antigens of Hemopoietic Stem Cells: The Expression of BAS, Thy-1, and H-2 Antigens on CFU-s

Ger van den Engh,
Jim Russell,
and Diane de Cicco

INTRODUCTION

Cell separation experiments with mouse marrow suspensions have yielded a fairly detailed picture of the pluripotent hemopoietic stem cell. The physical characteristics of the stem cell, such as buoyant density (14), sedimentation rate (13), and light scattering properties (12) lie within the same range as those of a medium- to large-sized lymphocyte. It may, therefore, be concluded that the gross morphology of the stem cell, with respect to its size, shape, and nucleus/cytoplasm ratio, resembles that of a lymphocyte. Studies in which the candidate stem cells have been identified in microscopic preparations have indeed shown cells that fit this description (1).

Despite its unique biological properties, the stem cell shares physical properties with many other cell types. It seems, therefore, unlikely that separation procedures that are based exclusively on physical differences between cells will permit the concentration of these rare cells to a sufficient degree. To arrive at a description of the stem cell that allows us to recognize this cell unequivocally from committed progenitor cells and lymphocytes, properties more closely related to cell function have to be included. Useful criteria for stem cell recognition may be found in the occurrence of specific membrane components. In immunology, the presence or absence of surface antigens is successfully used to classify functional subpopulations of lymphocytes. Those surface antigens that specifically mark stages of cell differentiation are often indicated by the term "differentiation antigens" (2).

Two types of hypotheses have been put forward concerning the role of differentiation antigens in stem cell differentiation. Davis (4) proposed that the stem cell, as the most undifferentiated cell of the hemopoietic series, expresses none of the differentiation antigens and may therefore be found among the "null" cells present in hemopoietic cell suspensions. An alternative possibility has been suggested by Till (10). The capacity of stem cells to differentiate into any of the maturation lines may be reflected in its membrane composition, in that the cell may have receptors for all differentiation lines. In this view, the stem cell is characterized by an abundance of surface markers and is far from being a null cell. The data presented in this paper show that stem cells do express unique surface antigens. Therefore, the use of the term null cell may be misleading if not incorrect.

The presence of antigens on the stem cell surface may be demonstrated by incubating mouse bone marrow cells with antisera of known specificity and complement and subsequently measuring

9

the suppression of the formation of spleen colonies by hemopoietic stem cells (CFU-s). The first evidence that CFU-s may express unique antigens was supplied by Golub (7), who showed that CFU-s share an antigen with brain tissue which is absent on other blood cells. In further studies it is shown that this brain-associated stem cell antigen (BAS) is lost during the first steps of differentiation and cannot be detected on committed progenitor cells (CFU-c) (6).

In this review, experiments are described that extend our knowledge on the tissue distribution of BAS and the mode of action of anti-BAS serum. Further, CFU-s are tested for the presence of the T-cell specific Thy-1 antigen. A third group of antigens, which may be useful in the recognition of CFU-s, are those belonging to the mouse major histocompatibility complex (H-2). These antigens are coded for by a complex of adjacent genes located on chromosome 17. By the use of specific antisera, three regions can be recognized: the K, I, and D regions. The K and D antigens are detectable on the surface of most, if not all cells. However, their density varies considerably (8). Most tissues express only traces of K and D antigens, whereas lymphocytes, particularly those from spleen and lymph nodes, express high amounts. The I region antigens have a restricted tissue distribution. They are found exclusively on B-cells and on monocytes. The character of the H-2 antigens and their assumed role in the immune response makes these antigens of particular interest.

MATERIALS AND METHODS

Mice

The $(C_{57}BL/Ka \times CBA/Rij)F_1$ (H-$2^{b/q}$), C57BL/Ka (H-2^b), and CBA/T$_6$ (H-2^k) mice were bred and maintained at the Radiobiological Institute TNO, Rijswijk. The B10.A (H-2^a) mice were bred at Purdue using stocks obtained from Jackson Laboratories, Bar Harbor, Me.

Antisera

The following antisera were supplied by Dr. D. Sachs: B10.A anti-B10, A.SW anti-A.TH, (B10.A × A/J) anti-B10.D$_2$. Dr. C. S. David supplied the serum A.TH anti-A.TL. Anti-mouse brain serum was prepared according to Golub (7) and was absorbed with liver cells, red blood cells, and thymocytes in order to obtain a specific anti-BAS activity. Anti-Thy 1.2 serum was prepared by immunizing AKR mice six times with C$_3$H thymocytes at weekly intervals.

Assay for Spleen Colony-forming Cells (CFU-s)

The spleen colony assay for pluripotent hemopoietic stem cells was performed as described by Till and McCulloch (11).

Assay for in vitro colony-forming cells (CFU-c)

The technique used was a modification of the culture system described by Bradley and Metcalf (3). Briefly, the cultures consisted of 35 mm Petri dishes containing 5×10^4 bone marrow cells in 1 ml Dulbecco's medium with 0.3 percent agar, to which a suitable growth stimulator (pregnant mouse uterus extract) was added (5). The dishes were kept at 37°C in a humidified incubator, which was gassed with a mixture of 10 percent CO_2 in air. Colonies of granulocytes and/or macrophages were scored after 7 days of culture.

Determination of Live/Dead Cell Ratio

The cytotoxic effect of the combined antiserum and complement treatment was determined either manually by counting the ratio of Trypan Blue-stained and unstained cells through a microscope or by using a cytoflow meter (Biophysics, cytofluorograph). In the latter method, dead cells are distinguished from live cells by their ability to incorporate propidium iodine; following the incorporation, they appear red when excited with blue layer light. The cell viability after antiserum treatment was expressed as a percentage of an untreated cell sample. If necessary, the number of dead cells were corrected for the number of dead cells found in the control sample.

Antiserum Treatment

Here, 10^6 cell in 0.1 ml of Hank's balanced salt solution (HBSS) containing 10 percent inactivated calf serum, were treated for 30 min on ice with appropriate dilutions of 0.1 ml of antiserum. Subsequently, 1 ml of a 1:5 dilution of guinea pig complement in HBSS was added and the suspensions were incubated for 60 min at 37°C. The cells were washed once in HBSS. The suspensions were assayed for CFU-s or CFU-c content and/or the live to the dead cell ratio was determined.

Papain Treatment of Antibody-coated Cells

Cells were suspended in HBSS with Hepes buffer (pH 6.8) to which papain (Worthington) and cysteine and 10 percent decomplemented calf serum had been added. The final concentration of papain was 1 mg/ml and the cysteine concentration was 0.1 mg/ml. The papain digestion was allowed to proceed for 60 min at 37°C.

RESULTS

Brain-associated Stem Cell Antigen (BAS)

Mouse brain tissue carries an antigen that is also present on hemopoietic stem cells. Treatment with rabbit anti-mouse brain sera abolishes the capacity of CFU-s to form spleen colonies. This is demonstrated in Figure 2-1. The anti-brain serum used in this experiment had been previously absorbed with mouse liver cells, red blood cells, and thymocytes. The toxicity of the antiserum to bone marrow cells, as measured by dye exclusion, is very low, yet at the higher antiserum concentration, CFU-s activity is almost completely abolished. The absence of toxicity to thymocytes indicates that the absorption procedure has effectively removed all antibody species directed against brain associated Thy-1, suggesting that BAS is a different antigenic membrane structure. In contrast to the suppressive effect of the antiserum on *in vivo* colony formation, *in vitro* colony formation by committed progenitor cells (CFU-c) is not noticeably reduced. This indicates that BAS is lost during the differentiation from CFU-s to CFU-c.

Further absorption studies provide additional proof that BAS is a rare antigen among hemopoietic

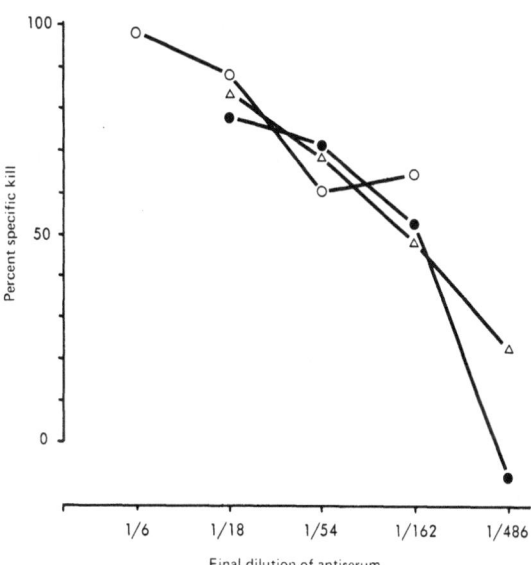

FIGURE 2-2. Effect of bone marrow absorption on CFU-s suppression by rabbit anti-mouse brain serum: Not absorbed with bone marrow cells (o—o), absorbed twice with bone marrow cells (△—△), absorbed four times with bone marrow cells (●—●).

cells. Figure 2-2 shows the effect of absorption with bone marrow cells on the anti-CFU-s titer of an anti-BAS serum. Up to four absorptions with bone marrow cells do not reduce the suppressive effect of the antiserum on *in vivo* colony formation. This finding is compatible with the idea that BAS is either expressed in very low amounts on marrow cells or is only present on a very rare cell type.

The anti-stem cell activity of anti-BAS serum has been measured by the capacity of this serum to suppress *in vivo* colony formation. Since many steps take place before a stem cell is expressed as a spleen colony, the suppression may not necessarily be due to the killing of CFU-s. The antibody could interfere in different ways with CFU-s function. It appears that *in vitro* incubation of antibody-treated cells with complement is not essential to obtain the suppression of *in vivo* colony formation. More detailed studies indicate that the anti-BAS activity may be a non-cytotoxic effect, suppressing CFU-s colony formation without killing the cells. Batches of anti-BAS serum have been obtained in which the CFU-s suppressive effect could be overcome by treating the antibody-coated cells with papain (to be published in detail elsewhere). Papain is known to cleave the F_c fragments from the antibody molecules. Antibody-treated cells, incubated first with complement and then treated with papain, remain viable and are able to form hemopoietic colonies in the spleen. This suggests that CFU-s suppression may occur at antibody levels

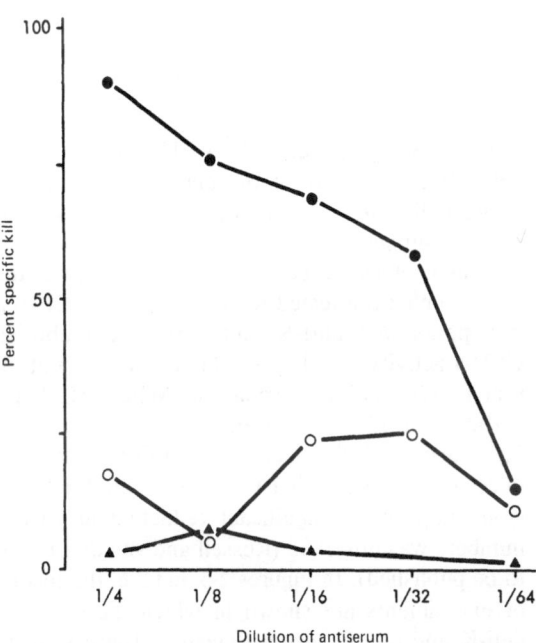

FIGURE 2-1. Suppression of CFU-s activity by rabbit anti-mouse brain serum. The antiserum is made specific for the brain-associated stem cell antigen by absorption with mouse liver, red blood cells, and thymocytes. Toxicity to thymocytes (▲—▲) and *in vitro* colony-forming cells (CFU-c) (o—o) is low; CFU-s (●—●) activity is abolished at high serum concentrations.

that are not yet cytotoxic. However, the F_c fragments appear to be necessary for suppression. Thus, a direct comparison between *in vivo* phenomena and *in vitro* cytotoxicity on various cell types should be undertaken with care. If possible, CFU-s antigens should be quantified in experiments in which complement-dependent toxicity toward CFU-s *in vitro* can be demonstrated.

Thy-1 Not Expressed on CFU-s

When CFU-s are tested for the presence of Thy-1, no detectable amounts of this antigen are found. Figure 2-3 shows that incubation of C_3H bone marrow cells with an AKR anti-C_3H thymocyte serum with a high titer against Thy 1.2 and complement does not reduce the number of CFU-s present.

Expression of H-2 Antigens on CFU-s

The CFU-s from mouse bone marrow have considerable amounts of H-2 antigens on their surface. Figure 2-4 shows that CFU-s activity is greatly reduced after incubation with an antiserum against the whole of the H-2 region (anti-$K^bI^bD^b$ tested in b mice). The suppression of CFU-s is comparable to the cytotoxic titer of the antiserum for mouse spleen cells. Since spleen cells express exceptionally high levels of H-2 antigens, this may indicate that CFU-s are also rich in surface H-2. In contrast,

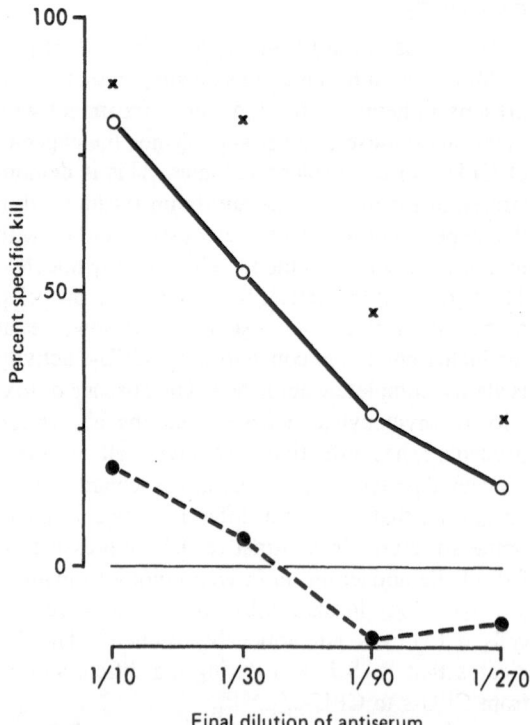

FIGURE 2-4. Suppression of CFU-s activity after incubation of $C_{57}BL/Rij$ (H-2^b) bone marrow cells with B10.D2 anti-B10 serum (anti-H-2^b) and complement. The anti-CFU-s activity of the antiserum is comparable to the toxicity to spleen cells. The CFU-c activity is affected to a much lesser extent. CFU-s (o—o), CFU-c (●---●), spleen cells (x).

in vitro colony formation by CFU-c is not noticeably affected at any of the antiserum concentrations, indicating that this cell type expresses far less H-2 antigens.

When antisera directed against subregions of the H-2 complex are tested for their effect on CFU-s, it appears that anti-K and anti-D sera abolish CFU-s activity but that anti-Ia serum does not affect *in vivo* colony formation. When EL-4 absorbed, anti-H-2^b was used against b mice, anti-K^dI^d against H-2^b mice, and anti-I^k against H-2^k mice, combinations that test specifically for Ia region antigens, no significant reduction in CFU-s numbers was observed (Russell and van den Engh, to be published). In Figures 2-5 and 2-6, the results of experiments are shown in which the effect of anti-K and anti-D sera was tested. These sera do suppress CFU-s. Again, the strength of the antisera with respect to CFU-s suppression is comparable to their toxicity toward splenic lymphocytes.

As has been demonstrated, the anti-CFU-s activity of antiserum is not necessarily due to cytotoxicity. Spleen cytotoxicity and the CFU-s suppressive effect of anti-H-2 sera may only be compared directly if both are equally dependent on

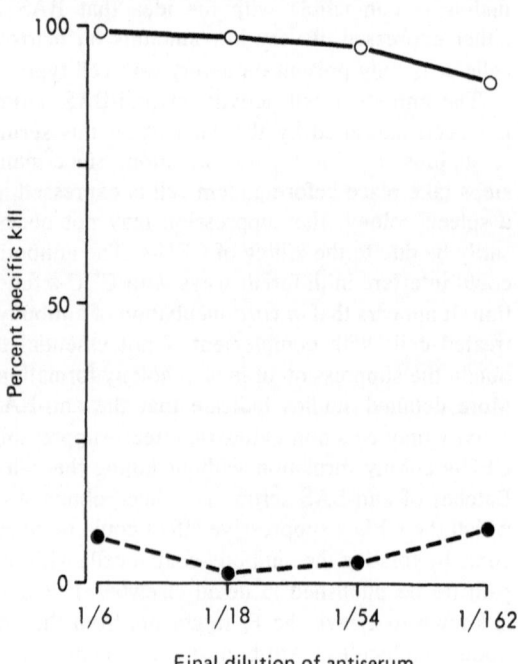

FIGURE 2-3. Incubation of C_3H bone marrow cells with AKR anti-C_3H thymocyte serum does not abolish CFU-s activity. A control experiment showed that this serum has anti-C_3H thymocyte activity: C_3H thymocytes (o—o), C_3H CFU-s (●---●).

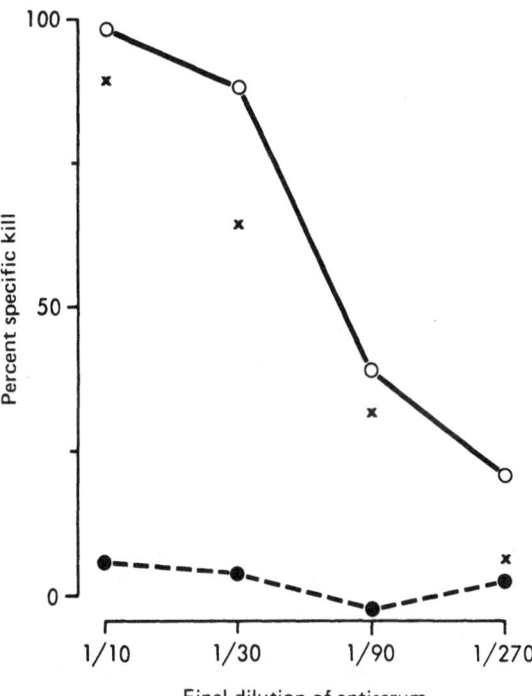

FIGURE 2-5. Suppression of CFU-s from DBA/2 (H-2^d) bone marrow after incubation with (B10.A × A/J) anti-B10.D2 serum (anti H-2K^d) and complement. Stem cells from H-2^b mice, which share an Ia antigen with H-2^d, are not suppressed by this antiserum. H-2^d (o—o), H-2^b (●-- -●), spleen cells (x).

TABLE 2-1 Rescue of Antiserum-treated Cells by Incubation with Papain Before and After Complement^a

| | ANTISERUM DILUTION | |
TREATMENT	1/10	1/30
Antiserum/complement	83^a	52^b
Antiserum/papain/complement	15	7
Antiserum/complement/papain	60	61

^{a}Bone marrow cells of C$_{57}$BL/Rij (H-2^b) mice were incubated for 30 min with anti H-2^b serum at 4°C. Subsequently, the cells were incubated with guinea pig complement for 60 min at 37°C. Thereafter the cells were treated with papain (60 min, 37°C) and injected into recipient mice. As a control, the sequence of papain and complement treatment was reversed.

^{b}Percent specific suppression of CFU-s activity.

in vitro complement treatment. Table 2-1 shows that by the removal of antibody from the cells at the end of the *in vitro* procedure, the CFU-s suppression by anti-H-2 sera can be made dependent on the presence of complement. If this procedure is followed, no shift in anti-CFU-s titers is observed, indicating that CFU-s suppression is related to *in vitro* cytotoxicity. Therefore, the similarity in titer between CFU-s suppression and the killing of splenic lymphocytes may indicate a similar density of surface antigens of both cell types.

The density of H-2 antigens on the CFU-s surface appears to be independent of its cell cycle characteristics. Figure 2-7 shows the anti-spleen colony effect of an anti-H-2 serum when tested against marrow of normal mice and marrow cells from mice 1 week after irradiation and bone marrow grafting. The majority of CFU-s in the latter are in active cell cycle, yet they are as sensitive to anti-H-2 treatment as the normal control cells. This suggests that the high H-2 density found on both CFU-s and splenic lymphocytes is not due to the fact that both cell types (in contrast to CFU-c) are in a resting state and consequently have a low turnover of membrane components. The fact that expression of surface antigens is independent of the proliferation rate suggests a high H-2 density for both cell types.

DISCUSSION

The expression of surface antigens on the hemopoietic stem cell was investigated by measuring the capacity of antisera with various specificities to abolish colony formation by CFU-s. The results for CFU-s and other cell types are summarized in Table 2-2. There is some uncertainty in the interpretation of these data. Cells with foreign antibodies on their membranes may exhibit atypical behavior when they are injected into a mouse. It is therefore

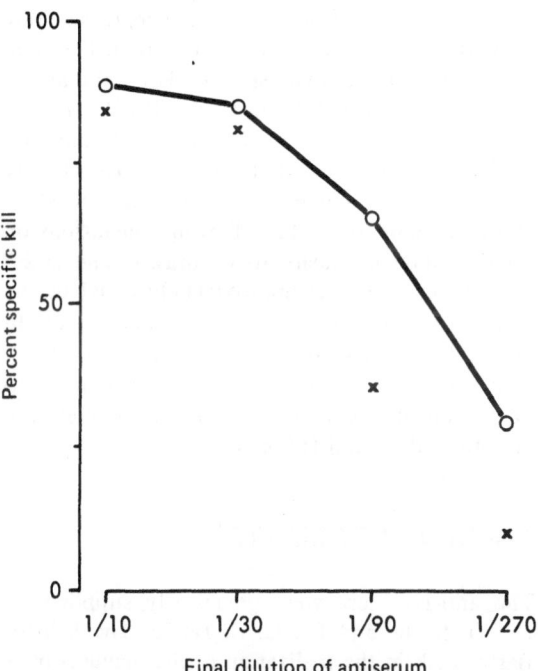

FIGURE 2-6. Suppression of CFU-s (o—o) from B10.A (H-2 D^d) bone marrow cells by A.SW anti-A.TH serum (anti-H-2 D^d) and complement. Spleen cells (x).

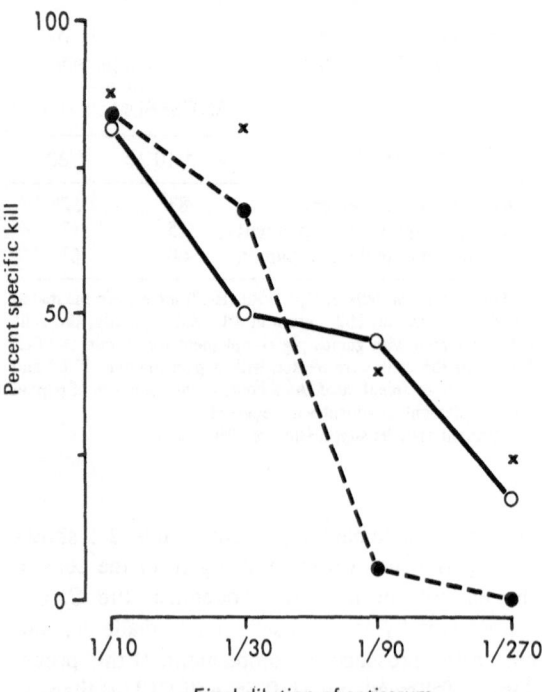

FIGURE 2-7. Suppression of normal and rapidly proliferation CFU-s from C₅₇BL/Rij (H-2ᵇ) mice by B10.A anti-B10 antiserum (anti-H-2ᵇ), in comparison to the complement-mediated toxicity of the same antiserum to H-2ᵇ spleen cells (x). Normal marrow (o—o), marrow-reconstituted mice (•---•).

conceivable that CFU-s inactivation may be achieved in other ways than through cell death. Since the specificities of the antisera used are defined in terms of cytotoxicity, some care should be taken in comparing the effects *in vivo* with cytotoxic titers.

In the case of the BAS antigen, it was found that the elimination of CFU-s may take place after the cells have been injected into mice and that CFU-s suppression can be caused by non-cytotoxic antibodies. Other experiments leading to the same conclusion have been reported (9). On the other hand, absorption studies indicate that the antigen is present in very small amounts in the hemopoietic tissues and it seems therefore unlikely that CFU-s

TABLE 2-2 Surface Antigens Present on Various Mouse Blood Cells as Detected by Cytotoxicity or Suppression of Colony Formation

BLOOD CELLS	H-2 K	Ia	D	THY-1	BAS
CFU-s	+	−	+	−	+
CFU-c	−	−	−	−	−
B-cell	+	+	+	−	−
T-cell	+	−	+	+	−

share the antigen with many other cell types. The capacity to react with anti-brain sera may be specific for CFU-s.

The suppressive effect of CFU-s activity by anti-H-2 sera was shown to be due to complement-mediated cytotoxicity. It seems therefore justified to compare CFU-s suppression *in vivo* with cytotoxic titers. If this is done, the level of H-2ᴷᴰ expression on CFU-s is comparable to that of splenic lymphocytes, and CFU-s may be characterized as having a high H-2ᴷᴰ density. This similarity between CFU-s and lymphocytes may be more than coincidental and may reflect the ontogenetic relationship between the two cell types.

No Ia (H-2ᴵ) and Thy-1 antigens, known markers of mature lymphocytes, could be demonstrated on the stem cell surface. The expression of these antigens must take place at a later stage of blood cell differentiation. The high surface density of H-2ᴷᴰ antigens on lymphocytes does not appear to be an exclusive property of these cells, since stem cells already have a high H-2ᴷᴰ density. The absence or low levels of H-2ᴷᴰ on CFU-c and red blood cells (8) suggests that H-2 transcription is lost during maturation in most of the blood cell lines. The high H-2 levels on lymphocytes may indicate a stem cell-like property of these cells.

SUMMARY

The surface antigens of hemopoietic stem cells were investigated using antisera to suppress colony formation by CFU-s *in vivo* as a criterion. The stem cells were found to express H-2ᴷ, H-2ᴰ, and an antigen associated with brain tissue (BAS). The differentiation antigens Ia (H-2ᴵ) and Thy-1 could not be demonstrated. The elimination of CFU-s activity by anti-H-2 sera could be shown to be caused by cytotoxic antibodies, thus allowing conclusions on relative antigen density to be drawn. The H-2ᴷᴰ density on CFU-s is approximately equal to the high density of those antigens on lymphocytes. In contrast, *in vitro* colony-forming cells (CFU-c), which are close relatives of CFU-s committed to maturation into granulocytes and monocytes, are not affected by anti-H-2 sera.

ACKNOWLEDGMENTS

The anti-H-2 sera were generously supplied by Dr. D. Sachs and Dr. C. S. David. The help of Barb Trask in the realization of the manuscript is greatly appreciated.

This investigation is part of a project program on the regulation of hemopoiesis, which is supported by FUNGO. The work at Purdue was supported by MSPHS grant CA 18641-02 and Regulation training grant 5T32GM 0706-02. Jim Russell and Diane de Cicco received travel grants from KWF.

REFERENCES

1. Bekkum, D.W. van, van Noord, M.J., Maat, B., and Dicke, K.A. Attempts at identification of the hemopoietic stem cell in the mouse. *Blood, 38*:547, 1971.
2. Boyse, E.A., and Old, L.J. Some aspects of normal and abnormal cell surface genetics. *Ann. Rev. Genetics, 3*:269, 1969.
3. Bradley, T.R., and Metcalf, D. The growth of mouse bone marrow cells *in vitro. Aust. J. exp. Biol. med. Sci., 44*:287, 1966.
4. Davis, S. Hypothesis: differentiation of the human lymphoid system based on cell surface markers. *Blood, 45*:871, 1975.
5. Engh, G.J. van den. Quantitative *in vitro* studies on stimulation of murine haemopoietic cells by colony stimulating factor. *Cell Tissue Kinet., 7*:537, 1974.
6. Engh, G.J. van den, and Golub, E.S. Antigenic differences between hemopoietic stem cells and myeloid progenitors. *J. exp. Med., 139*:1621, 1974.
7. Golub, E.S. Brain-associated stem cell antigen: An antigen shared by brain and hemopoietic stem cells. *J. exp. Med., 142*:1200, 1972.
8. Klein, J. *Biology of the Mouse Histocompatibility-2 Complex*. Berlin: Springer, 1975.
9. Kuznetsky, R.D., Adler, S., and Trobaugh, F.E. Anti-mouse brain anti-serum: Characteristics and spectrum of activity against hemopoietic cells. *Exp. Hemat., 4 (suppl)*:60, 1976.
10. Till, J.E. Regulation of hemopoietic stem cells. In A. B. Cairnie, P. K. Lala, and D. G. Osmond, eds., *Stem Cells of Renewing Cell Populations*. New York: Academic Press, 1976.
11. Till, J.E., and McCulloch, E.A. A direct measurement of the radiation sensitivity of normal mouse bone marrow cells. *Radiat. Res., 14*:213, 1961.
12. Visser, J., and Engh, G. van den. Manuscript in preparation.
13. Visser, J., Engh, G. van den, Williams, N., and Mulder, D. Physical separation of the cycling and non-cycling compartments of murine haemopoietic stem cells. In S. J. Baum and G. D. Ledney, eds., *Experimental Haematology Today*. New York: Springer, 1977, pp. 21–27.
14. Worton, R.G., McCulloch, E.A., and Till, J.E. Physical separation of hemopoietic stem cells from cells forming colonies in culture. *J. Cell. Physiology, 74*:171, 1969.

3

The Regulation of Hemopoiesis: Effect of Thymosin or Thymocytes in a Diffusion Chamber

S. J. Sharkis, A. Ahmed,
L. L. Sensenbrenner,
W. W. Jedrzejczak,
A. L. Goldstein,
and K. W. Sell

INTRODUCTION

Investigation of the mechanisms regulating the proliferation and differentiation of hemic tissue remains an important area of research despite the many accomplishments in this field. Both humoral and cellular systems have been implicated in many of the maturational pathways, from the pluripotent hemopoietic stem cell (HSC) to mature cellular elements of the blood. Thus, hormones, such as erythropoietin, are known to interact with erythroid precursor elements in culture to eventually differentiate into circulating red blood cells (RBC) (1,19). More recently, we have described a system (11,17,18) in which cells carrying the well-known surface antigen associated with lymphocytes (the theta- or THY-1 antigen) may regulate the normal proliferation of stem cells and erythroid precursors in the anemic W/W^v mouse. Thus, bone marrow treated *in vitro* with alloantisera (CBA thymocytes given to AKR mice) plus complement and transplanted into anemic mice produces sufficient numbers of macroscopic colonies on the surface of spleens, but these colonies are predominantly committed granulocyte precursor elements with no potential to produce stem cells (17,18).

It is the purpose of this review to determine if a humoral factor can restore the capacity of bone marrow depleted of the theta-sensitive regulatory cell (TSRC) to cure W/W^v anemia and to further determine if this factor has any effect on proliferation of stem cells. We, therefore, examined the possible effect of thymosin or thymocytes in a cell-impermeable diffusion chamber on stem cell proliferation and erythroid maturation in the transplanted W/W^v anemic host.

MATERIALS AND METHODS
Animals

Male WBB_6F_1 and female anemic (W/W^v) and hematologically normal coisogeneic littermate (+/+) mice, obtained from Jackson Laboratories at 6 to 8 weeks of age were used in these studies. All animals were given Purina chow and water *ad libitum*.

Preparation of Cell Inocula

Donor W/W^v and +/+ mice were killed by cervical dislocation and marrow was collected by flushing cold RPMI 1640 media through femurs with a syringe and needle. Cell counts and percent viabilities (all suspensions were at least 95 percent viable just prior to injection) were performed by routine methods. Whole thymocytes were obtained from +/+ or W/W^v donors by mincing thymic tissue with a fine forceps and a rubber spatula. Single-cell suspensions were obtained by passing the suspension through a syringe and 25 gauge needle several times.

Reconstitution of Erythropoiesis and the Spleen Colony Assay Determination

Marrow cells from +/+ donors were treated *in vitro* in some experiments with anti-Thy 1.2 serum plus rabbit complement. Details of this treatment are described elsewhere (11). For studies that required evaluation of erythroid differentiation of transplanted stem cells, 10^7 viable +/+ bone marrow cells treated *in vitro* with anti-Thy 1.2 serum plus complement to deplete TSRC were injected into groups of five W/W^v recipients. For spleen colony determinations, 10^5 cells, treated with antiserum or normal mouse serum as a control plus complement, were injected into groups of either ten lethally irradiated +/+ mice (1,050 rad, 137 rad/min from dual ^{137}Cs sources) or five nonirradiated W/W^v recipients. Seven days later, the spleens were removed, placed in fixative, and examined for the appearance and number of macroscopic colonies.

Effect of *In Vitro* Incubation of Thymosin with W/W^v Bone Marrow

For one experimental design, both W/W^v and +/+ donor bone marrow cells at a concentration of 10^6/ml were incubated with 10 μg of thymosin (fraction 5) or 1 percent bovine serum albumin (BSA) as a control at 37°C for 1 hr. The cells were washed, and 10^5 viable cells were injected into irradiated +/+ recipients for the CFUs assay, and 10^7 cells into W/W^v recipients for reconstitution of the erythroid compartments of the anemic host.

In Vivo Thymosin Therapy

Thymosin was given to W/W^v recipient mice for 6 weeks following a marrow graft of 10^7 TSRC-depleted cells. For reconstitution of anemic recipients, each animal received 500 μg of fraction 5 thymosin intraperitoneally three times a week in a volume of 0.5 ml. For spleen colony assay, 10^5 TSRC-depleted cells were given followed by three doses of thymosin on alternate days (700 μg/dose) and the W/W^v mice were sacrificed 7 days after marrow grafting. Experiments for the determination of spleen colonies used normal mouse serum-treated cells given as a control for the antiserum treatment; whole thymocytes (10^7) given to compare with the effect of thymosin, and 1 percent BSA was used as a control for the injection of thymosin.

Diffusion Chamber Studies

To parallel the *in vivo* thymosin studies, we attempted to determine if thymocytes in the diffusion chamber would elaborate a humoral factor and if this factor would interact with TSRC-depleted +/+ bone marrow to cure the anemia of W/W^v recipients. Therefore, 24 hr prior to the intravenous injection of 10^7 anti-Thy 1.2 serum and complement-treated +/+ marrow cells, 2×10^6 +/+ thymocytes were placed in diffusion chambers, and two chambers were placed in the peritoneal cavity of W/W^v recipients. Empty chambers were implanted as a control. The details of chamber filling and implantation have been described (16). Recipient mice did not receive cyclophosphamide prior to chamber implantation.

Hematological Values

Blood was obtained via the retro-orbital sinus and collected in heparinized capillary tubes from lightly anaesthesized mice. Hematocrit values and RBC counts were determined by routine methods, at least biweekly to see if the anemic W/W^v hosts had been cured.

RESULTS

In Vitro Incubation of W/W^v Bone Marrow with Thymosin

McCulloch et al. (13) established the absence of a normal stem cell in W/W^v mice, but the defect presumably is quantitative, since these mice do maintain a reduced but significant number of circulating blood cells. Restoration of normal erythropoiesis in these mice require an interaction between the stem cell and the TSRC (11). It, therefore, became of interest to see if the defective stem

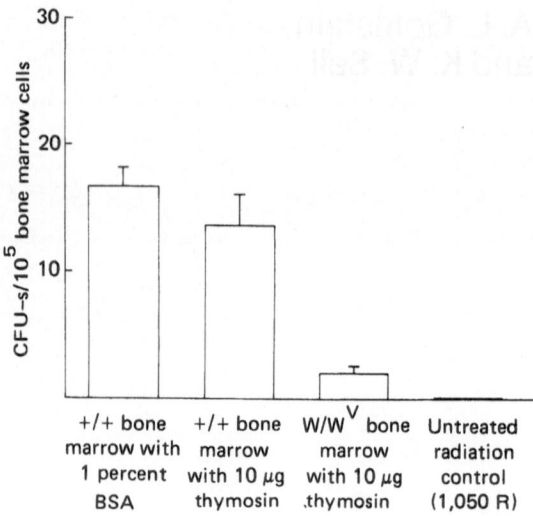

FIGURE 3-1. The effect of *in vitro* incubation of W/W^v or +/+ bone marrow cells on the potential of these cells to produce macroscopic colonies in lethally irradiated +/+ mice. Mean ± SEM of 10 recipients in each group.

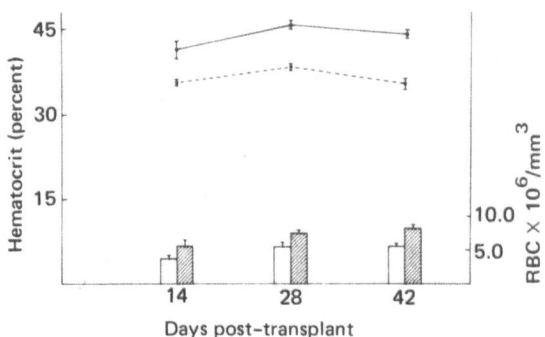

FIGURE 3-2. The effect of *in vitro* incubation of W/Wᵛ (●------●) or +/+ (●———●) bone marrow cells with thymosin on the potential of these cells to increase hematocrit values of transplanted W/Wᵛ recipients. Hatched bars, RBC values of W/Wᵛ recipients of thymosin-treated W/Wᵛ bone marrow. Mean ± SEM for five recipients.

cell could be induced to produce normal RBC by generating TSRC in the W/Wᵛ marrow. This was attempted by incubating W/Wᵛ bone marrow cells with thymosin. We tested (a) the capacity of such incubated marrow to produce CFUs in lethally irradiated recipients and (b) the ability of thymosin-treated W/Wᵛ marrow to cure anemic recipients. As can be seen in Figure 3-1, incubation of W/Wᵛ bone marrow cells with thymosin produces fewer than three colonies, compared to the 14 colonies formed with +/+ bone marrow. This suggests that thymosin may not generate TSRC in W/Wᵛ bone marrow. Apparently, W/Wᵛ marrow incubated with thymosin is ineffective in producing normal erythroid cells in W/Wᵛ recipients as well (Figure 3.2).

The Effect of Thymosin or Whole Thymocytes on the Colony-Forming Capacity of TSRC Depleted +/+ Bone Marrow

The removal of the TSRC from +/+ bone marrow had no effect on the production of CFU (Table

3-1) in nonirradiated W/Wᵛ mice. If thymosin is given to the recipient of anti-Thy 1.2 serum plus complement-treated +/+ bone marrow, a fourfold augmentation in the number of macroscopic colonies (51.2 ± 3.8) is observed. Interestingly, this same augmentation was observed when whole thymocytes were given with the TSRC-depleted marrow cells (52.0 ± 6.9 colonies). These data suggest that thymosin, like whole thymocytes, provides a factor that modifies spleen colony production.

The Effect of Thymosin In Vivo on Reconstitution of W/Wᵛ Mice Given TSRC-Depleted Marrow Grafts

Two groups of five W/Wᵛ mice each, transplanted with 10⁷ TSRC-depleted bone marrow cells, received either 500 μg of thymosin three times weekly for six weeks or 1 percent BSA as a control. In the group receiving thymosin following the TSRC-depleted graft (Figure 3-3), increases in hematocrit values occurred consistent with a cure of the anemia, whereas in the animals receiving BSA following the TSRC-depleted graft, hematocrit values never rose above 40 percent. The level of hematocrit values in the group receiving thymosin did not reach that of donor +/+ mice (48.0 ± 1.0), but this may be related to the dose of thymosin employed.

The Effect of Thymocytes in Diffusion Chambers on the Reconstitution of Circulating Erythroid Cells in W/Wᵛ Recipients of TSRC-Depleted Grafts

Increases in hematocrit values following thymosin and anti-Thy 1.2 serum plus complement-treated +/+ bone marrow grafts suggested that a humoral factor could effect replacement of whole thymocytes and restore normal erythropoiesis in anemic recipients. Experiments were performed to examine the effect of placing thymocytes in the diffusion chamber so that only humoral factors and

TABLE 3-1 Effect of Thymocytes or Thymosin on Macroscopic Colony Formation in Nonirradiated W/Wᵛ Recipient Mice

BONE MARROW	TREATMENT OF MARROW[a]	THYMOSIN[b]	THYMOCYTES	MACROSCOPIC COLONIES
W/Wᵛ	NMS+C'	–	–	0.0
–	–	–	W/Wᵛ	0.0
+/+	NMS+C'	–	–	12.8 ± 1.7
+/+	NMS+C'	+	–	57.75± 4.3
–	–	–	+/+	0.5 ± 0.3
+/+	NMS+C'	–	+/+	20.8 ± 2.2
+/+	αθS+C'	–	–	13.4 ± 3.2
+/+	αθS+C'	–	+/+	52.0 ± 6.9
+/+	αθS+C'	+	–	51.2 ± 3.8

[a]NMS+C', normal AKR mouse serum + complement', αθS+C', anti-Thy 1.2 antiserum + complement.
[b]700 μg of thymosin given i.p. three times prior to sampling for CFUs.

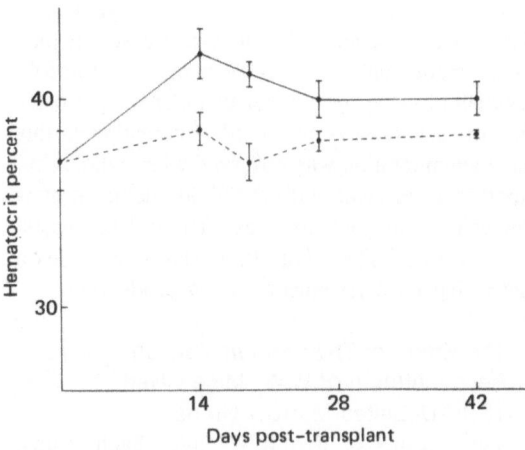

FIGURE 3-3. The effect on hematocrit values of W/Wᵛ recipients given anti-Thy 1:2 serum plus complement-treated +/+ bone marrow, followed by 500 μg of thymosin three times per week for 6 weeks (●————●) or 1 percent as a central BSA (●------●). Each point represents the mean ± SEM for five animals.

not cells could pass through the chamber and enter the circulation of the host. The results of implantation of the chambers with whole thymocytes are seen in Figure 3-4. Once again, hematocrit values increased in the recipients of +/+ thymocytes in the chamber, so that by 6 weeks, they are 44.4 ± 0.2 percent; in the animals receiving empty chambers, hematocrit values did not rise above 40 percent.

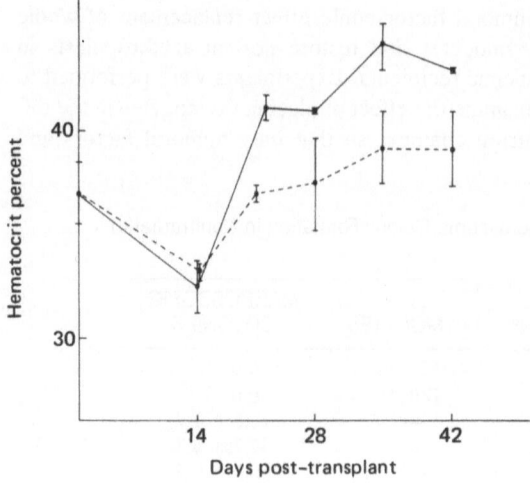

FIGURE 3-4. Hematocrit values of W/Wᵛ recipients receiving TSRC-depleted bone marrow grafts and thymocytes (●————●) or empty diffusion chambers (●------●). Each point represents the mean ± SEM for five animals. $\alpha\theta^s$ + c¹, anti-theta serum plus complement.

DISCUSSION

The W/Wᵛ anemic mouse is a product of genetic mutations, which produce phenotypic expression with pleiotropic effects. The mouse has a hypoplastic bone marrow, presumably a result of a defective stem cell (15) and a circulating macrocytic anemia, and there are other blood cell deficiencies, such as a reduction in circulating granulocytes (5) and reduced numbers of megakaryocytes (5,6). We have recently described a defective or absent population of cells in this mouse that is present in the bone marrow, spleen, and thymus of the normal littermate, the theta-sensitive regulatory cell (TSRC). The TSRC appears to be required for normal differentiation of stem cells into erythroid precursors and for self-renewal of stem cells (11,17,18).

The effect of addition of thymocytes to TSRC-depleted +/+ bone marrow grafts to produce increased numbers of spleen colonies and to stimulate production of normal erythroid cells in the transplanted W/Wᵛ mouse prompted us to examine the effect of a thymic humoral factor on this repopulation. Thymosin has been shown (2) to increase the number of Thy 1.2-positive cells when incubated in vitro with bone marrow. Quantitatively, the W/Wᵛ mouse has in its thymus adequate numbers of theta-positive cells, but preliminary evidence from our laboratory indicates that the population of these cells may be altered. With the use of the fluorescence activated cell sorter, it was demonstrated that the distribution of brightly stained Thy 1.2-positive cells of the W/Wᵛ thymus shifts, to become less fluorescent on the cell membrane when compared with the fluoresceinated antiserum-stained littermate thymocytes (Sharkis et al., unpublished observations). Separation studies are warranted to examine this phenomena further. Since thymosin can generate Thy 1.2-positive cells in vitro, and the W/Wᵛ bone marrow has sufficient numbers of stem cells to protect a lethally irradiated mouse (10), it was felt that perhaps thymosin might induce the formation of TSRC, which, in turn, could generate normal stem cell proliferation from W/Wᵛ bone marrow. The result of incubation of W/Wᵛ bone marrow with thymosin (a) to generate sufficient colonies in +/+ mice and (b) to prevent W/Wᵛ anemia, met with complete failure (Figures 3-1 and 3-2). Two possible hypotheses can be advanced: (i) the thymosin does not act upon the defective stem cell of W/Wᵛ mice to produce sufficient TSRC or (ii) even if TSRC are generated, the defective stem cell of the host W/Wᵛ bone marrow cannot interact with these TSRC.

The effect of thymosin on colony formation in the W/Wᵛ recipient is of interest. We have shown previously (16,17) that the W/Wᵛ recipient of an anti-Thy 1.2 serum plus complement-treated marrow graft together with whole thymocytes from the +/+ donor will produce increased numbers of colonies on the surface of recipient spleens. The action of thymosin was similar to that seen with whole thymocytes. The pleuripotential stem cell is usually not in cell cycle (G_0). It is intriguing to speculate that the stem cells incubated with thymosin *in vitro* or interacting with TSRC *in vivo* are somehow placed into cell cycle and produce larger numbers of CFUs in the deficient W/W^v environment. Thymosin has been shown to increase cyclic guanosine 5' monophosphate (cGMP) levels in murine thymocytes (14) and cGMP has been shown to increase the number of CFUs in the cell cycle (4). Frindel and Croizat (8) have observed that, in thymectomized C_3H mice, implantation of diffusion chambers containing thymocytes significantly increases the number of CFUs in the cell cycle compared with thymectomized animals receiving empty chambers ·in which no demonstrable CFUs in the cycle were observed. Thymic-dependent antigens have been used to increase the number of CFUs in the cycle, which suggests that a product of the T-lymphocytes might trigger the stem cells into entering the cycle (9).

It has been suggested that, in addition to the defective stem cell of W/W^v mice, these mice appear to demonstrate a maturational defect in the erythroid cell line (15). The effect of thymic hormone (thymosin) on the cure of W/W^v mice given TSRC-depleted marrow grafts might be directly on a committed erythroid precursor and be separate from the augmentation effect of the hormone on colony formation. The W/W^v mouse does not respond to exogenous erythropoietin (3,12), making it unlikely that the effect might be due to an erythropoietin-like substance as a contaminant of the thymosin. Lymphocytes have recently been shown to have both suppressor effects (7) and enhancing effects (20) on CFU-e production, suggesting alternate control mechanisms for erythroid differentiation. The mechanism of action of thymosin on erythroid proliferation is unknown and will be the subject of further study.

SUMMARY

The results presented in this communication suggest the following: (i) Incubation of W/W^v bone marrow with thymosin does not generate significant numbers of colonies in lethally irradiated recipients. (ii) Incubation of W/W^v bone marrow with thymosin fails to correct the anemia of W/W^v recipients of such treated marrow. (iii) Injection of thymosin into W/W^v recipients of anti-Thy 1.2 serum plus complement-treated bone marrow increases macroscopic colony formation and increases hematocrits of these recipients. (iv) Thymocytes in chamber also increases hematocrit values of anemic recipients when given with anti-Thy 1.2 serum plus complement-treated +/+ bone marrow grafts.

ACKNOWLEDGMENTS

We wish to thank Ms. D. Boulware and Ms. B. Hooke for excellent technical assistance. We also gratefully acknowledge the helpful editorial assistance of Ms. A. Fields.

This work was supported in part by the Naval Medical Research and Development Command, Work Unit No. MR000.01.01.1159. The opinions or assertions contained herein are the private ones of the authors and are not to be construed as official or reflecting the views of the U.S. Navy Department or the naval service at large. Support was also in part from Public Health Service research grant CA06973, the National Cancer Institute, National Institutes of Health, Bethesda, Maryland. The experiments reported herein were conducted according to the principles set forth in the "Guide for the Care and Use of Laboratory Animals," Institute of Laboratory Animal Resources, National Research Council, DHEW Pub. No. (NIH) 74-23.

REFERENCES

1. Axelrad, A.A., McLeod, D.L., Shreeve, M.M., and Heath, D.S. Properties of cells that produce erythrocytic colonies *in vitro*. In Robinson, W., ed., et al. *Hematopoiesis in Culture*. Washington, D.C.: U.S. Government Printing Office, 1974, p. 266.
2. Bach, J.F., Dardenne, M., Goldstein, A.L., Guha, A., and White, A. Appearance of T-cell markers in bone marrow rosette forming cells after incubation with thymosin, a thymic hormone. *Proc. Nat. Acad. Sci. (USA)*, 68:2734, 1971.
3. Bernstein, S.E., Russell, E.S., and Keighley, G. Two hereditary mouse anemias (Sl^d/Sl_d and W/W^v) deficient in response to erythropoietin. *Ann N.Y. Acad. Sci.*, 149:475, 1968.
4. Byron, J. Molecular basis for the triggering of hematopoietic stem cells into DNA synthesis. In Robinson, W., ed., *Hemopoiesis in Culture*. Washington, D.C.: U.S. Government Printing Office, 1974, p. 91.
5. Chervenick, P.A., and Boggs, D.R. Decreased neutrophils and megakaryocytes in anemic mice by donor cells. *Transplantation*, 22:42, 1969.
6. Ebbe, S., Phalen, E., and Stohlman, F. Abnormalities of megakarocytes in W/W^v mice. *Blood*, 42:857, 1969.
7. Hoffman, R., Zanjani, E.D., Viva, J., Zalusky, R., Lutton, J.D., and Wasserman, L.R. Diamond-Black-

fan syndrome. Lymphocytes mediated suppression of erythropoiesis. *Science, 193*:899, 1976.

8. Frindel, E., and Croizat, H. The relationship between CFU kinetics and the thymus. *Ann. N.Y. Acad. Sci., 249*:468, 1975.

9. Frindel, E., Leuchers, E., and Davies, A.J.S. Thymus dependency of bone marrow stem cell proliferation in response to certain antigens. *Exp. Hematol., 4*:275, 1976.

10. Harrison, D.E. Life sparing ability (in lethally irradiated mice) of W/W^v mouse marrow with no macroscopic colonies. *Rad. Res., 52*:553, 1972.

11. Jedrzejczak, W.W., Sharkis, S.J., Ahmed, A., Sell, K.W., and Santos, G.W. Theta-sensitive cell and erythropoiesis: Identification of a defect in W/W^v anemic mice. *Science, 196*:313, 1977.

12. Keighley, G.H., Lowy, P., Russell, E.S., and Thompson, M.W. Analysis of erythroid hemeostatic mechanisms in normal and genetically anemic mice. *Brit. J. Hematol., 12*:461, 1966.

13. McCulloch, E.A., Siminovitch, L., and Till, J.E. Spleen colony formation in anemic mice of genotype W/W^v. *Science, 144*:844, 1964.

14. Naylor, P.H., Sheppard, H., Thurman, G.B., and Goldstein, A.L. Increase in cyclic GMP induced in murine thymocytes by thymosin fraction 5. *Biochem. Biophys. Res. Commun., 73*:843, 1976.

15. Russell, E.S. Problems and potentialities in the study of genic action in the mouse. In Brudette, W.J., ed., *Methology in Mammalian Genetics*, p. 217. San Francisco: Holden Day, Inc., 1963.

16. Sensenbrenner, L.L., Steele, A.A., and Santos, G.W. Recovery of hematologic competence without engraftment following attempted bone marrow transplantation for aplastic anemia. Report of a case with diffusion chamber studies. *Exp. Hematol., 5*:51, 1977.

17. Sharkis, S.J., Jedrzejczak, W.W., Ahmed, A., McKee, A., Santos, G.W., and Sell, K.W. Characterization of the theta-sensitive lymphoid cell responsible for restoration of normal erythropoiesis in the W/W^v anemic mouse. *Fed. Proc., 36*:1088, 1977.

18. Sharkis, S.J., Jedrzejczak, W.W., Ahmed, A., Santos, G.W., McKee, A., and Sell, K.W. Theta sensitive regulatory cell (TRSC) and hematopoiesis regulation of the differentiation of transplanted stem cells in W/W^v and normal mice. Submitted for publication, 1977.

19. Stephenson, J.R., Axelrad, A.A., McLeod, D.L., and Shreeve, M.M. Induction of colonies of hemoglobin-synthesizing cells by erythropoietin in·vitro. *Proc. Nat. Acad. Sci. (USA), 68*:1542, 1971.

20. Torok-Storb, B., Storb, R., and Weiden, P. The effects of lymphocytes from non-transfused dogs on the growth of erythroid colonies (EC) from DLA identical littermates. *Exp. Hematol. (Suppl.), 5*:97, 1977.

4

Anti-CFU-s Activity of Rabbit Anti-mouse Brain Serum: Mechanism of Action

Solomon S. Adler,
Richard D. Kuznetsky,
and Frank E. Trobaugh, Jr.

INTRODUCTION

In 1972, Golub (14) reported and others have subsequently confirmed (12,20,22,24) that antiserum raised against mouse brain tissue when incubated with mouse marrow cells prevents the pluripotent stem cells (CFU-s) from forming surface spleen nodules in lethally irradiated mice. The anti-CFU-s activity is not complement dependent (12,20,28). Van den Engh and Golub (28) have shown that anti-mouse brain serum does not interfere with the formation of granuloid/macrophage colony-forming units (CFU-c) *in vitro*. We have confirmed this and also found that rabbit anti-mouse brain serum (RAMBS) impairs the colony-forming capacity of spleen cells to the same extent as that of marrow cells (unpublished) and cannot inhibit the growth of erythroid colony-forming unit (CFU-e) cells *in vitro* (1).

In this review, we will define the basic characteristics of our RAMBS. We will also describe how the splenic colony-forming capacity of RAMBS-treated marrow cells can be rescued. The latter finding may provide insight into the mechanism of action of RAMBS.

MATERIALS AND METHODS

Animals

Female CAF$_1$ (Balb/c × A/He) mice were purchased from Cumberland View Farms, Clinton, Tenn. The mice used for experimental purposes were 12 to 20 weeks old. For the chromosomal analysis study, we used 11- to 23-week-old CBA/H-T6T6 female donor mice and 11- to 20-week-old CBA/Ca female recipient mice. The donors and some of the recipients were bred in our institution from stock originally obtained from Jackson laboratories (Bar Harbor, Me.), whereas the remaining recipients were purchased from Jackson Laboratories. The mice were housed 10 or less per cage and received *ad libitum* Wayne Lab-Blox® feed and tapwater adjusted to pH 3.2 to prevent early irradiation deaths.

New Zealand white female rabbits (Thompson Research Foundation, Monee, Ill.) were housed separately and received *ad libitum* Wayne Rabbit Ration® feed and tapwater.

Irradiation

Total body irradiation was delivered with a ^{137}Cs Gammacell 40® irradiation unit (Atomic Energy of Canada Ltd., Ottawa, Canada) at a rate of 127.2 rad/min. The CFU-s assay mice received 750 or 800 rad whereas the carrageenan-treated mice for RAMBS-r-treated CFU-s rescue experiments received 900 or 950 rad.

23

Preparation of Sera

Rabbit anti-mouse brain serum was produced by a method modified from Golub (13). Here, 1 ml of a suspension containing a minced CAF_1 mouse brain in phosphate-buffered saline was added to 1 ml of Freund's complete adjuvant, mixed well, and injected intramuscularly, in two divided doses, into a rabbit on days 0, 7, and 61. On day 68, the serum was harvested, inactivated at 56°C for 30 min, and stored at −20°C until use.

Control rabbit serum (CRS) was obtained from either untreated or Freund's adjuvant-treated rabbits and heat-inactivated at 56°C for 30 to 45 min. Since no difference was noted between the two sources of CRS, the type will not be specified. Sera were diluted with Hanks' balanced salt solution (HBSS) and sterilized by filtration thru 0.45 μm Millipore membranes (Millipore Corp., Bedford, Mass.).

Preparation of IgG Fraction from Sera

The IgG fraction was separated from serum by DEAE column chromatography using Whatman® DE52 DEAE advanced ion exchange cellulose and 0.02 M phosphate buffer at pH 7.5. Samples were dialyzed overnight against HBSS at 4°C. The IgG specificity was confirmed by immunoelectrophoresis against goat anti-rabbit whole serum. The sera were stored at −20°C until use.

Complement

Guinea pig (Microbiological Associates, Bethesda, Maryland) and rabbit (GIBCO, Grand Island, New York) complement solutions were absorbed with Difco Noble agar (10), using 100 mg of agar/ml of complement. The complement was stored at −70°C. Guinea pig complement was diluted 1:4 and rabbit complement 1:10 with HBSS prior to use.

Preparation of Tissues for Cytotoxicity Studies, Absorption Procedures, and Hemagglutination Assays

To prepare cell suspensions for the cytotoxicity studies, HBSS containing 2 percent heat-inactivated fetal calf serum (HBSS-FCS) was used throughout. Thymic, splenic, or lymph node tissue was mashed through a series of three stainless steel wire meshes, the finest of which had 200 openings/linear inch, to obtain a suspension of single cells. Femoral marrow cells were harvested and suspended by repeated aspiration and flushing through a 22-gauge needle inserted into the marrow cavity. Each cell suspension was washed at least twice by centrifuging at 1,000 rpm for 10 min at 4°C.

To prepare thymic or splenic tissue for absorption, the mashed tissues were suspended in FCS-free HBSS and washed three times. Brain tissue (0.4- to 0.5-g samples) used for absorption was minced with forceps, suspended in cold HBSS, mixed either by agitation or by repeated aspiration through a Pasteur pipette, and washed at least twice with HBSS by centrifuging at 3,000 rpm for 10 min at 4°C.

Packed red blood cells (RBC) used for absorption procedures and hemagglutination assays were washed four times with 0.85 percent saline; the buffy coat was removed after each wash. A 2 percent RBC suspension in 0.85 percent saline was prepared for the hemagglutination assays.

Hemagglutination Assay

Aliquots of 0.1 ml of serially diluted (0.85 percent saline) serum were mixed in 10×75 mm glass tubes with 0.05 ml of the RBC suspension, incubated at 37°C for 15 min, and centrifuged in an Adams® Serofuge for 1.5 min. Hemagglutination titers are reported as the reciprocals of the highest dilutions yielding macroscopic agglutination.

Cytotoxicity Assay

Aliquots of 0.1 ml of serially diluted HBSS-FCS serum were mixed in 10×75 mm plastic or glass tubes with 0.1 ml of the test cell suspension (5.0×10^6 cells/ml) and incubated for 30 min in an ice bath, with gentle mixing at 15 and 30 min. Diluted guinea pig complement (0.1 ml) was added to each tube. After mixing, the tubes were incubated at 37°C for 30 min, mixed at 15 min, and then refrigerated at 4°C until use (30). Aliquots of 0.1 ml of a freshly prepared 0.16 percent Trypan Blue solution in isotonic saline were added to each tube about 6 min prior to reading (5). From each test cell suspension, 200 cells were evaluated under $400 \times$ magnification. The following cells were scored as dead: cells unstained but badly deformed, cells with only the nucleus stained, cells with both the cytoplasm and nucleus stained, and cell ghosts. Cells in large clumps were not counted. The cytotoxicity index (CI) at a particular dilution was calculated using the following equation:

$$CI = \frac{\% \text{ dead with RAMBS} - \% \text{ dead with CRS}}{100 - \% \text{ dead with CRS}} \times 100$$

A positive CI was at least 15 percent (18). Cytotoxicity titers are reported as the reciprocals of the highest final dilutions resulting in a positive CI.

Absorption Procedures

Red cell-absorbed CRS (CRS-r), which was used in some experiments, was prepared by absorption of CRS with two separate samples of CAF_1 mouse packed RBC in a CRS:RBC volume ratio of 4:1.

The initial absorption was at 37°C for 40 min, and the second at 4°C for 45 min. The hemagglutination titer after absorption was always less than 1:8. There was no inhibition of CFU-s by either unabsorbed CRS or CRS-r.

Red cell-absorbed antiserum (RAMBS-r) was prepared by absorption with three separate samples of packed RBC in a RAMBS:RBC volume ratio of 2.7:1. The initial absorption was at 37°C for 40 min and the subsequent two at 4°C for 60 min. After the third absorption, the hemagglutination activity was usually absent or only weakly positive at 1:8. If necessary, further absorptions were performed at room temperature for 30 min in approximately the same volume ratio until hemagglutination no longer occurred.

Spleen tissue-absorbed antiserum was prepared by absorbing a 1.0 ml sample of RAMBS (diluted 1:16 with HBSS) three times with 20×10^6 splenocytes at 4°C for 45 min with continuous mixing on an Ames® aliquot mixer. The CI against splenocytes was less than 15 percent at a final dilution of 1:48.

Thymus-absorbed antiserum (RAMBS-RT) was prepared as follows from RAMBS-r. Thymocytes from at least 15 mouse thymus glands were divided into three or four aliquots, and 1.0 ml of RAMBS-r was absorbed with each, at 0 or 4°C for 30 to 45 min. Because of the expense involved, no further absorption was performed if the CI was less than 50 percent against thymocytes at a final dilution of 1:48.

Brain-absorbed RAMBS was absorbed three times with 0.5-g brain tissue samples at 0°C for 60 min in a volume (ml) to mass (g) ratio of 1:1 and twice with 0.4-g brain tissue samples at 4°C for 40 min in a volume (ml) to mass (g) ratio of 0.75:1.

Assay for Anti-CFU-s Activity of RAMBS

Marrow cells from the femurs of at least three mice were suspended in HBSS and adjusted to 10×10^6 cells/ml. One volume of this suspension was incubated with one volume of diluted serum (RAMBS or CRS) at 0°C for 30 min, with gentle mixing at 15 and 30 min. In the studies utilizing complement, four volumes of diluted complement were added after the initial incubation. This mixture was incubated at 37°C for 60 min, with gentle mixing at 20 and 40 min. Both incubations were performed in those experiments where complement was not used. Cell suspensions were adjusted to 1×10^5 nucleated cells/ml, and 0.5-ml samples were injected into the tail veins of mice within 3 hr after lethal total body irradiation. The mice were sacrificed on day 8 or 9, their spleens fixed in Bouin's solution, and surface colonies were counted under 2.5× magnification.

Carrageenan- and Heparin-rescue Studies

We first investigated an uncharacterized potassium carrageenan (lot number 71C-90119; Sigma Chemical Co., St. Louis, Mo.); however, because it was highly toxic for mice when combined with irradiation, we also investigated iota carrageenan (courtesy of Marine Colloids, Park Ridge, Ill.). Poly-2-vinylpyridine-N-oxide (PVPNO) was purchased from Polysciences (lot number 275-125; Washington, Pa.). These compounds were assayed for endotoxin by *Limulus* lysate gelation, which can be used to detect endotoxin at a concentration of 0.5 μg/ml. Iota carrageenan and PVPNO were negative; potassium carrageenan, because of its very viscous nature in solution, interfered with the interpretation of the test. For the rescue experiments, potassium carrageenan was suspended in a concentration of 1 to 2 mg/ml and iota carrageenan at 0.5 mg/ml in 0.9 percent preservative-free, sterile, nonpyrogenic saline and dissolved in a hot water bath. The PVPNO was dissolved in a concentration of 3 to 6 mg/ml in HBSS.

Assay mice were injected intravenously with 1 to 1.2 mg of potassium carrageenan or 0.25 mg of iota carrageenan one day before they were lethally irradiated; they were injected with 5×10^4 RAMBS-r ($\pm$C)- or CRS-r ($\pm$C)-treated nucleated marrow cells within 3 hr after irradiation. Surface spleen colonies were counted 8 days later. In PVPNO studies, 3 or 6 mg of PVPNO (0.5 ml, s.c., in each of two dorsal sites) was administered 24 hr prior to injection of carrageenan.

It is important to note several additional points with respect to the carrageenan-rescue studies. The dosage combination of carrageenan and irradiation was found to be critical and probably mouse-strain dependent. In some potassium carrageenan-treated groups, an increase in irradiation exposure from 900 to 950 rad produced a marked increase in mortality during the first 4 days after irradiation, and for unknown reasons, mice transplanted with marrow cells (either CRS- or RAMBS-treated) were more susceptible to this than were lethally irradiated mice treated with carrageenan but with no marrow cell transplants. Preliminary studies to define the optimal dose combination of potassium carrageenan and irradiation in CAF$_1$ mice required about 500 mice and many separate protocols, making these studies costly and time-consuming; therefore, we did not attempt to define the optimal conditions for the CBA/Ca mice used for the chromosomal studies, which were performed only to confirm rescue of the RAMBS-r–treated donor cells. As will be noted in the section on results, the percent rescue in these studies was less than that in the basic rescue studies. This was probably because the conditions studied were not optimal, and the dose of iota carrageenan used was about one-fourth that of the potassium compound used in the basic studies. The smaller dose was necessitated by the potent stimulation of endogenous colony formation by iota carrageenan. However, iota carrageenan had the advantages of being endotoxin-free and less toxic to mice.

For the study designed to evaluate the capacity of heparin to rescue RAMBS-r–treated marrow cells, mice were injected intravenously with 65 units heparin (Panheprin®; Abbot Laboratories, North Chicago, Ill.) in 0.5 ml HBSS just prior to cell injection.

Statistical Methods

The data are presented as means ± one standard error of the mean (SEM). For evaluation of the percentage of colonies rescued by carrageenan, the mean number of endogenous surface spleen colonies in mice treated with carrageenan was subtracted from the mean number of colonies in the experimental groups.

The unpaired Student's t test was used to determine whether statistically significant differences ($p < 0.05$) existed between groups.

RESULTS

General Characteristics of RAMBS

Unabsorbed RAMBS had a hemagglutination titer against CAF_1 mouse red blood cells of 512. The cytotoxic effects of RAMBS against mouse tissues, the anti-CFU-s effects of variously absorbed RAMBS, and the anti-CFU-s potency of diluted RAMBS-r are shown in Table 4-1, Table 4-2, and Figure 4-1, respectively. The IgG fraction of RAMBS-r has substantial anti-CFU-s activity (Fig-

TABLE 4-1 Cytotoxic Effect of RAMBS on CAF_1 Tissues

RAMBS ABSORPTION:	CYTOTOXICITY TITER[a]			
	Thymus	Spleen	Marrow	Lymph Node
—	192	96	192	192
RBC	96	48	12	ND[b]
RBC + thymus	<12	<24	<12	<24
Brain	<24	<24	<12	ND[b]

[a]The cytotoxicity titer of RAMBS against a cell suspension is reported as the reciprocal of the highest final dilution resulting in a cytotoxicity index ≥ 15 percent.
[b]ND, not done.

ure 4-2). We did not evaluate other immunoglobulin fractions for anti-CFU-s activity.

Rescue of CFU-s in RAMBS-treated Marrow Cells

Experiment 1 In lethally irradiated mice, RAMBS-r–treated marrow cells yielded 11.7 percent as many surface spleen colonies as did CRS-r–treated marrow cells; however, in host mice injected i.v. with 1 to 1.2 mg potassium carrageenan 24 hr before irradiation and transplantation, RAMBS-r–treated marrow cells yielded 72 percent as many surface spleen colonies as the control (Figure 4-3). Carrageenan treatment had no effect on the numbers of surface spleen colonies formed from CRS-r–treated marrow cells (Figure 4-3). The percentage of colonies rescued (percent rescue) by carrageenan (car) was 68.4 percent and was calculated using the following equation:

$$\text{Percent rescue} = \frac{(\text{RAMBS-r}_{car} - \text{CRS-r-I}_{car}) - (\text{RAMBS-r})}{(\text{CRS-r}_{car} - \text{CRS-r-I}_{car}) - (\text{RAMBS-r})} \times 100$$

TABLE 4-2 Number of Surface Spleen Colonies Formed from 5×10^4 Marrow Cells Treated with RAMBS or Tissue-Absorbed RAMBS

EXPERIMENT	SERUM[a]	TISSUE(S) USED FOR ABSORPTION	NUMBER OF ASSAY MICE	MEAN NUMBER OF SPLEEN COLONIES/ 5×10^4 NUCLEATED CELLS ± SEM	PERCENT OF CONTROL	p VALUE[b]	NUMBER OF REPLICATIONS
1	CRS	—	19	13.9±0.8	—	—	2
	RAMBS	—	19	0.5±0.2	3.4	<0.001	
	RAMBS	RBC	20	1.7±0.3	11.8	<0.001	
2	CRS	RBC	10	15.8±1.1	—	—	1
	RAMBS	RBC	11	2.8±0.6	17.8	<0.001	
	RAMBS	RBC+thymus	10	1.2±0.2	7.6	<0.001	
3	CRS	—	8	15.0±0.8	—	—	1
	RAMBS	—	8	0.3±0.2	1.7	<0.001	
	RAMBS	Spleen	8	1.5±0.3	10.0	<0.001	
4	CRS	—	10	26.9±1.8	—	—	1
	RAMBS	Brain	11	27.7±1.5	103.1	>0.500	

[a]Unabsorbed CRS was diluted 1:8 or 1:16, RBC-absorbed CRS 1:4, and RAMBS 1:8 or 1:16 prior to incubation with marrow cells.
[b]The p values for experiments represent the comparison between unabsorbed or absorbed RAMBS and the CRS control.

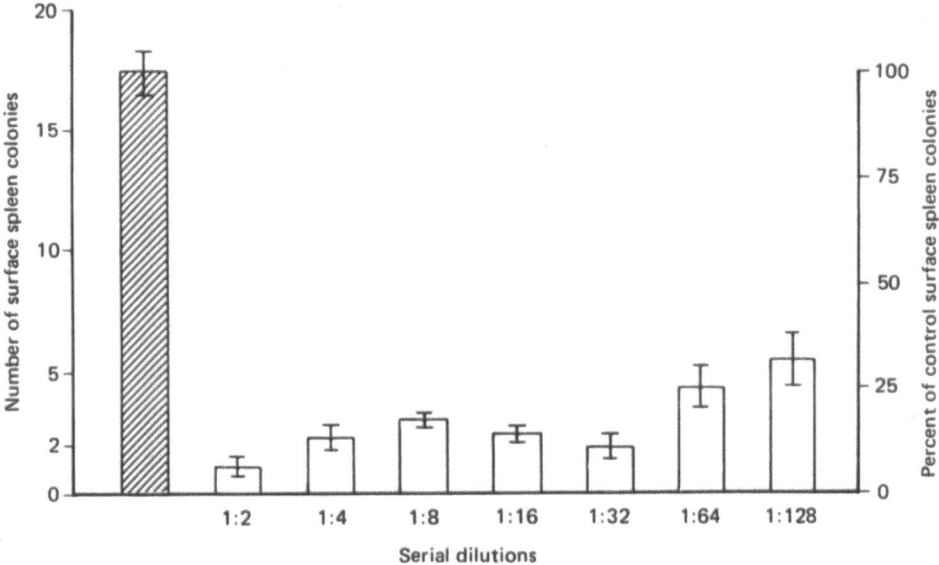

FIGURE 4-1. Number of surface spleen colonies formed from 5×10^4 marrow cells treated with serial dilutions of RAMBS-r in the presence of complement. The data are presented as means ± standard error (SEM). There were nine to 10 assay mice in each RAMBS-r group. Dilutions of RAMBS-r (open histograms) during the primary incubation, i.e., prior to the addition of complement, are indicated. The number of spleen colonies formed after treatment with CRS-r (cross-hatched histogram) is the mean of three separate assays (nine to 10 mice/assay) using dilutions at 1:2, 1:4, and 1:8 during the primary incubation. The numbers of spleen colonies for these three individual determinations were 14.7 ± 1.4, 18.7 ± 1.3, and 18.7 ± 1.7, respectively. For all dilutions of RAMBS-r, $p < 0.001$.

where RAMBS-r_{car} denotes the number of colonies (18.6 ± 1.4) formed in potassium carrageenan-treated hosts transplanted with 5×10^4 RAMBS-r–treated nucleated marrow cells; CRS-r-I_{car}, the number of colonies (1.1 ± 0.3) in potassium carrageenan-treated hosts transplanted with 5×10^4 CRS-r–treated and irradiated (I) nucleated marrow cells (i.e., the control for endogenous colony stimulation); RAMBS-r, the number of colonies (2.8 ± 0.4) in hosts transplanted with 5×10^4 RAMBS-r–treated nucleated marrow cells; and CRS-r_{car}, the number of colonies (25.4 ± 1.2) in carrageenan-treated hosts transplanted with 5×10^4 CRS-r–treated nucleated marrow cells.

Treatment of mice with PVPNO 24 hr before potassium carrageenan injection and 48 hr before lethal irradiation and transplantation of 5×10^4 RAMBS-r–treated nucleated marrow cells decreased the number of colonies from 17.5 ± 1.3 to 12.9 ± 0.6 ($0.005 > p > 0.001$). The percent rescue effected by carrageenan after treatment of the host mice with PVPNO(P) was 46.0, calculated from the following equation:

Percent rescue =

$$\frac{(\text{RAMBS-}r_{pcar} - \text{CRS-r-}I_{pcar}) - (\text{RAMBS-}r_p)}{(\text{CRS-}r_{pcar}) - \text{CRS-r-}I_{pcar} - (\text{RAMBS-}r_p)} \times 100$$

where RAMBS-r_{pcar} denotes the number of colonies (12.9 ± 0.6) in PVPNO- and carrageenan-treated hosts transplanted with 5×10^4 RAMBS-r–treated nucleated marrow cells; CRS-r-I_{pcar}, the number of colonies (0.2 ± 0.2) in PVPNO- and car-

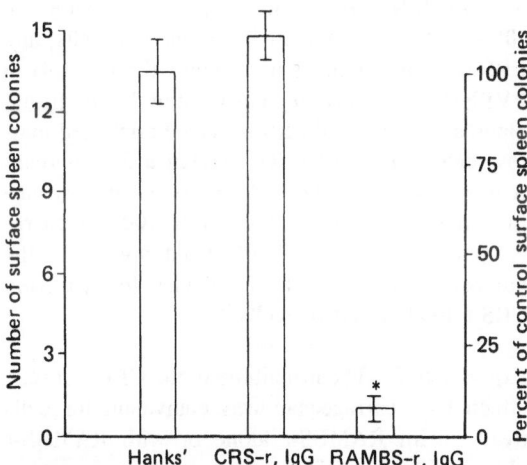

FIGURE 4-2. Number of surface spleen colonies formed from 5×10^4 marrow cells treated with the IgG fraction of CRS-r or RAMBS-r in the absence of complement. There were 10 assay mice in each group. The IgG fractions of CRS-r and RAMBS-r were diluted 1:4 and 1:8 in HBSS, respectively, during the primary incubation. Mean ± SEM; asterisk, $p < 0.001$.

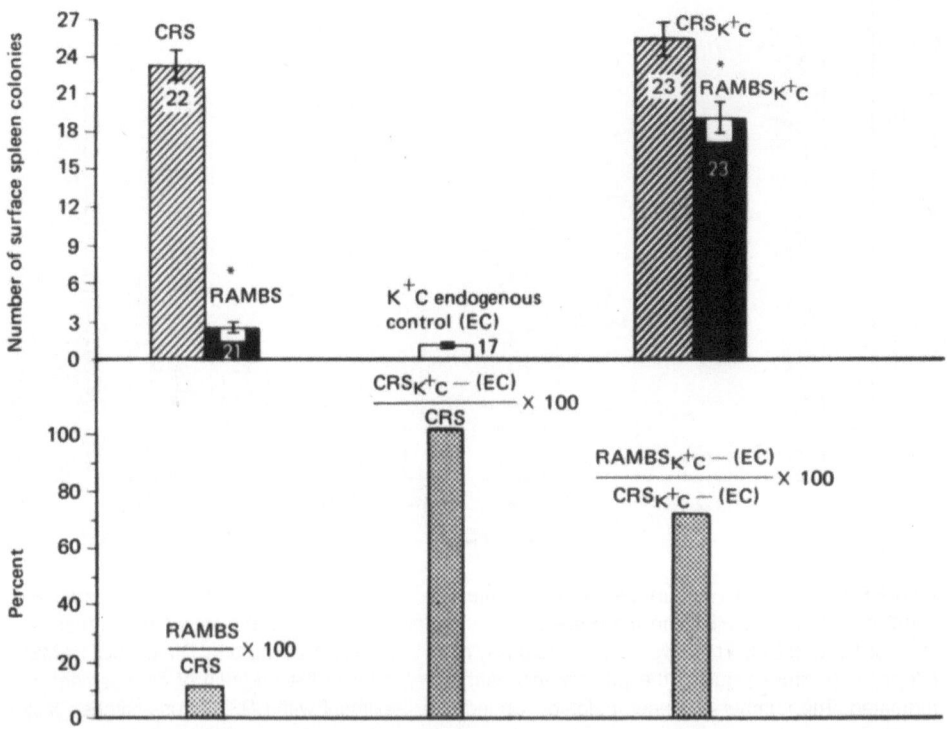

FIGURE 4-3. Effect of potassium carrageenan (K⁺ C) treatment of CAF₁ mice on the number of surface spleen colonies formed by 5×10^4 marrow cells treated with CRS-r or RAMBS-r. The data are pooled from two replications. The endogenous control represents the number of spleen colonies (1.1 ± 0.3) formed in carrageenan-treated hosts transplanted with 5×10^4 CRS-r–treated and irradiated nucleated marrow cells. Mean $\pm$ SEM; asterisk, $p < 0.001$.

rageenan-treated hosts transplanted with 5×10^4 CRS-r–treated and irradiated (I) nucleated marrow cells (i.e., the control for endogenous colony stimulation); RAMBS-r_p, the number of colonies (2.3 ± 0.3) in PVPNO-treated hosts transplanted with 5×10^4 RAMBS-r–treated nucleated marrow cells; and CRS-r_{pcar}, the number of colonies (25.1 ± 1.4) in PVPNO- and carrageenan-treated hosts transplanted with 5×10^4 CRS-r–treated nucleated marrow cells. Thus, PVPNO effected a 32.7 percent reduction in the number of CFU-s rescued by carrageenan $[(68.4 - 46.0)/68.4] \times 100$. In carrageenan-treated hosts, PVPNO did not affect the number of surface spleen colonies formed from CRS-r–treated marrow cells.

Experiment 2 The magnitude of the CFU-s rescue effected by carrageenan was equivalent for cells treated with RAMBS-r alone or with RAMBS-r plus complement (Figure 4-4).

Experiment 3 In both of the above experiments, we found that carrageenan yielded only 1 to 1.5 endogenous surface spleen colonies per mouse (Figure 4-3 and 4-4). To exclude definitively the possibility that we were only inducing endogenous

colony formation with the combination of carrageenan treatment and RAMBS-r–treated cell injection, rather than actually rescuing the RAMBS-r–treated cells, we designed a study in which we identified the origin of the surface spleen colonies cytogenetically. Donor marrow cells bearing T6T6 chromsomes were treated with RAMBS-r ($\pm$ complement) and injected into lethally irradiated CBA/Ca hosts injected (i.v.) with 0.25 mg of iota carrageenan before irradiation. The chromosomal analyses convincingly demonstrated that the rescued colonies were of exogenous origin (Table 4-3). Note that in these studies a larger number of CFU-s escaped RAMBS-r inactivation than in our other studies; this most likely resulted from our crossing a mouse strain barrier, since our RAMBS was raised against CAF₁ mouse brain but used here to inactivate cells from CBA/H-T6T6 mice. In these studies also, the percent rescue with iota carrageenan was less than that in the basic studies using potassium carrageenan; the reasons for this are discussed in the section on materials and methods.

Experiment 4 Animals injected with 65 units of heparin (i.v.) just prior to the injection of 5×10^4 RAMBS-r–treated marrow cells had only 1.7 ± 0.4

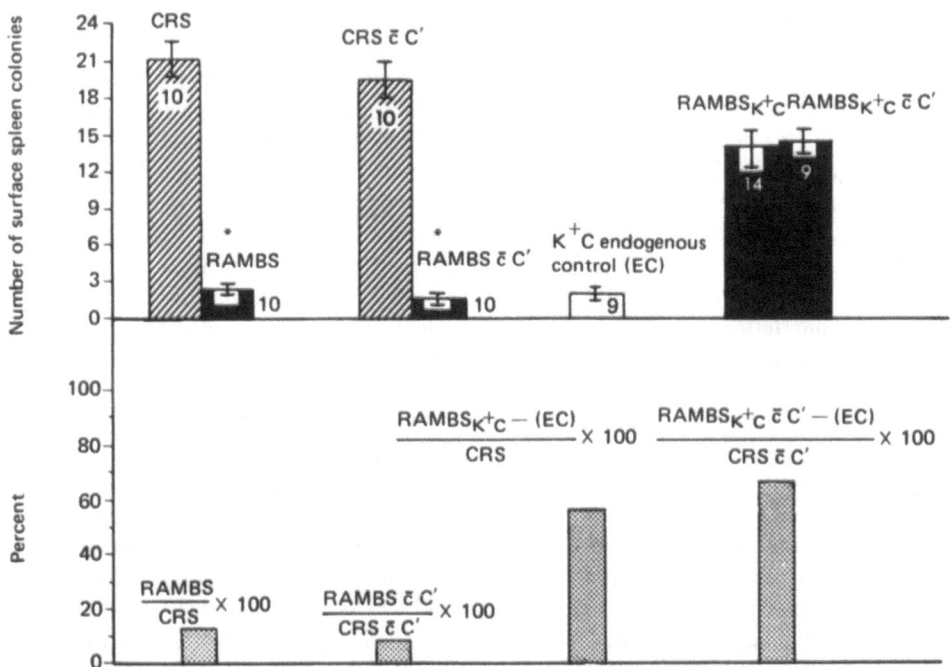

FIGURE 4-4. Effect of potassium carrageenan (K⁺ C) treatment of CAF₁ mice on the number of surface spleen colonies formed by 5 × 10⁴ marrow cells treated with CRS-r (with or without complement) or RAMBS-r (with or without complement). C′, complement. The endogenous control represents the number of spleen colonies (1.9 ± 0.6) formed in carrageenan-treated hosts. Mean ± SEM; asterisk, $p < 0.001$.

surface spleen colonies, whereas heparin-treated mice injected with CRS-r–treated marrow cells had 19.9 ± 1.0 colonies. Thus, rescue was not effected by heparin.

DISCUSSION

Our RAMBS was raised in a fashion adopted from that of Golub (13). Its hemagglutination and cytotoxicity characteristics and anti-CFU-s activity after appropriate absorption are similar to those reported by other investigators (12,13,14,15,24,28); thus, we can assume that we are dealing with a very similar antiserum.

In these studies, we have shown that carrageenan treatment of mice can rescue the spleen colony-forming capacity of RAMBS-r–treated CFU-s. By exploiting this observation and the observations of others on the activities of the carrageenans, we may be able to shed some light on the mechanism of RAMBS anti-CFU-s activity. Carrageenans are

TABLE 4-3 Effect of Iota Carrageenan on the Number of Surface Spleen Colonies and the Origin of the Colonies Formed by 5 × 10⁴ Marrow Cells Treated with RAMBS-r with or without Complement

TREATMENT		NUMBER OF SURFACE SPLEEN COLONIES	CHROMOSOMAL ANALYSES				
T6/T6 donor cell treatment	CBA/Ca host mice treatment	Mean ± SEM	Number of mice studied	Number of colonies studied	Number of cells studied/ colony	Number of T6/T6 donor cell colonies	Number of CBA/Ca host colonies
1. RAMBS	Saline	3.9 ± 0.7 (N = 9)	6	15	20	15	0
2. RAMBS	Iota carrageenan	7.8 ± 0.8 (N = 8)	5	17	10–30	16	1
3. RAMBS + complement	Iota carrageenan	7.3 ± 0.9 (N = 6)	3	17	10–30	17	0

high molecular weight sulfated polygalactoses obtained from red seaweed. They have a multiplicity of actions (2-4,6-9,11,16,17,19,21,25-27,29), which can be attributed to interference with one or more of the following three physiological systems: (i) complement (4,11,29); (ii) coagulation (16,17,26); and (iii) mononuclear phagocyte system (2,8,9,19,25). It is likely that the carrageenan rescue RAMBS-r–treated cells by their inhibition of one or more of these systems.

If carrageenans effected rescue of RAMBS-r–treated CFU-s by interfering with the activation of the mouse complement system, we would not expect carrageenans to rescue cells treated *in vitro* with RAMBS-r plus rabbit complement; however, we were able to rescue such CFU-s, making this alternative less attractive.

If RAMBS-r–coated CFU-s initiated intravascular coagulation and subsequent fibrin clot deposition, these opsonized cells could be trapped in organs with large vascular beds, such as the liver and the lungs. One might then argue that carrageenan, by inhibiting the coagulation process, permits the cells to circulate and seed the spleen. The failure of heparin, a much more potent anticoagulant than carrageenan (16,17), to rescue RAMBS-r–coated CFU-s mitigates against this alternative.

Thus, the third alternative (i.e., interference with the mononuclear phagocyte system) is the most attractive explanation for the capacity of carrageenans to effect the rescue of RAMBS-r–treated CFU-s. To support further this contention we found that PVPNO, a macrophage-stabilizing agent (2,23), reduced by about one-third the number of RAMBS-r–treated CFU-s rescued by carrageenan.

If we knew which macrophages (extrasplenic, e.g., from the liver and lungs, or splenic) were inactivated or killed by carrageenan, we could infer which macrophages are responsible for the inhibition of spleen colony formation by RAMBS-r–treated CFU-s. Theoretically, extrasplenic macrophages could act by preventing splenic seeding of RAMBS-r–opsonized CFU-s; splenic macrophages might kill opsonized CFU-s directly or impair their interaction in the splenic microenvironment. In preliminary studies, we found no decrease in the number of mouse spleen cells capable of ingesting latex particles 24 hr after treatment with 1 mg potassium carrageenan (i.v.) and 2 hr after lethal total body irradiation (unpublished); we did not study latex ingestion by liver or lung macrophages. Sawicki and Catanzaro (25) injected calcium carrageenan (i.p.) into guinea pigs and demonstrated by histochemical staining that Kupffer cells take up more carrageenan than splenic, marrow, and lymph node macrophages; lung macrophages were not studied. Moreover, cytotoxic histological changes

were evident only in the Kupffer cells. Their study suggests that Kupffer cells are more susceptible than splenic macrophages because more of the compound reaches the liver, an organ with a richer blood supply, than the spleen. In addition, the liver may clear much of the circulating carrageenan from the blood and thus prevent sufficient amounts from reaching other organs. Although we injected carrageenan i.v., it is likely that carrageenan is similarly distributed in our mice. If this were so, then we might speculate that it is primarily the extrasplenic macrophages (Kupffer cells and possibly lung macrophages) that are responsible for the inactivation of RAMBS-r–treated CFU-s.

In conclusion, these studies indicate that the anti-CFU-s activity of RAMBS-r is mediated *neither* by direct cytotoxicity, since carrageenan treatment of host mice effected rescue of CFU-s treated *in vitro* with RAMBS-r plus complement, *nor* by activation of the coagulation cascade by the opsonized CFU-s, since treatment of host mice with carrageenan effected the rescue of CFU-s treated with RAMBS-r, whereas heparin did not. Rather, it is likely that RAMBS-r–treated CFU-s are prevented from forming spleen colonies by an *in vivo* macrophage-mediated process, perhaps by inhibition of splenic seeding.

SUMMARY

Antiserum raised in rabbits against mouse brain tissue (RAMBS) and incubated with mouse hemopoietic cells markedly interferes with their ability to form surface spleen colonies upon injection into lethally irradiated mice. Granuloid (CFU-c) or erythroid (CFU-e) progenitor cell growth *in vitro* was not inhibited by RAMBS. In addition, the activity of RAMBS is not dependent on *in vitro* incubation with complement.

The hemagglutination, cytotoxicity, anti-CFU-s, and absorption characteristics of the RAMBS used in these studies are similar to those of anti-mouse brain sera used by other investigators. To elucidate the mechanism of action of RAMBS, we performed studies designed to alter the fate of RAMBS-treated CFU-s. Here, CAF_1 mouse marrow cells were incubated with red cell-absorbed RAMBS (RAMBS-r) at 0°C for 30 min, then with or without complement at 37°C for 60 min; 5×10^4 treated cells were injected (i.v.) into lethally irradiated mice (900 or 950 rad ^{137}Cs), and macroscopic spleen colonies were counted on day 8 or day 9. In regular assay, mice RAMBS-r–treated cells yielded about 12 percent of the number of colonies formed by cells treated with control rabbit serum (CRS), whereas in assay, mice injected with 1 to 1.2 mg potassium carrageenan (i.v.) 24 hr prior to lethal irradiation the number of colonies resulting from RAMBS-r–treated cells increased to 72 percent ($p < 0.001$) of control. Potassium carrageenan treatment even increased the number of surface spleen colonies formed by marrow cells treated with RAMBS-r plus complement from 8 to 65 percent of control ($p < 0.001$), suggesting that RAMBS-r is not cytotoxic to CFU-s even in the presence of complement. Potassium carrageenan treatment alone or in combination with irradiated cell injection yielded only one endogenous colony, on average, and did not affect the number of colonies formed by CRS-r–treated marrow cells. To demonstrate definitively that carrageenan rescues RAMBS-r–treated CFU-s, rather than merely stimulating endogenous colony formation, donor CBA/H-T6T6 marrow cells bearing marker chromosomes were treated with RAMBS-r with and without complement and transplanted into syngeneic CBA/Ca host mice that had been treated with iota carrageenan. The chromosomal analyses of cells in the spleen colonies demonstrated their exogenous origin, and hence, rescue.

Heparin, 65 U (i.v.), given just prior to the injection of RAMBS-r–treated cells did not effect rescue, implying that it is not the anticoagulant action of carrageenan that effects rescue. To define further the mechanism by which carrageenan effects rescue and thereby the mechanism by which RAMBS-r treatment inactivates CFU-s, we attempted to block carrageenan rescue by treating mice (prior to the administration of carrageenan) with poly-2-vinylpyridine-N-oxide, a macrophage-stabilizing agent. This treatment decreased the number of colonies rescued by carrageenan by about one-third ($0.005 > p > 0.001$), supporting the thesis that it is the effect of carrageenan on macrophages that effects rescue.

ACKNOWLEDGMENTS

We are grateful to Art Conti, Max Dansbie, Salah Husseini, and Dan May for their skillful technical help; to Linda Bielitzki and Michele Kuznetsky for their excellent secretarial help; to Marilyn Uhl for her excellent art work; to Alice Gillman-Sacks, Dr. Sheldon Dray's Laboratory, Department of Microbiology, University of Illinois Medical Center, Chicago, Ill., for supplying the goat anti-rabbit whole serum; to Scott E. Rangus, Marine Colloids, Inc., for supplying iota carrageenan; and to Dr. John Robinson at Loyola University Medical Center, Maywood, Ill., for performing the endotoxin assays.

This work was supported in part by research grant #CA04144 from the National Cancer Institute, Bethesda, Maryland; by training grant # 5 T01 HL06021 from the National Heart, Lung and Blood Institute, Bethesda, Maryland; and by research grants from the Leukemia Research Foundation, Chicago, Illinois.

Dr. Adler was the recipient of an NIH research fellowship, #6F22-CA9452, National Cancer Institute, Bethesda, Maryland when this work was performed.

REFERENCES

1. Adler, S.S., Kuznetsky, R., and Trobaugh, F.E., Jr. Additional characteristics of rabbit anti-mouse brain antiserum (Abst.) *Am. Soc. Hemat.*, 19th Annual Meeting, 1976.
2. Allison, A.C., Harington, J.S., and Birbeck, M. An examination of the cytotoxic effects of silica on macrophages. *J. Exp. Med. 124*:141, 1966.
3. Bice, D., Schwartz, H.J., Lake, W.W., and Salvaggio, J. The effect of carrageenan on the establishment of delayed hypersensitivity. *Int. Arch. Allergy Appl. Immunol. 41*:628, 1971.
4. Borsos, T., Herbert, J.R., and Crisler, C. The interaction between carrageenan and the first component of complement. *J. Immunol, 94*:662, 1965.
5. Boyse, E.A., Old, L.J., and Chouroulinkow, I. Cytotoxic test for demonstration of mouse antibody. *Methods. Med. Res., 10*:39, 1964.
6. Calne, R.Y., Wall, W.J.P., and Wilkins, D.C. Inhibition of rejection of canine renal allografts by treatment with sulphated polysaccharides, Promethazine Hydrochloride and Azathioprine. *IRCS Med. Sci., 3*:556, 1975.
7. Calne, R.Y., Wall, W.J.P., and Wilkins, D.C. The individual and combined roles of carrageenan, promethazine hydrochloride and azathioprine as immunosuppressants in dogs with renal allografts. *IRCS Med. Sci., 4*:19, 1976.
8. Catanzaro, P.J., Schwartz, H.J., and Graham, R.C., Jr. Spectrum and possible mechanism of carrageenan cytotoxicity. *Am. J. Pathol., 64*:387, 1971.
9. Chaouat, G., and Howard, J.G. Influence of reticuloendothelial blockade on the induction of tolerance and immunity of polysaccharides. *Immunology, 30*:221, 1976.
10. Cohen, A., and Schlesinger, M. Absorption of guinea pig serum with agar. A method for elimination of its cytotoxicity for murine thymus cells. *Transplantation, 10*:130, 1970.
11. Davies, G.E. Inhibition of complement by carrageenan: Mode of action, effect on allergic reactions and on complement of various species. *Immunology, 8*:291, 1965.
12. Filppi, J.A., Rheins, M.S., and Nyerges, C.A. Antigenic cross-reactivity among rodent brain tissues and stem cells. *Transplantation, 21*:124, 1976.
13. Golub, E.S. Brain-associated θ antigen: Reactivity of rabbit anti-mouse brain with mouse lymphoid cells. *Cell. Immunol., 2*:353, 1971.

14. Golub, E.S. Brain-associated stem cell antigen: An antigen shared by brain and hemopoietic cells. *J. Exp. Med., 136*:369, 1972.

15. Golub, E.S. Brain-associated erythrocyte antigen: An antigen shared by brain and erythrocytes. *Exp. Hematol., 1*:105, 1973.

16. Hawkins, W.W., and Leonard, V.G. Antipeptic and antithrombotic properties of carrageenan. *J. Lab. Clin. Med., 60*:641, 1962.

17. Hawkins, W.W., and Leonard, V.G. The antithrombotic activity of carrageenan in human blood. *Can. J. Biochem. Physiol., 41*:1325, 1963.

18. Hellström, K.E. Cytotoxic effect of isoantibodies of mouse tumor cells in vitro. *Transplant. Bull., and 6*:411, 1959.

19. Ishizaka, S., Otani, S., and Morisawa, S. Effects of carrageenan on immune responses. I. Studies on the macrophage dependency of various antigens after treatment with carrageenan. *J. Immunol., 118*:1213, 1977.

20. Kuznetsky, R.D., Adler, S., and Trobaugh, F.E., Jr. Anti-mouse brain antiserum: Characteristics and spectrum of activity against hemopoietic cells. *Exp. Hematol. (Suppl.), 4*:60, 1976.

21. Lotzová, E., Gallagher, M.T., and Trentin, J.J. Involvement of macrophages in genetic resistance to bone marrow grafts. Studies with two specific anti-macrophage agents, carrageenan and silica. *Biomedicine, 22*:387, 1975.

22. Monette, F.C., Eichacker, P.Q., Byrt, W., Garver, R.I., Gilio, M.J., and DeMello, J.B. An immunologic approach to cell cycle analysis of the stem cell. In Baum, S.J., and Ledney, G.D., eds., *Experimental Hematology Today*. New York: Springer-Verlag, 1977, pp. 11–19.

23. Rios, A., and Simmons, R.L. Poly-2-vinylpyridine-N-oxide reverses the immunosuppressive effects of silica and carrageenan. *Transplantation, 13*:343, 1972.

24. Rodt, H., Thierfelder, S., and Eulitz, M. Anti-lymphocytic antibodies and marrow transplantation. III. Effect of heterologous anti-brain antibodies on acute secondary disease in mice. *Eur. J. Immunol., 4*:25, 1974.

25. Sawicki, J.E., and Catanzaro, P.J. Selective macrophage cytotoxicity of carrageenan *in vivo. Int. Arch. Allergy Appl. Immunol., 49*:709, 1975.

26. Schwartz, H.J., and Kellermeyer, R.W. Carrageenan and delayed hypersensitivity II. Activation of Hageman Factor by carrageenan and its possible significance. *Proc. Soc. Exp. Biol. Med., 132*:1021, 1969.

27. Thomson, A.W., and Horne, C.H.W. Toxicity of various carrageenans in the mouse. *Br. J. Exp. Pathol., 57*:455, 1976.

28. Van den Engh, G.J., and Golub, E.S. Antigenic differences between hemopoietic stem cells and myeloid progenitors. *J. Exp. Med., 139*:1621, 1974.

29. Ward, P.A., and Cochrane, C.G. Bound complement and immunologic injury of blood vessels. *J. Exp. Med., 121*:215, 1965.

30. Zimpel, H., Stark, R., and Thiele, H.G. An approach to avoiding errors in cytotoxic testing. *J. Immunol. Methods, 3*:411, 1973.

PART II

Physiology of Committed Stem Cells

D. Metcalf

5

The Control of Neutrophil and Macrophage Production at the Progenitor Cell Level

D. Metcalf

INTRODUCTION

This chapter will briefly review the current knowledge on the regulation of the formation of neutrophils and monocyte-macrophages. Recent advances in this field have originated almost entirely from the introduction, in 1966, of methods for cloning the progenitor cells of these populations in semisolid *in vitro* cultures (3,31).

In appropriately prepared agar or methylcellulose cultures of bone marrow cells from most species, it is possible to grow colonies of neutrophils and/or macrophages (see review, reference 45). Evidence has accumulated that the cells initiating these colonies (granulocyte-macrophage colony-forming cells, GM-CFC or CFU-c) are the specific progenitor cells, or committed stem cells, for granulopoiesis and monocyte-macrophage formation. Subsequent work has developed comparable *in vitro* colony-forming systems for erythroid, eosinophil, megakaryocytic, and mixed hemopoietic populations. Superficially similar cloning systems for B- and T-lymphocytes differ in that the colonies are probably formed by relatively mature members of these hemopoietic populations.

GRANULOCYTE-MACROPHAGE COLONY-FORMING CELLS

Hemopoietic colony-forming cells are an important, but difficult to define class of hemopoietic cell, since they are the first cells in the sequence from stem cell to mature end cell that are committed to a specific pathway of differentiation, e.g., granulopoiesis. Furthermore, from the colony-forming cell stage onwards, cell proliferation and differentiation within a hemopoietic series is almost wholly controlled by one or more of a number of highly specific hemopoietic regulatory factors.

Proliferation of progenitor cells eventually leads to the formation of the morphologically recognizable members in each hemopoietic class. These events appear to be faithfully reproduced during hemopoietic colony formation *in vitro*. Thus, in granulocyte-macrophage colony formation, colonies come to be composed of enlarging populations of cells morphologically identifiable as myeloblasts, myelocytes, metamyelocytes, polymorphonuclear cells, promonocytes, monocytes, and macrophages. The ability of cells to mature almost completely during colony growth *in vitro* allows the use of agar cultures as a convenient and versatile system for analyzing the proliferation and differentiation of neutrophils and macrophages under controlled conditions.

The most striking finding concerning granulocyte-macrophage (GM) progenitor cells is the extreme heterogeneity in the size and shape of the GM colonies generated by these cells in semisolid cultures (44,45). It can be deduced from this that the ancestral (colony-forming) cells of polymorphs and macrophages must be a quite heterogeneous group of cells. Whereas GM progenitor cells are no doubt interrelated, it is clear that they tend to be grouped into distinct subpopulations of cells. Recognition that polymorphs and macrophages, like lymphocytes, occur in functionally heterogeneous subpopulations in the peripheral blood has been belated, but there is now increasing evidence for such heterogeneity (83). Production of these heterogeneous end cells appears to be achieved by the maintenance of distinct progenitor cell compartments. As shall be discussed later, the various progenitor cells are often selectively responsive to variants of the basic regulatory factor controlling cell proliferation.

In the mouse, analysis has defined at least five subpopulations of neutrophil and/or macrophage progenitor cells (Table 5-1). Since these cells are all the progeny of multipotential hemopoietic stem cells, they clearly must have some type of interrelationship. However, the nature of this relationship has not yet been clarified. From the failure of *in vitro* suicide with [³H]TdR to influence the relative proportions of granulocytic and/or macrophage colony-forming cells (43), it seems improbable that the subpopulations are simply different cell cycle stages of a single, colony-forming cell population. One radical hypothesis has proposed that *all* hemopoietic progenitor cells are of a single type and that the pathway of differentiation entered is determined by the stage in the cell cycle at which a particular regulator makes effective contact with the cell (18). This proposition is extremely improbable, but it is difficult to disprove experimentally.

The spectacular clonal purity of most hemopoietic colonies and a variety of colony transfer studies have made it clear that once a hemopoietic cell is committed to a particular pathway of differentiation, it cannot revert to another type of differentiation (45). Thus, in cultures capable of supporting neutrophil, eosinophil, and megakaryocyte proliferation, colonies are never composed of mixtures of more than one of these cells, even though adequate concentrations of regulators for all three types of cell are bathing all the colony cells in the culture.

Neutrophil-macrophage colonies are an exception to this rule, since most macrophage colonies initially contain a neutrophilic population and, for a major portion of the life-span of the colony, both populations coexist. Single-cell cloning experiments with both monkey and mouse bone marrow cells have shown that mixed GM colonies are clones initiated by single progenitor cells (34,64).

TABLE 5-1 Types of Granulocyte (Neutrophil) and Macrophage (GM) Progenitor Cells in the Mouse

PROGENITOR CELL	TYPE OF COLONY FORMED	LOCATION OF CELLS	FREQUENCY	CHARACTERISTICS	KEY REFERENCES
Multiple G-CFC	Multiple granulocytic	Bone marrow Spleen Peripheral blood	In marrow, 10-20/ 10^5 cells	Form only neutrophils Requires GM-CSF concentrations 10- to 20-fold higher than other cells.	47
G-CFC	Granulocytic	Bone marrow Spleen Peripheral blood	In marrow, 20-30/ 10^5 cells	Form only neutrophils Slowly sedimenting Require high GM-CSF concentrations	47
GM-CFC	Mixed granulocytic-macrophage	Bone marrow Spleen Peripheral blood	In marrow, 200-300/10^5 cells	The most common type of GM-CFC, initially forming neutrophils, then macrophages	45,47,48
M-CFC	Macrophage	Bone marrow Spleen Peripheral blood	In marrow, 20-30/ 10^5 cells	Form only macrophages Rapidly sedimenting cells, responsive to low GM-CSF concentrations	47
M-CFC	Macrophage	Peritoneal cavity	After thioglycollate, 5-10/10^2 cells	Form only macrophages Can survive in absence of GM-CSF but proliferation dependent on GM-CSF.	38,81

The conclusion is that neutrophils and macrophages are closely related and share common ancestral cells.

In erythropoiesis, the cells forming multiple (burst) erythroid colonies (BFU-e) are almost certainly the ancestors of the cells forming the small single erythroid colonies (CFU-e) (28). Colony formation by BFU-e requires at least a 10-fold higher concentration of erythropoietin than colony formation by CFU-e (21). By analogy, it is tempting to speculate that the cells forming multiple granulocytic colonies might also be the immediate ancestors of the cells forming the more numerous GM colonies. Formation of multiple granulocytic colonies also requires GM-CSF concentrations 10- to 20-fold higher than those needed to stimulate the formation of other GM colonies (47). Despite these intriguing similarities, the proposed interrelationship is unlikely, since multiple granulocytic colonies never develop macrophage populations, as would be expected if the cells forming multiple colonies generated GM-CFC. Similarly recloning of multiple granulocytic colonies has failed to demonstrate the presence in such colonies of other types of GM-CFC (Metcalf, unpublished data).

At the other end of the spectrum, it is likely that at least one interrelationship can be demonstrated between these various GM-CFC. The peritoneal cells forming pure macrophage colonies (M-CFC) seem to be relatively mature members of the monocyte-macrophage family (38). Furthermore, M-CFC can be obtained from many GM colonies grown *in vitro*. It is likely, therefore, the M-CFC are the immediate progeny of GM-CFC.

Larger, more rapidly sedimenting GM-CFC tend to form macrophage colonies and to be more responsive to stimulation by lower concentrations of GM-CSF than the more slowly sedimenting GM-CFC that preferentially form granulocytic colonies (47). Whereas most GM-CFC are able to form both types of progeny cell, an almost continuous spectrum of progenitor cells appears to exist, whose different thresholds of responsiveness to GM-CSF largely account for the well-known observation that high concentrations of GM-CSF tend to stimulate granulocytic colony formation, whereas low concentrations stimulate the macrophage colony formation.

In the adult mouse bone marrow, GM progenitor cells have been found to have a heterogeneous buoyant density and to sediment with a peak velocity of 4.5 mm/hr (27,47,53). From identification of these cells in monkey bone marrow and correlative studies of electronmicroscopic and light microscopic morphology with the frequency of colony-forming cells in fractionated marrow populations from a number of species, it is apparent that most GM-CFC in adult marrow are unremarkable, medium-sized cells with a slightly eccentric round nucleus and basophilic, but agranular cytoplasm (15,64). The morphology of GM-CFC is, however, not constant. Since most are in active cell cycle, the cells will enlarge as the S phase of the cell cycle proceeds. A pronounced increase in size in mouse bone marrow GM-CFC has also been observed following the injection of endotoxin (51) and following whole body irradiation (52). In the fetus, GM-CFC are clearly different in morphology from GM-CFC in adult marrow, being rapidly sedimenting light density cells, presumably with the appearance of large blast cells (33).

There are some distinct strain differences in the relative frequency of the various subtypes of GM-CFC in the marrow, with $C_{57}BL$ and CBA mice serving as prototypes for two distinct patterns of colony growth. In cultures of $C_{57}BL$ marrow cells stimulated by high concentrations of mouse lung-conditioned medium GM-CSF, about 10 percent of the colonies are multiple granulocytic colonies, 20 to 30 percent are granulocytic, and 10 percent are macrophagic. In sharp contrast, CBA marrow cells form no multiple granulocytic colonies but do form a high proportion of macrophagic colonies. Most of the mice examined had GM-CFC populations clearly falling into one type or the other (Table 5-2), but the genetic basis for this has yet to be established. When $C_{57}BL$ marrow cells are incubated with serum inhibitors present at high levels in CBA or BALB/c mice and the marrow cells are then cultured, the pattern of colonies produced resembles that produced by CBA marrow cells (9,49). It is possible that the growth pattern of colonies produced by CBA and similar marrow cells reflects their prior exposure *in vivo* to this type of serum inhibitor.

Although it was initially suggested that some or many GM colonies might have been generated by stem cells (CFU-s), a large body of information (review, reference 45) now indicates that most GM-CFC are the immediate progeny of CFU-s. Correlation analyses of GM-CFC and CFU-s content of individual spleen colonies also indicated a relatively close relationship between these two cell types (22).

In a study in this laboratory, individual spleen colonies were examined for their content of cells capable of forming colonies *in vitro*. It was observed that when spleen colonies contained GM-CFCs, more than one subtype of GM-CFC was usually present. In the experiment shown in Table 5-3, where 20, 12-day-old , $C_{57}BL$ spleen colonies were analyzed, 18 of 19 contained GM-CFC and seven of 19 contained all four types of GM-CFC populations present in the marrow. In addition, 10

TABLE 5-2 Classification of Mouse Strains According to Whether Their Bone Marrow Cells Generate Mainly Granulocytic or Mainly Macrophage Colonies *In Vitro*[a]

MAINLY GRANULOCYTIC COLONIES (C_{57}BL TYPE)		MAINLY MACROPHAGE COLONIES (CBA TYPE)	
C_{57}BL/6	C_{57}BL/10	CBA	ATH
NZB Ig[b]	NZB Choc	BALB/c	SJL × CBA
AKR/J	A/J	DBA/1	C_3H He
SJL × B_{10} A (2R)	129J	ATL	BALB/Igb N2
AKR-2b	B_{10}Br	NZW	BALB/Igf
θ × AKR-2b	C_{57}BL/6 IgeN14	BALB/c H-2b N4	BALB/H-2k
SJL × ATH	NZB H-2b N10	ATH × ATL	BALB F_{13}
B_{10}A (2R)	C_{57}BL/61geN2	BALB × NZB	BALB IgC N7
NZB Choc N_{14}	B_{10}D2	BALB/c H-2k N4	
C_{57}BL × NZB	AQR/SF	C_3H	
B_{10}A	NZC	ATL	
NZB	NZB H-2k N8	ASW	
θ	B_{10}D$_2$Ige	CBA T_6T_6	
	SJL		
	SJL × BALB/c		
	NZO		

[a]All cultures are stimulated by supramaximal concentrations of GM-CSF from mouse lung-conditioned medium. For each strain, replicate cultures were prepared of marrow cells from a pool of three 3-month-old mice. Colonies were typed on day 7 of incubation.

of 19 also contained eosinophil CFC and one of 19, megakaryocyte CFC. In this example, morphological analysis showed that most spleen colonies contained a mixed population of cells, but several of them would have appeared as pure erythroid colonies in conventional sections of the irradiated spleen. Data of this type not only emphasizes the early heterogeneity of apparently pure spleen colonies but also constitutes direct proof that the various subpopulations of GM-CFC are related and can originate from a single CFU-s.

A somewhat similar heterogeneity of GM-CFC has recently been demonstrated in human marrow populations. Rapidly sedimenting GM-CFC were shown to form macrophage colonies preferentially as in the mouse and were found to respond to lower

TABLE 5-3 Analysis of Individual Spleen Colonies for Hemopoietic Progenitor Cells[a]

COLONY	TOTAL COLONY CELLS (× 10^6)	PERCENT COLONY CELLS E	G	Other	TOTAL GM-CFC/ SPLEEN COLONY	PERCENT COLONIES IN AGAR Multiple G	G	Mix	M	TOTAL EO-CFC/ SPLEEN COLONY	TOTAL MEG-CFC/ SPLEEN COLONY
1	0.02	—	—	—	181	0	89	11	0	0	0
2	0.24	—	—	—	1,070	4	96	0	0	176	0
3	0.68	—	—	—	487	0	37	37	26	85	0
4	5.8	78	11	11	3,931	0	68	24	8	0	0
5	1.48	34	56	10	5,137	2	71	27	0	685	0
6	1.56	85	9	6	2,223	2	77	17	4	0	0
7	6.2	78	12	10	—	—	—	—	—	—	—
8	1.92	96	1	3	13,653	1	40	59	0	82	0
9	3.0	37	52	11	1,694	0	75	0	25	0	0
10	1.48	83	11	6	4,525	11	83	3	5	306	0
11	1.52	32	50	18	3,603	0	66	34	0	310	232
12	6.0	81	15	4	6,105	1	89	10	0	0	0
13	8.0	71	23	6	5,460	0	72	28	0	280	0
14	1.88	77	13	10	2,397	4	49	37	10	131	0
15	2.28	91	4	5	0	0	0	0	0	0	0
16	6.2	78	18	4	9,407	10	48	26	16	0	0
17	7.2	87	7	6	1,811	0	47	37	16	0	0
18	8.6	—	—	—	2,910	0	44	22	34	342	0
19	4.0	76	16	8	8,928	1	78	14	7	216	0
20	7.2	76	14	10	1,277	0	71	0	29	0	0

[a]Individual 12-day-old spleen colonies in irradiated C_{57}BL mice injected with 75,000 C_{57}BL marrow cells; G-M colony formation stimulated by 0.1 ml mouse lung GM-CSF in 20-fold excess concentration; eosinophil and megakaryocyte colony formation stimulated in parallel cultures by 0.2 ml pokeweed mitogen stimulated C_{57}BL spleen-conditioned medium (57,59). E, erythroid; G, neutrophilic; Mix, mixed neutrophil-macrophage; M, macrophage.

concentrations of GM-CSF than the more slowly sedimenting GM-CFC that form mainly granulocytic colonies (36).

The existence of different subpopulations of GM-CFC in the adult marrow strongly suggests that this arrangement is advantageous for polymorph and macrophage production. It can be expected that, in certain disease states or conditions, changes will be found to involve preferentially one or the other of these populations. Little information exists on this question at present and future studies on GM-CFC levels in various diseases or in response to various procedures should document which subpopulations are involved in these responses.

CONTROL BY GM-COLONY-STIMULATING FACTOR (GM-CSF)

The survival and proliferation of GM-CFC in culture requires the continuous presence of adequate concentrations of a factor variously termed GM-colony-stimulating factor (GM-CSF), colony-stimulating factor (CSA), or macrophage-granulocyte inducer (MGI) (41). Removal of developing neutrophil or macrophage colonies to cultures lacking GM-CSF causes an immediate cessation of proliferation, and usually, the early death of colony cells (46,68).

A sigmoid dose-response relationship exists between GM-CSF concentration and the number of colonies developing. This dose-response relationship forms the basis for the use of marrow cultures to bioassay GM-CSF concentrations (80), and the cultures can detect $10^{-11}M$ of the regulator.

Although GM-CSF has now been purified, the exact mechanism of action of this regulator on GM colony-forming cells and their progeny is not known. Some facts have been established:

1. Whereas GM-CSF is necessary for *in vitro* survival of GM-CFC (41), it is not necessary for the survival of M-CFC (81), and GM-CSF appears therefore not to be simply a survival factor.
2. When target GM-CFC are not in cycle, incubation with GM-CSF initiates cellular DNA synthesis within 3 hr (62).
3. When GM-CSF concentrations are increased, there is a progressive shortening of mean colony cell cycle time (48).
4. In the presence of high concentrations of GM-CSF, differentiation to post-mitotic metamyelocytes and polymorphs appears to be delayed (48).
5. With high concentrations of GM-CSF, a high proportion of granulocytic colonies develops, and conversely, with low GM-CSF concentra-

tions, most colonies are composed of macrophages (44,48). This well-known association probably does not indicate that low concentrations of GM-CSF can force the common ancestral GM-CFC to enter a pathway of macrophage formation. Because of the heterogeneity in responsiveness of GM-CFC discussed above, it is more likely that, at low GM-CSF concentrations, only the rapidly sedimenting, macrophage-forming GM-CFC can be activated (47).

6. Since GM-CSF selectively increases RNA synthesis in granulocytic cells within 10 min, even in post-mitotic polymorphs (5), a capacity of GM-CSF to activate certain facets of the functional activity of even mature polymorphs and macrophages is suggested. In support of this, GM-CSF has been shown to increase protein synthesis in such cells (6) and to increase prostaglandin E production by mature macrophages (37).
7. Purified GM-CSF can stimulate the proliferation of all GM-CFC, and the same molecule stimulates granulopoiesis and monocyte-macrophage formation (6).
8. No action has been noted of GM-CSF on other hemopoietic populations or on non-hemopoietic cells (48).

The above information can be summarized by saying that GM-CSF is a specific proliferative stimulus for granulopoietic and macrophage-forming cells that is required for each cell division. Although it is not exactly clear whether GM-CSF can influence the pattern of differentiation of GM-CFC or the rate at which the progeny colony cells can differentiate, GM-CSF clearly can influence RNA synthesis rates and probably the levels of certain functional activities in mature cells. This dual action of GM-CSF in stimulating both proliferation and cellular synthesis is essentially identical to the double action of erythropoietin in stimulating erythropoiesis and hemoglobin synthesis.

NATURE AND SOURCES OF GM-CSF

Although it is clear that the existence of heterogeneous subpopulations of GM-CFC with varying threshholds of responsiveness to GM-CSF permits subtle fluctuations in GM-CSF-stimulated polymorph and macrophage formation, further complexities introduce even more variations. Analysis has shown that more than one molecular form of GM-CSF exists (1,7,8,24,74,77,79) and that some GM-CFC respond only or preferentially to certain types of GM-CSF. Thus for example, human urine GM-CSF is unable to stimulate the proliferation of multiple granulocytic GM-CFC or the slowly sedimenting granulocytic CFC (47); pregnant uterus CSF in the presence of red cell lysates preferentially stimulates colony formation by GM-CFC with a buoyant density of 1.080 g/cm³ (85), and yolk sac GM-CSF tends to stimulate only the formation of macrophage colonies regardless of the concentration used and is also relatively inefficient in stimulating the proliferation of fetal GM-CFC (35).

At this stage, the heterogeneity of the various GM-CSF and the interrelationships between them have not yet been fully explored. The GM-CSF seem at present to belong to one or other of two broad groups—those tending to stimulate granulocytic colony formation, e.g., the GM-CSF produced by mouse lung tissue in vitro, and those tending to stimulate macrophage colony formation, e.g., the GM-CSF produced by mouse yolk sac or pregnant uterus.

In the context of the present discussion, the heterogeneous nature of the various GM-CSF that might possibly impinge on target progenitor cells in the bone marrow considerably increases the flexibility of the control systems and gives these systems the capacity to regulate the production of the extremely variable numbers of polymorphs and monocyte-macrophages according to circumstances.

The GM-CSF is readily detectable in the serum, tissues, and urine in concentrations significantly higher than those required to stimulate the proliferation of granulocytes and macrophages in vitro (75). Serum levels of GM-CSF fluctuate sharply under conditions known to involve altered rates of polymorphs and monocyte production. Thus within minutes of the injection of endotoxin or bacterial antigens into animals, serum GM-CSF levels rise to levels approaching 100-fold higher than those in normal serum (42), and high serum levels are sustained during the acute phases of viral, bacterial, and protozoal infections (50,56,82). Serum and urine levels of GM-CSF fluctuate in cyclic neutropenia in man and dogs (14,23,66), and serum GM-CSF levels are elevated by whole-body irradiation (26,67) or the administration of cytotoxic drugs (73).

This indirect evidence is consistent with the possibility that GM-CSF functions in vivo as a genuine humoral regulator of polymorph and monocyte production. It has been suggested, however, that normal serum GM-CSF levels are too high for this factor to serve as a satisfactory regulator. This objection seems highly theoretical and takes no account of possible biological filters or dampening systems operating in vivo to screen mouse target cells from GM-CSF. Furthermore, the demonstration that some GM-CFC require extremely high GM-CSF levels to be activated (47) provides a rational explanation for GM-CSF fluctuations that seem large.

A similar set of theoretical objections can be raised concerning the lipid or lipoprotein inhibitors present in serum (9,20,49). These block the action of GM-CSF in vitro and, as mentioned above, when incubated with marrow cells, force the colony cells subsequently generated by these cells to undergo premature macrophage transformation (9,49). Levels of these serum inhibitors are high in BALB/c or CBA mice, and in cultures containing inhibitors at the concentrations present in such sera, no concentration of GM-CSF can stimulate colony formation (49). It can be questioned, therefore, how GM-CSF is able to stimulate granulopoiesis in BALB/c or CBA mice. Although there is no solution to this question at present, again it may be that filters exist in vivo, e.g., vascular endothelium, to protect target GM-CFC and their progeny from such inhibitors.

Two forms of GM-CSF have been purified by acceptable biochemical criteria (7,77). The GM-CSF synthesized in vitro by mouse lung tissue has a molecular weight of 23,000 by SDS polyacrylamide gel electrophoresis (7); it is a neuraminic acid-containing glycoprotein with a carbohydrate content of approximately 10 percent (A. W. Burgess and D. Metcalf, unpublished data). Incubation with neuraminidase did not alter the in vitro activity of the molecule (76). The other form of GM-CSF so far purified is that present in L-cell conditioned medium (77). This GM-CSF is also a neuraminic acid-containing glycoprotein, but with a molecular weight of approximately 70,000. Treatment of this molecule with mercaptoethanol suggests that it might be composed of two equal subunits of 35,000.

Measurements using gel filtration tend to overestimate molecular weights, particularly when impure materials are being analyzed. With this reservation, size estimates for other GM-CSF are 45,000 for the GM-CSF in human urine (79), 40,000 for that in human placenta-conditioned medium (8), and 40,000 for the GM-CSF in mouse spleen-conditioned medium (55). Conversely, the size estimate for embryo cell-conditioned medium GM-CSF is 60,000 to 70,000 (24) and a similar estimate has been obtained for the GM-CSF produced in vitro by mouse yolk sac cells (A. W. Burgess and G. R. Johnson, unpublished data).

Given the known inaccuracies associated with these various estimates, it is possible that two broad groups of GM-CSF have been documented: (a) a group with molecular weights in the 25,000 to 40,000 range and (b) a group in which the molecular weight appears to be about twice this—60,000 to 70,000. Again, analogies with erythropoietin are of interest. Only two forms of erythropoietin have been examined in any detail. Both are neuraminic acid-containing glycoproteins; the erythropoietin in human urine has a molecular weight of 39,000 and that in anemic sheep plasma a molecular weight of 46,000 (60).

Despite a number of studies in the past 10 years, it is still quite unclear which cells in the body produce GM-CSF, since GM-CSF can be extracted from all tissues in concentrations higher than that present in the serum (75) and can be released and/or produced by many tissues *in vitro*, including a number of continuous cell lines. Two broad alternatives exist: (a) GM-CSF is potentially able to be produced by most cells in the body under certain conditions or (b) certain cells common to all tissues, e.g., macrophages or lymphocytes, are the sole source of GM-CSF. Although cloned continuous cell lines, such as L-cells or melanoma cells, can produce GM-CSF, this should be discounted from the discussion because of the possibility that such cells may be abnormally derepressed and not exhibiting normal functional activity.

Logic suggests that a production site for GM-CSF close to most target GM-CFC, e.g., by the cells in the bone stroma, would be an effective control system, and in fact, evidence has been produced that bone stromal cells can produce significant amounts of GM-CSF (10), production varying significantly during such situations as post-irradiation regeneration (11).

However, such tissues as the lung or heart also have an extraordinary capacity to produce GM-CSF *in vitro* (57,59), and it is difficult to envisage that such tissues would normally be closely involved in the regulation of granulopoiesis.

ROLE OF MACROPHAGES AND LYMPHOCYTES

A large literature exists on the production of GM-CSF by monocyte-macrophages (12,17,62,66) and by activated lymphocytes (13,40,57,59,69,70,71,72). Proponents of production by these tissues have asserted that, for example, macrophages are *the* only source of GM-CSF—an extraordinary conclusion based on essentially no evidence other than that macrophages *can* produce GM-CSF. In truth, many of the experiments using macrophages and lymphocytes were initiated simply because of the ready availability of blood as a source of cells and the current interest of other workers in products of activated lymphocytes. The evidence that macrophages and activated lymphocytes can produce GM-CSF is quite conclusive, and a number of intriguing biological aspects are raised by these ob-

servations. However, from the point of view of the control of polymorph and macrophage production, it is far from clear whether macrophages or lymphocytes are important tissue sources of GM-CSF, nor is it clear whether different tissues fluctuate in their importance as sources under certain conditions or whether some types of GM-CSF are activated or transformed by other tissues.

Because GM-CSF production *in vitro* has been shown to occur in cultures of a wide variety of tissues, the demonstration that activated lymphocytes also produce GM-CSF is, in itself, of no special interest. However, under certain conditions, such activated lymphoid populations are also able to produce the specific factors to stimulate eosinophil, megakaryocyte, and erythroid proliferation. Indeed, the only available method for producing the eosinophil and megakaryocyte factors from normal tissues is to stimulate mouse spleen cells by the addition of pokeweed mitogen (57,59).

From the evidence with phytohemagglutinin-stimulated cultures, it seems clear that mitogen-stimulated lymphocytes can produce GM-CSF (13,70,71). In the more versatile and interesting pokeweed mitogen-stimulated system the evidence is less clear. T-lymphocytes are certainly necessary for factor production, since spleen cells from *nu/nu* (congenitally hypothymic) mice are virtually inactive, whereas cells from *nu/nu* mice grafted with thymus tissue have normal functional activity (Table 5-4). However, removal of adherent cells either by Petri dish adherence or by adherence to carbonyl ion grossly reduces the capacity of spleen cells to condition medium (Table 5-5), indicating that adherent cells are involved in the production of active conditioned medium. It is still uncertain which of the two interacting cells—T-lymphocytes or adherent (? macrophage) cells—actually produces the various stimulating factors.

Studies using metabolic inhibitors have shown that the production of medium with all four colony-stimulating activities depends on early DNA synthesis and can also be blocked by inhibitors of protein synthesis (D. Metcalf, unpublished data). In pokeweed mitogen-stimulated spleen conditioned medium, four separate factors with colony-stimulating activity for neutrophil-macrophages (GM-CSF) are produced: eosinophils (EO-CSF), megakaryocytes (MEG-CSF), and erythroid cells (E-CSF) (55). All appear to be glycoproteins, with molecular weights of 40,000 or less, and because of their general similarity, they are difficult to separate from one another using the conventional methods of protein chemistry.

The production of EO-CSF by pokeweed mitogen-stimulated spleen cells is of particular interest in view of the *in vivo* evidence that eosinophil pro-

TABLE 5-4 Failure of Spleen Cells from *nu/nu* (Athymic) Mice to Condition Medium and Subsequent Medium Conditioning by Thymus Grafting

	COLONY-STIMULATING ACTIVITY[b]			
SOURCE OF SPLEEN CELLS[a]	Neutrophil-macrophage	Erythroid	Eosinophil	Megakaryocyte
C$_{57}$BL	73 ± 11	4 ± 2	3 ± 2	5 ± 2
BALB/c	91 ± 8	4 ± 1	5 ± 2	7 ± 2
nu/nu	0	0	0	0
nu/nu grafted with BALB/c thymus	112 ± 17	7 ± 4	8 ± 1	10 ± 2

[a]The *nu/nu* congenitally athymic mice were back-crossed to BALB/c mice. All mice were held under specific pathogen free conditions, and spleen cells tested at the age of 2 months. Pools of two to three mice were used in each group. Suspension cultures contained 2×10^6 spleen cells/ml, plus 0.05 ml of 1:15 dilution pokeweed mitogen (Gibco)/ml of 1640 medium containing 5 percent tested human plasma. Conditioned media harvested at 7 days and 0.2 ml added to cultures of 75,000 C$_{57}$BL marrow cells for neutrophil-macrophage, eosinophil, and megakaryocyte colony formation (57,59) or 20,000 12-day-old CBA fetal liver cells for erythroid colony formation (34).

[b]Mean number of colonies stimulated by 0.2 ml of conditioned medium ± standard deviations.

duction depends on stimulation by a factor released by activated lymphocytes (2). It seems reasonable to assume, as a working hypothesis, that EO-CSF is this lymphocyte-derived factor. Although there is no *in vivo* evidence linking megakaryocyte or neutrophil-macrophage production to lymphocyte function, it seems improbable that only one of four chemically similar colony-stimulating factors produced by lymphoid populations under the same culture conditions should have an *in vivo* function, while the other three with essentially similar *in vitro* activity should merely be artifacts.

In this context, it is of some interest that recent experiments have suggested links between T-lymphocytes and erythropoiesis. Goodman and Shinpock (19) showed that the injection of T-lymphocytes increased erythroid colony formation in irradiated animals injected with parental bone marrow cells, and it has also been shown that the injection of theta-positive cells can cure the anemia of stem cell-depleted W^v mice (84).

At this state of knowledge, it seems unwise to reject the possibility that the production of hemopoietic factors by activated lymphoid populations may actually occur *in vivo* under certain conditions. One intriguing question posed by these observations is whether each lymphocyte has the capacity to produce all four factors simultaneously or whether separate subsets of lymphocytes exist, each capable of producing only a single factor.

Associated with the proposal that macrophages might be a significant source of GM-CSF with a positive feedback stimulating further macrophage production has been considerable speculation that macrophages and polymorphs might also produce inhibitory factors and thus represent an inhibitory feedback control for cell production.

The search for inhibitors of polymorphs and macrophage origin has been marred by the same type of wishful thinking associated with the attempts to show that macrophages were the only source of GM-CSF. It has proved relatively easy

TABLE 5-5 Effect of Removal of Adherent Cells on Ability of C$_{57}$BL Spleen Cells to Condition Medium

		COLONY-STIMULATING ACTIVITY[b]			
EXPERIMENT	TYPE OF CELLS[a]	Neutrophil-macrophage	Erythroid	Eosinophil	Megakaryocyte
I	Unfractionated	114 ± 8	28 ± 5	4 ± 2	4 ± 1
	Non-Adherent	2 ± 1	0	0	0
II	Unfractionated	94 ± 2	10 ± 1	5 ± 3	4 ± 2
	Non-Adherent	0	0	0	0

[a]Suspension cultures contained 2×10^6 C$_{57}$BL unfractionated or non-adherent cells/ml, plus 0.05 ml 1:15 dilution pokeweed mitogen (GIBCO)/ml of 1640 medium containing 5 percent tested human plasma. Non-adherent cells separated by incubation with carbonyl iron. Conditioned media were harvested at 7 days, and 0.2 ml of the media was added to cultures of 75,000 C$_{57}$BL marrow cells for neutrophil-macrophage, eosinophil, or megakaryocyte colony formation (57,59) or 20,000 12-day-old CBA fetal liver cells for erythroid colony formation (34).

[b]Mean number of colonies stimulated by 0.2 ml of conditioned medium ± standard deviations.

to design experiments in which the addition of polymorphs inhibits colony formation, particularly in cultures of human marrow cells (29). Analysis of this system suggests that polymorph-derived products can have a moderately inhibitory influence on GM-CSF-producing cells but appear not to be cytotoxic for GM-CFC (4). It has also been shown that excessive numbers of macrophages can inhibit GM-colony formation (32) and that macrophages, particularly after stimulation by GM-CSF, produce prostaglandin E, an inhibitor of GM-colony formation (37).

Although this may indeed be important evidence for feedback control of polymorph and macrophage production, there has been a remarkable reluctance to determine whether these cells are the *only* source of such inhibitors and whether the inhibitory effects are in any way specific for polymorph and macrophage production. Limited tests with prostaglandin E have, in fact, shown it to be inhibitory for lymphoid proliferation (37), and on this ground alone, the factor would appear to be an unsatisfactory candidate for a specific inhibitor of macrophage production *in vivo*. The serum inhibitors of GM-colony formation are equally unsatisfactory candidates, since they too appear to inhibit a wide range of hemopoietic cells *in vitro* (49).

GM-CSF AND MYELOID LEUKEMIA

Apart from the general importance of characterizing the action of GM-CSF for an understanding of the control mechanisms regulating polymorph and macrophage production, the importance of GM-CSF has been sharply increased by the observation that the leukemic cells in acute and chronic myeloid leukemia appear to retain full responsiveness to, and dependency on, GM-CSF. At least *in vitro*, leukemic cells are quite unable to proliferate in the absence of adequate concentrations of GM-CSF (58). A similar observation has been made for mouse myeloid leukemic cells, and the cells have also been shown to respond to stimulation by purified GM-CSF (16,30,45,54). Formal proof that human leukemic cells are responding to the GM-CSF in white cell-conditioned or placental-conditioned media cannot be obtained until this factor is purified from these human sources, but no doubt exists regarding the dependency of the leukemic cells.

In one mouse model system, addition of GM-CSF (MGI) permitted or forced a population of blast cells to differentiate to apparently normal polymorphs and macrophages. The ability to respond by differentiation was genetically determined in the leukemic population and not all members of the leukemic population were capable of responding (39).

In the case of human leukemia, it is also clear that intrinsic defects exist in the leukemic population, particularly in patients with acute myeloid leukemia (45,61,65). Thus, although proliferation in the leukemic population is GM-CSF-dependent, the differentiation that occurs in normal GM populations proliferating under the influence of GM-CSF either does not occur in the leukemic cells or is bizarre and abnormal.

Despite this evidence for intrinsic defects in human myeloid leukemic populations, leukemia is essentially a disease of proliferative competition between normal and leukemic populations. The fact that leukemic cells remain dependent on the normal proliferative stimulus GM-CSF is an extremely encouraging observation that opens the possible use of GM-CSF analogues in the control of leukemic proliferation.

GENERAL COMMENT

Faced with a heterogeneity of molecular forms of GM-CSF and great uncertainty concerning the cellular origin of these regulator molecules, some have questioned whether GM-CSF really is the *in vivo* regulator of polymorph and macrophage production and whether any progress has been made in the last 50 years since the bewildering days when the injection of almost any substance appeared to affect blood polymorph levels.

My personal view is that the introduction of semisolid cloning systems has had a dramatic impact on the analysis of factors controlling polymorph and macrophage production. It is now possible in defined culture conditions [even under serum-free conditions (25)] to add minute concentrations of a purified glycoprotein and predictably stimulate both neutrophil and macrophage production. This system is, in fact, the best available biological model for the general study of cell proliferation and differentiation because of the unique advantages of a system in which two such different end cells as polymorphs and macrophages are produced by a common ancester cell in response to stimulation by a single regulator molecule. Quite clearly, we are still a long way from understanding how polymorph and macrophage production is controlled *in vivo*, but it is encouraging to find that most substances or procedures, shown in earlier years to influence granulopoiesis, have since been shown to elicit major changes in GM-CSF levels. The implication is strong that these heterogeneous substances work by activating the GM-CSF final pathway.

43

The unexpected finding clearly emerging from the recent *in vitro* studies is the extraordinary heterogeneity of granulocyte and macrophage progenitor populations. If this is to be put to use, the conclusion is inescapable that end-cell polymorphs and macrophages must have at least a similar degree of functional heterogeneity; in other words, that distinct subpopulations of these end cells must exist. Preliminary evidence that this is, in fact, so is now appearing and should be actively extended (83). Although it is unwelcome news to find that any cell system is much more complex than had previously been expected, at least it is encouraging that *in vitro* cloning systems offer the perfect technical approach to analyzing such heterogeneity at the level of individual progenitor cells.

SUMMARY

Polymorph and monocyte-macrophage production from the progenitor cell (granulocyte-macrophage colony-forming cells and colony-forming unit-culture) stage onwards appears to be stimulated by the glycoprotein (granulocyte-macrophage colony stimulating factor). This regulator occurs in more than one molecular form, some forms tending to stimulate preferentially polymorph production, and others macrophage production. There is a corresponding heterogeneity in the progenitor cell compartment, suggesting both a capacity for subtle modulation in the production of these cells and the existence of heterogeneous subsets of end-cell polymorphs and monocyte-macrophages.

ACKNOWLEDGMENT

This work was supported by the Carden Fellowship Fund of the Anti-Cancer Council of Victoria, the National Health and Medical Research Council, Canberra, and the National Cancer Institute, Washington, Contract No. NOI-CB-33854.

REFERENCES

1. Austin, P.E., McCulloch, E.A., and Till, J.E. Characterisation of the factor in L-cell conditioned medium capable of stimulating colony formation by mouse marrow cells in culture. *J. Cell. Physiol.*, 77:121, 1971.
2. Basten, A., and Beeson, P.B. Mechanism of eosinophilia. II. Role of the lymphocyte. *J. Exp. Med.*, 131:1288, 1970.
3. Bradley, T.R., and Metcalf, D. The growth of mouse bone marrow cells *in vitro*. *Aust. J. Exp. Biol. Med. Sci.*, 44:287, 1966.
4. Broxmeyer, H.E., Baker, F.L., and Galbraith, P.R. *In vitro* regulation of granulopoiesis in human leukemia: Application of an assay for colony-inhibiting cells. *Blood*, 47:389, 1976.
5. Burgess, A.W., and Metcalf, D. The effect of colony-stimulating factor on the synthesis of ribonucleic acid by mouse bone marrow cells in vitro. *J. Cell. Physiol.*, 90:471, 1977.
6. Burgess, A.W., and Metcalf, D. Colony-stimulating factor and the differentiation of granulocytes and macrophages. In S. J. Baum and G. D. Ledney, eds., *Experimental Hematology Today*. New York: Springer-Verlag, 1977, pp. 135–146.
7. Burgess, A.W., Camakaris, J., and Metcalf, D. The purification and properties of colony stimulating factor from mouse lung conditioned medium. *J. Biol. Chem.*, 252:1988, 1977.
8. Burgess, A.W., Wilson, E.M.A., and Metcalf, D. Stimulation by human placental conditioned medium of hemopoietic colony formation by human marrow cells. *Blood*, 49:573, 1977.
9. Chan, S.H. Influence of serum inhibitors on colony development *in vitro* by bone marrow cells. *Aust. J. Exp. Biol. Med. Sci.*, 49:553, 1971.
10. Chan, S.H., and Metcalf, D. Local production of colony-stimulating factor within the bone marrow. Role of nonhematopoietic cells. *Blood*, 40:646, 1972.
11. Chan, S.H., Metcalf, D. Local and systemic control of granulocytic and macrophage progenitor cell regeneration after irradiation. *Cell Tissue Kinet.*, 6:185, 1973.
12. Chervenick, P.A., and LoBuglio, A.F. Human blood monocytes: Stimulators of granulocyte and mononuclear formation *in vitro*. *Science 178*:164, 1972.
13. Cline, M.J., and Golde, D.W. Production of colony-stimulating activity by human lymphocytes. *Nature (Lond), 248*:703, 1974.
14. Dale, D.C., Brown, C.H., Carbone, P., and Wolf, S.M. Cyclic urinary leukopoietic activity in gray collie dogs. *Science, 173*:152, 1971.
15. Dicke, K.A., Van Noord, M.J., Maat, B., Schaefer, U.W., and Van Bekkum, D.W. Identification of cells in primate bone marrow resembling the hemopoietic stem cell in the mouse. *Blood, 42*:195, 1973.
16. Fibach, E., and Sachs, L. Control of normal differentiation of myeloid leukemic cells. VII. Induction of differentiation to mature granulocytes in mass culture. *J. Cell. Physiol., 86*:221, 1975.
17. Golde, D.W., and Cline, M.J. Identification of the colony-stimulating cell in human peripheral blood. *J. Clin. Invest., 51*:2981, 1972.
18. Goldwasser, E. Erythropoietin and the differentiation of red blood cells. *Fed. Proc., 34*:2285, 1975.
19. Goodman, J.W., and Shinpock, S.G. Influence of thymus cells on erythropoiesis of parenteral marrow in irradiated hybrid mice. *Proc. Soc. Exp. Biol. Med., 129*:417, 1968.
20. Granstrom, M. Conditions influencing inhibitors to the colony stimulating factor (CSF). *Exp. Cell. Res., 87*:307, 1974.
21. Gregory, C. J. Erythropoietin sensitivity as a differentiation marker in the hemopoietic system: Studies of three erythropoietic colony responses in culture. *J. Cell. Physiol., 89*:289, 1976.

22. Gregory, C.J., McCulloch, E.A., and Till, J.E. Erythropoietic progenitors capable of colony formation in culture. State of differentiation. *J. Cell. Physiol.*, *81*:411, 1973.

23. Guerry, D., Adamson, J.W., Dale, D.C., and Wolff, S.M. Human cyclic neutropenia: Urinary colony-stimulating factor and erythropoietin levels. *Blood*, *44*:257, 1974.

24. Guez, M., and Sachs, L. Purification of the protein that induces cell differentiation to macrophages and granulocytes. *FEBS Letters*, *37*:149, 1973.

25. Guilbert, L.J., and Iscove, N.N. Partial replacement of serum by selenite, transferrin, albumin and lecithin in haemopoietic cell cultures. *Nature (Lond)*, *263*:594, 1976.

26. Hall, B.M. The effects of whole body irradiation on serum colony stimulating factor and in vitro colony-forming cells in the bone marrow. *Brit. J. Haematol.*, *17*:553, 1969.

27. Haskill, J.S., McNeill, T.A., and Moore, M.A.S. Density distribution analysis of *in vivo* and *in vitro* colony-forming cells in bone marrow. *J. Cell. Physiol.*, *75*:167, 1970.

28. Heath, D.S., Axelrad, A.A., McLeod, D.L., and Shreeve, M.M. Separation of the erythropoietin-responsive progenitors BFU-e and CFU-e in mouse bone marrow by unit gravity sedimentation. *Blood*, *47*:777, 1976.

29. Heit, W., Kern, P., Kubanek, B., and Heimpel, H. Some factors influencing granulocytic colony formation *in vitro* by human white blood cells. *Blood*, *44*:511, 1974.

30. Ichikawa, Y. Further studies on the differentiation of a cell line of myeloid leukemia. *J. Cell. Physiol.*, *76*:175, 1970.

31. Ichikawa, Y., Pluznik, D. H., and Sachs, L. *In vitro* control of the development of macrophage and granulocyte colonies. *Proc. Nat. Acad. Sci. (USA)*, *56*:488, 1966.

32. Ichikawa, Y., Pluznik, D. H., and Sachs, L. Feedback inhibition of the development of macrophage and granulocyte colonies. I. Inhibition by macrophages. *Proc. Natl. Acad. Sci. (USA)*, *58*:1480, 1967.

33. Johnson, G.R., and Metcalf, D. Characterisation of mouse fetal liver granulocyte-macrophage colony-forming cells (GM-CFC) by velocity sedimentation at unit gravity. *Exp. Hematol.*, *6*:246, 1978.

34. Johnson, G.R., and Metcalf, D. Pure and mixed erythroid colony formation *in vitro* stimulated by spleen conditioned medium with no detectable erythropoietin. *Proc. Natl. Acad. Sci. (U.S.A.)*, (in press) 1978.

35. Johnson, G.R., and Metcalf, D. Sources and nature of granulocyte-macrophage colony-stimulating factor in fetal mice. *Exp. Hematol.*, (in press) 1978.

36. Johnson, G.R., Dresch, C., and Metcalf, D. Heterogeneity of human neutrophil, macrophage and eosinophil progenitor cells demonstrated by velocity sedimentation separation. *Blood*, (in press) 1977.

37. Kurland, J., and Moore, M.A.S. Inhibition of normal and neoplastic hemopoietic proliferation by a diffusible factor from activated macrophages. In S. J. Baum and G. D. Ledney, eds. *Experimental Hematology Today*. New York: Springer-Verlag, 1977, pp. 51–62.

38. Lin, H.S., and Stewart, C.C. Peritoneal exudate cells. I. Growth requirement of cells capable of forming colonies in soft agar. *J. Cell. Physiol.*, *83*:369, 1974.

39. Lotern, J., and Sachs, L. Different blocks in the differentiation of myeloid leukemic cells. *Proc. Natl. Acad. Sci. (USA)*, *71*:3507, 1974.

40. McNeill, T.A. Release of bone marrow colony-stimulating activity during immunological reactions *in vitro*. *Nature New Biology*, *244*:175, 1973.

41. Metcalf, D. Studies on colony formation *in vitro* by mouse bone marrow cells. II. Action of colony-stimulating factor. *J. Cell. Physiol.*, *76*:89, 1970.

42. Metcalf, D. Acute antigen-induced elevation of serum colony stimulating factor (CSF) levels. *Immunology*, *21*:427, 1971.

43. Metcalf, D. Effect of thymidine suiciding on colony formation *in vitro* by mouse hematopoietic cells. *Proc. Soc. Exp. Biol. Med.*, *139*:511, 1972.

44. Metcalf, D. Regulation by colony-stimulating factor of granulocyte and macrophage colony formation *in vitro* by normal and leukemic cells. In R. Baserga, ed., *Control of Proliferation in Animal Cells*. New York: Cold Spring Harbor Laboratory, 1974, pp. 887–905.

45. Metcalf, D. *Hemopoietic Colonies. In vitro Cloning of Normal and Leukemic Cells*. Berlin: Springer-Verlag, 1977.

46. Metcalf, D., and Foster, R. Behavior on transfer of serum stimulated bone marrow colonies. *Proc. Soc. Exp. Biol. Med.*, *126*:758, 1967.

47. Metcalf, D., and MacDonald, H.R. Heterogeneity of *in vitro* colony- and cluster-forming cells in the mouse marrow. Segregation by velocity sedimentation. *J. Cell. Physiol.*, *85*:643, 1975.

48. Metcalf, D., and Moore, M.A.S. Regulation of growth and differentiation in haemopoietic colonies growing in agar. In G. E. W. Wolstenholme, ed., *Haemopoietic Stem Cells CIBA Foundation Symposium 13*. Amsterdam: Associated Scientific Publishers, 1973, pp. 157–182.

49. Metcalf, D., and Russell, S. Inhibition by mouse serum of hemopoietic colony formation *in vitro*. *Exp. Hematol.*, *4*:339, 1976.

50. Metcalf, D., and Wahren, B. Bone marrow colony-stimulating activity of sera in infectious mononucleosis. *Brit. Med. J.*, *3*:99, 1968.

51. Metcalf, D., and Wilson, J. Endotoxin-induced size change in bone marrow progenitors of granulocytes and macrophages. *J. Cell. Physiol.* *89*:381, 1976.

52. Metcalf, D., Johnson, G.R., and Wilson, J. Radiation-induced enlargement of granulocytic and macrophage progenitor cells in the mouse bone marrow. *Exp. Hematol.* (in press) 1977.

53. Metcalf, D., Moore, M.A.S., and Shortman, K. Adherence column and buoyant density separation of bone marrow stem cells and more differentiated cells. *J. Cell. Physiol.*, *78*:441, 1971.

54. Metcalf, D., Moore, M.A.S., and Warner, N.L. Colony formation *in vitro* by myelomonocytic leukemic cells. *J. Natl. Cancer Instit.*, *43*:983, 1969.

55. Metcalf, D., Russell, S., and Burgess, A.W. Production of hemopoietic stimulating factors by pokeweed mitogen-stimulated spleen cells. *Transplant. Proceedings* (in press), 1977.

56. Metcalf, D., Chan, S.H., Gunz, F.W., Vincent, P., and Ravich, R.B.M. Colony-stimulating factor and inhibitor levels in acute granulocytic leukemia. *Blood,* 38:143, 1971.

57. Metcalf, D., MacDonald, H.R. Odartchenko, N., and Sordat, B. Growth of mouse megakaryocyte colonies *in vitro. Proc. Nat. Sci. (USA),* 72:1744, 1975.

58. Metcalf, D., Moore, M.A.S., Sheridan, J.W., and Spitzer, G. Responsiveness of human granulocytic leukemic cells to colony-stimulating factor. *Blood,* 43:847, 1974.

59. Metcalf, D., Parker, J., Chester, H.M., and Kincade, P.W. Formation of eosinophilic-like granulocytic colonies by mouse bone marrow cells *in vitro. J. Cell. Physiol.,* 84:275, 1974.

60. Miyake, T., Kung, C.K-H., and Goldwasser, E. The purification of human erythropoietin. *J. Biol. Chem.* (in press), 1977.

61. Moore, M.A.S. *In vitro* studies in myeloid leukemias. In F. J. Cleban, D. Crowther, and J. S. Malpas, eds. *Advances in Acute Leukemia.* Amsterdam: North-Holland, 1975, pp. 161–227.

62. Moore, M.A.S., and Williams, N. Physical separation of colony-stimulating cells from *in vitro* colony-forming cells in hemopoietic tissue. *J. Cell. Physiol.,* 80:195, 1972.

63. Moore, M.A.S., and Williams, N. Functional morphologic and kinetic analysis of the granulocyte-macrophage progenitor cell. In W. A. Robinson, ed., *Hemopoiesis in Culture.* Washington, D.C., DHEW Publication No. 74-205, 1973, p. 17–27.

64. Moore, M.A.S., Williams, N., and Metcalf, D. Purification and characterisation of the *in vitro* colony-forming cell in monkey hemopoietic tissue. *J. Cell. Physiol.,* 79:283, 1972.

65. Moore, M.A.S., Williams, N., and Metcalf, D. *In vitro* colony formation by normal and leukemic human hematopoietic cells: Characterisation of the colony-forming cells. *J. Natl. Cancer Instit.,* 50:603, 1973.

66. Moore, M.A.S., Spitzer, G., Metcalf, D., and Penington, D.G. Monocyte production of colony-stimulating factor in familial cyclic neutropenia. *Brit. J. Haematol.,* 27:47, 1974.

67. Morley, A.A., Quesenberry, P.J., Bealmear, P., Stohlman, F., and Wilson, R. Serum colony-stimulating factor levels in irradiated germfree and conventional CFW mice. *Proc. Soc. Exp. Biol. Med.,* 140:478, 1972.

68. Paran, M., and Sachs, L. The continuous requirement for inducer for the development of macrophage and granulocyte colonies. *J. Cell. Physiol.,* 72:247, 1968.

69. Parker, J.W., and Metcalf, D. Production of colony-stimulating factor in mixed leucocyte cultures. *Immunology,* 26:1039, 1974.

70. Parker, J.W., and Metcalf, D. Production of colony-stimulating factors in mitogen-stimulated lymphocyte cultures. *J. Immunol.,* 112:502, 1974.

71. Ruscetti, F.W., and Chervenick, P.A. Regulation of the release of colony-stimulating activity from mitogen-stimulated lymphocytes. *J. Immunol., 114*:1513, 1975.

72. Ruscetti, F.W., Cypess, R.H., and Chervenick, P.A. Specific release of neutrophilic- and eosinophilic-stimulating factors from sensitized lymphocytes. *Blood, 47*:757, 1976.

73. Shadduck, R.K., and Nagabhushananm, N.G. Granulocyte colony-stimulating factor. I. Response to acute granulocytopenia. *Blood, 38*:559, 1971.

74. Sheridan, J.W., and Metcalf, D. Studies on the bone marrow colony stimulating factor (CSF). Relation of tissue CSF to serum CSF. *J. Cell. Physiol., 80*:129, 1972.

75. Sheridan, J.W., and Stanley, E.R. Tissue sources of bone marrow colony stimulating factor. *J. Cell. Physiol., 78*:451, 1971.

76. Sheridan, J.W., Metcalf, D., and Stanley, E.R. Further studies on the factor in lung-conditioned medium stimulating granulocyte and monocyte colony formation *in vitro. J. Cell. Physiol., 84*:147, 1974.

77. Stanley, E.R., and Heard, P.M. Factors regulating macrophage production and growth: Purification and some properties of the colony stimulating factor from medium conditioned by mouse L cells. *J. Biol. Chem.* (in press), 1977.

78. Stanley, E.R., Cifone, M., Heard, P.M., and Defendi, V. Factors regulating macrophage production and growth. Identity of colony-stimulating factor and macrophage growth factor. *J. Exp. Med., 143*:631, 1976.

79. Stanley, E.R., Hansen, G., Woodcock, J., and Metcalf, D. Colony-stimulating factor and the regulation of granulopoiesis and macrophage production. *Fed. Proc., 34*:2272, 1975.

80. Stanley, E.R., Metcalf, D., Maritz, J.S., and Yeo, G.F. Standardised bioassay for bone marrow colony-stimulating factor in human urine: Levels in normal man. *J. Lab. Clin. Med., 79*:657, 1972.

81. Stewart, C.L., Lin, H-S., and Adles, C. Proliferation and colony-forming ability of peritoneal exudate cells in liquid culture. *J. Exp. Med., 141*:1114, 1975.

82. Trudgett, A., McNeill, T.A., and Killen, M. Granulocyte-macrophage precursor cell and colony-stimulating factor responses of mice infected with *Salmonella typhimurium. Infect. Imm.,* 8:450, 1973.

83. Walker, W.S. Functional heterogeneity of macrophages in the induction and expression of acquired immunity. *J. Reticuloendothel. Soc., 20*:57, 1976.

84. Wiktor-Jedrzejczak, W., Sharkis, S., Ahmed, A., and Sell, K.W. Theta-sensitive cell and erythropoiesis. Identification of a defect in W/W^v anemic mice. *Science* (Wash. D.C.) *196*:313, 1977.

85. Williams, N., and van den Engh, G.J. Separation of subpopulations of *in vitro* colony-forming cells from mouse marrow by equilibrium density centrifugation. *J. Cell. Physiol., 86*:237, 1975.

6

The Mononuclear Phagocyte and Its Regulatory Interactions in Hemopoiesis

Jeffrey I. Kurland

INTRODUCTION

The hemopoietic system exists as a unique hierarchy of pluripotent, committed and maturing cell populations, which continually differentiate into a diverse number of cell lineages. The system remains in exquisite homeostatic balance, allowing for steady-state production of mature end-stage cells, yet retains a rapid and specific responsiveness to altered demand for cell production in any one of the various cell lines. The existence of *in vivo* and *in vitro* clonal assay systems for detection of early hemopoietic stem cells and progenitor cells has permitted a detailed study of stem cell organization and regulation.

The *in vitro* agar culture assay system developed by Bradley and Metcalf (2) led to the recognition of a class of hemopoietic progenitor cells (CFU-c)* restricted to granulocyte-macrophage differentiation. Essential to the cloning technique is the presence of regulatory macromolecules, which have been operationally termed colony-stimulating factors (CSF). A sigmoidal dose-response relationship exists between CSF concentration and the number of colonies developing in marrow culture, and this provides a sensitive bioassay system capable of detecting as little as 10^{-11} to 10^{-12} M CSF/ml (42). The action of CSF *in vitro* is complex and it cannot be considered simply as an inducer of differentiation, since it is necessary for every cell division and differentiation step in the granulocyte-macrophage lineage *in vitro* (29). *In vitro* colony formation by stimulated peritoneal macrophages (CFU-PM) has been described (22,24) and appears to depend upon a factor that is similar, if not identical, to CSF (42).

Recognition that CSF is principally produced by monocytes and macrophages (5,11,21,32) and that it acts to promote granulopoiesis, as well as to increase monocyte production and macrophage proliferation, introduces the problem of mechanisms designed to counterbalance this positive feedback drive. A number of mechanisms have been revealed in *in vitro* studies and many, if not all, may be of physiological significance *in vivo*.

DUALISM BETWEEN CSF AND SYNTHETIC PROSTAGLANDINS

Conceptually, one could imagine two ways in which the proliferative stimulus of CSF might be limited. The first is by a direct action on the CSF molecule, resulting in either the masking or de-

---

*The term colony-forming unit (CFU-) denotes the existence of a hemopoietic cell on the basis of its ability to generate colonies by successive cell divisions.

struction of its biological activity, and second, by some action on the progenitor cells that results in their decreased responsiveness to stimulation by CSF. It has recently been found that the E-series prostaglandins (PGE), PGE_1, and PGE_2 (but not $PGF_{2\alpha}$ or $PGF_{1\alpha}$) profoundly suppress CFU-c proliferation *in vitro* (22). This inhibition can be effectively prevented by incubation of bone marrow cells with the synthetic prostaglandin antagonist, SC-19220 (40), as shown in Figure 6-1, thereby indicating the existence of prostaglandin receptors on the granulocyte-macrophage progenitor cell (19). Of particular importance, however, is the manner in which PGE inhibits the clonal proliferation of both CFU-c and CFU-PM. In the presence of PGE, these hemopoietic cells require an eight- to ninefold greater concentration of CSF to proliferate to the

same extent as in the absence of PGE (22). Therefore, inhibition by PGE operates by decreasing hemopoietic progenitor cell responsiveness to the stimulatory activity of CSF. In this regard, a physiological antagonism is indicated, since an increase in the concentration of available CSF can counteract the effectiveness of PGE-mediated inhibition.

Cell type specificity of the CSF-PGE dualism has been documented by comparisons with other clonogenic hemopoietic cells, including a B-lymphocyte (CFU-BL) and a murine myelomonocytic leukemia (WEHI-3), each of which proliferates in semisolid medium without an exogenous source of CSF (22). Though PGE can inhibit the clonal proliferation of these cells, CSF is without effect in reducing the inhibition. Thus, the antagonism between CSF and PGE is operative only within the

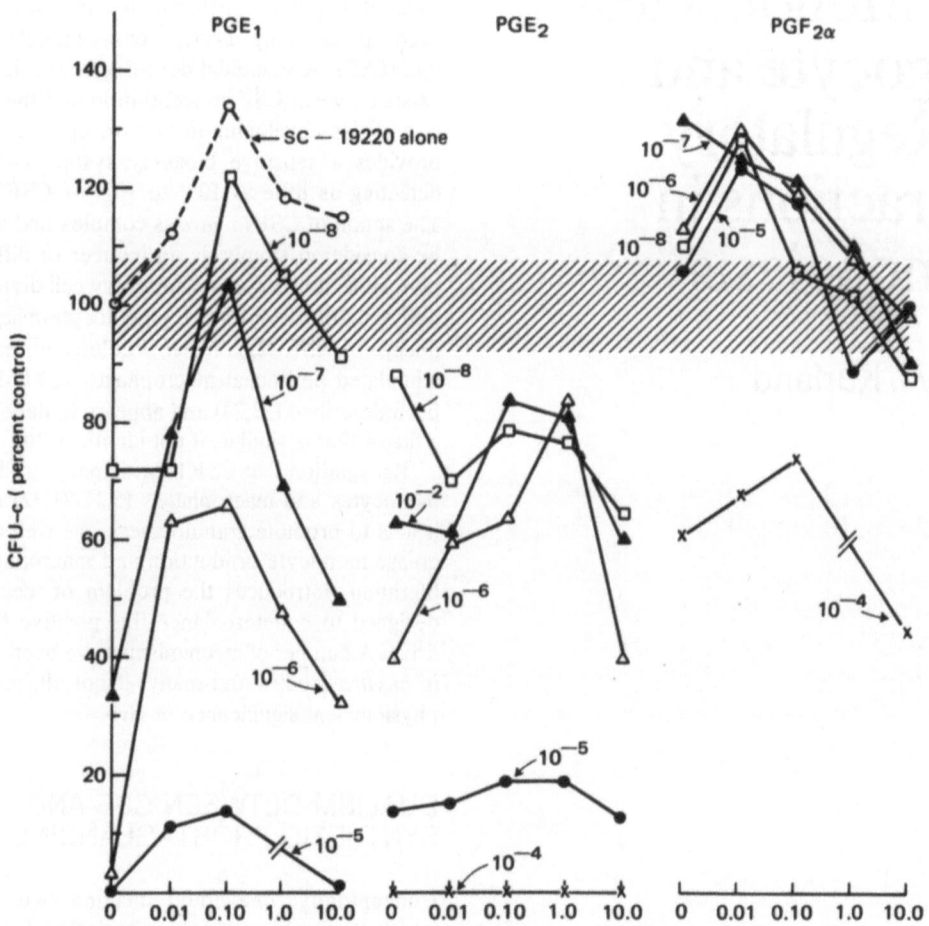

FIGURE 6-1. Effects of the prostaglandin (PGE) antagonist SC-19220 on mouse bone marrow c cultured in the absence and in the presence of PGE. 7.5×10^4 $C_{57}BL/6$ bone marrow cells were cultured in the presence of CSF provided by conditioned medium from the murine myelomonocytic leukemic (WEHI-3) cell line. Various molar concentrations of PGE_1, PGE_2, and $PGF_{2\alpha}$ were added to the cultures either in the absence or in the presence of various concentrations of SC-19220. Control colony formation ± standard deviation in the absence of PGE and SC-19220 is shown by the cross-hatching.

two hemopoietic progenitor compartments (CFU-c and CFU-PM), in which CSF is required as a proliferative stimulus (Table 6-1).

CYCLIC 3′,5-ADENOSINE MONOPHOSPHATE IN THE MEDIATION OF PGE ACTIVITY

Examination of various mammalian tissues reveals that the E-series prostaglandins possess an infinity for a receptor, are capable of activating membrane-associated adenylate cyclase, and can thereby increase intracellular levels of cyclic 3′,5′-adenosine monophosphate (cAMP). The extremely small incidence of CFU-c in hemopoietic tissues, such as bone marrow, spleen, or blood, does not permit a direct analysis of the ability of PGE to elevate cAMP levels within the CFU-c. However, by the use of a number of pharmacological agents that modulate the activity of the enzyme responsible for cAMP degradation (cAMP phosphodiesterase), there is indirect evidence that cAMP mediates granulocyte-macrophage progenitor cell inhibition by PGE (19). Theophylline, an inhibitor of the phosphodiesterase, markedly potentiates the inhibition of the progenitor cells by PGE_1, whereas imidazole, an activator of cAMP phosphodiesterase, effectively reverses CFU-c inhibition by PGE_1. This dualistic action by the two phosphodiesterase modulators suggests that the PGE acts on adenylate cyclase rather than on phosphodiesterase. Furthermore, carbamylcholine, $PGF_{2\alpha}$, and imidazole, agents reported to increase intracellular levels of cyclic 3′,5′-guanosine monophosphate (cGMP), augment the stimulatory actions of CSF, thereby suggesting that the CSF-PGE antagonism may extend to an intracellular dualism between cGMP and cAMP (19).

REGULATION OF GRANULOPOIESIS AND MACROPHAGE PRODUCTION BY MACROPHAGE-DERIVED PGE

Though synthetic prostaglandin can suppress CFU-c proliferation, it is important, from a physiological standpoint, to determine whether PGE is synthesized and released by cells of the hemopoietic system. By employing two-dimensional cell separation procedures of cell density and adherence, we have shown that the principal PGE-producing cells in human peripheral blood and mouse peritoneal exudates are the light density and adherent monocyte and macrophage (18), which possess histochemical affinity for cytoplasmic nonspecific α-naphthyl acetate esterase (46) and Neutral Red (44). In contrast,

TABLE 6-1 Influence of CSF and PGE_1 on Hemopoietic Cells Capable of Clonal Proliferation in Soft-agar Medium

CELL TYPE	CSF DEPENDENCE[a]	PGE SENSITIVITY[b]
CFU-c	+++	++
CFU-PM	++++	++
CFU-BL	−	+++
CFU/WEHI-3	−	++++

From Kurland and Moore (22). Reproduced by permission from *Experimental Hematology*.
[a]Denotes the requirement for CSF to promote maximum colony formation.
[b]Denotes the molar concentration of PGE required to significantly inhibit colony formation.

PGE is undetectable in culture supernatants prepared from the non-adherent components of mouse peritoneal exudates, as well as from pure populations of normal human peripheral blood granulocytes (nonadherent, density > 1.070 g/cm³) and lymphocytes (non-adherent, density < 1.070 g/cm³). These findings, therefore, suggest that in addition to being able to produce the myelopoietic stimulator, CSF, both monocytes and macrophages also possess the capacity to effectively inhibit the proliferation of both CFU-c and CFU-PM by the elaboration of PGE.

Experiments were carried out to establish whether such a proposed cellular interaction between monocytes or macrophages and the granulocyte-macrophage progenitor cell did, in fact, exist. A two-layer, semisolid culture system was established to permit quantitative analysis of the effects of populations of adherent monocytes or macrophages on the clonal proliferation of the hemopoietic progenitor cell. Direct cell contact interactions were prevented by a cell-free soft agar layer separating the adherent macrophages from the overlying bone marrow cells, thereby ensuring that diffusible molecules of monocyte or macrophage origin would be analyzed.

Such studies have shown that mouse peritoneal macrophages, by their production of CSF, stimulate the proliferation of mouse bone marrow CFU-c. Similarly, human monocytes are capable of stimulating human bone marrow CFU-c. The magnitude of stimulation (or the concentration of CSF) increases with the number of monocytes or macrophages present in the culture underlayer. However, colony formation declines to a point at which no detectable proliferation is observed when greater than optimal numbers of macrophages are present. This ability of the macrophage to both stimulate and inhibit the proliferation of CFU-c suggests the production of opposing regulators, with inhibition favored at high macrophage concentrations (21).

The influence of monocyte-derived CSF and PGE on the clonal proliferation of human CFU-c is shown in Figure 6-2(A). Non-adherent light density (< 1.070 g/cm³) human bone marrow cells were cultured over agar underlayers containing various numbers of adherent human peripheral blood monocytes (density < 1.070 g/cm³). In those cultures containing a monocyte-free underlayer, no colony formation was observed, but in the presence of increasing numbers of adherent monocytes, CFU-c were induced to proliferate. Above a monocyte concentration of 2.3×10^5 monocytes/culture, no further increase in net colony formation was observed. Addition of the prostaglandin synthetase inhibitor, indomethacin (8), into the monocyte underlayers resulted in a linear stimulation of CFU-c proliferation over the concentration range of monocytes tested, and no diminution of colony formation was observed at high monocyte numbers (17). Indomethacin had no direct effect on CFU-c and did not stimulate colony formation in the absence of monocytes. The potentiating effect of indomethacin occurred only in the presence of appreciable numbers of monocytes ($\geq 2.3 \times 10^5$/culture). In the absence of indomethacin, these numbers of monocytes elaborated significant levels of PGE into the medium of replicate cultures as determined by radioimmunoassay [Figure 6-2(B)]. Levels of prostaglandin first became detectable at a concentration of 2.3×10^5 monocytes, the optimal monocyte number above which no further increase in CFU-c stimulation occurred. The levels of PGE continued to increase as a linear function of the monocyte concentration. The inability of monocytes to synthesize PGE in the presence of indomethacin permitted colony formation to occur only in proportion to the underlying adherent monocyte population.

These findings indicate that the proliferation of the committed granulocyte-macrophage stem cell is regulated by both positive and negative feedback controls, involving monocyte-derived CSF and PGE, respectively. The accumulation of monocyte-derived PGE limits the numbers of CFU-c that undergo proliferation in response to a given concentration of monocyte-derived CSF and therefore results in an underestimate of the incidence of potentially clonable committed stem cells. When the constitutive synthesis of PGE is inhibited, the magnitude of CFU-c proliferation is determined only by the net level of CSF, which increases as a linear function of the monocyte concentration (17).

The ability of the mononuclear phagocyte to regulate *in vitro* myelopoiesis by the elaboration of two factors with opposing activities suggested that the macrophage may be able to modulate the synthesis of the indomethacin-sensitive inhibitory factor in response to the local concentration of CSF.

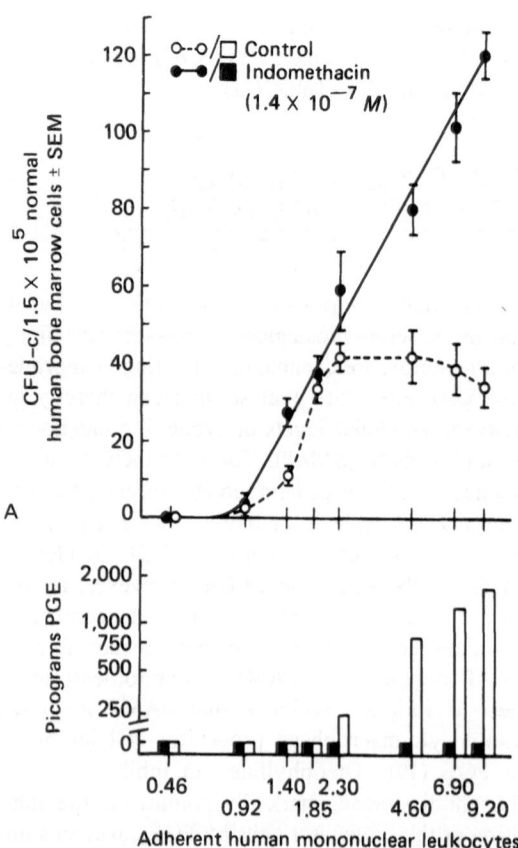

FIGURE 6-2. Modulation of human monocyte production of colony-stimulating factor (CSF) and Prostaglandin E (PGE) by indomethacin. (A) Various numbers of adherent human monocytes (density < 1.070 g/cm³) were overlayered with 1.0 ml of 0.5 percent agar in the absence and in the presence of $1.4 \times 10^{-7}M$ indomethacin. Then 1 ml of target cell suspensions of 1.5×10^5 normal human bone marrow cells in 0.3 percent soft agar were added above the adherent monocyte underlayers, and the number of colonies containing more than 40 cells in the bone marrow overlayers were determined after 10 days of culture. Quadruplicate cultures were scored for each point and the results are expressed as the number of CFU-c ($\pm$ SEM) that began to proliferate to form colonies. (B) Replicate liquid cultures containing the identical number of adherent monocytes as in the agar underlayers were used to obtain supernatant media to determine PGE production for each monocyte concentration. From Kurland, Bockman, Broxmeyer, and Moore (17). Reproduced by permission of *Science*. Copyright 1978 by the American Association for the Advancement of Science. *199*:552, 1978.

The experimental approach to this query of the homeostatic potential of the macrophage, shown in Figure 6-3, was based upon the species specificity of CSF action. Normal human bone marrow cell cultures were stimulated by the presence of human CSF elaborated by 1×10^6 normal human peripheral blood leukocytes suspended in an agar under-

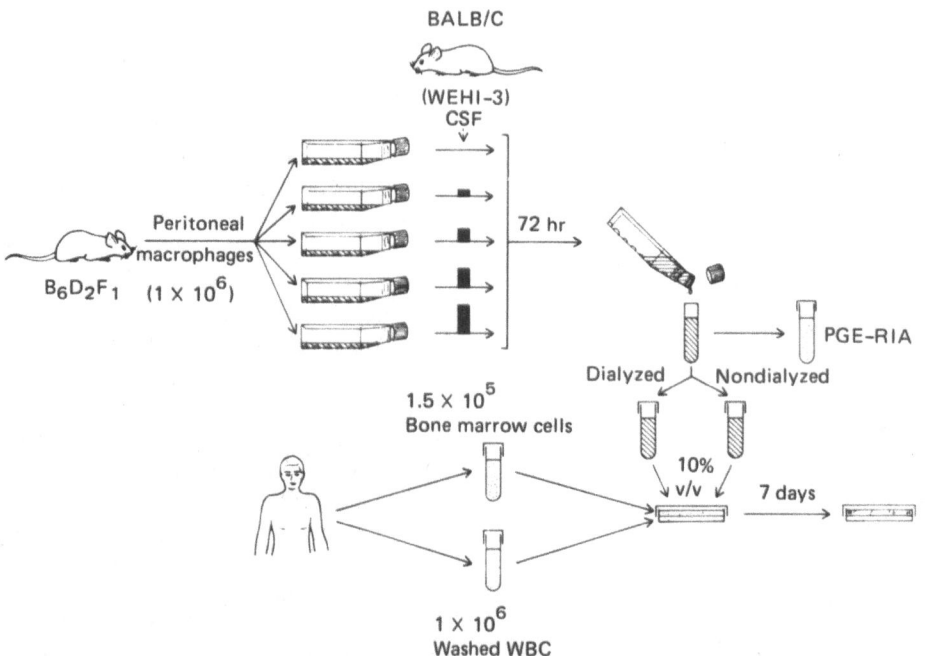

FIGURE 6-3. Experimental approach to test the ability of CSF to induce the generation of both CFU-c inhibitory activity and PGE by mouse peritoneal macrophages. Supernatant media from liquid cultures of 1×10^6 adherent $B_6D_2F_1$ peritoneal macrophages/ml incubated in the absence and in the presence of increasing concentrations of murine CSF provided by WEHI-3–conditioned medium were collected after 48 hr and assayed for PGE by radioimmunoassay. One-half of the murine macrophage supernatants were dialyzed against five changes of phosphate buffered saline containing 5.0 percent FCS for 3 days, sterilized by Millipore filtration (0.45 μ pore size), and 0.1 ml of each dialyzed and nondialyzed sample was added to soft agar cultures of 1.5×10^5 normal nonadherent human bone marrow cells (density < 1.070 g/cm³) stimulated by human leukocyte feeder layers.

layer, and the numbers of CFU-c that underwent clonal proliferation to form colonies is shown by the shaded area in Figure 6-4,A. Cell-free supernatants from 48-hr liquid cultures of 1×10^6 murine peritoneal macrophages/ml were prepared in the absence and presence of a potent source of murine CSF, provided by conditioned medium from a murine myelomonocytic leukemic cell line (WEHI-3) (5). Neither the individual presence of the murine macrophage supernatant nor the murine CSF to which the macrophages were exposed had any effect on the human CSF-stimulated CFU-c proliferation when tested at 10 percent (v/v) in soft agar cultures. However, the peritoneal macrophage supernatant prepared in the continued presence of the murine CSF profoundly inhibited the proliferation of human CFU-c stimulated by the human leucocyte-derived CSF [Figure 6-4(A)]. The human CFU-c inhibition was proportional to the concentration of murine CSF present during the active generation of the murine macrophage supernatant. After simple dialysis, these same macrophage supernatants lost all detectable inhibitory activity

against human bone marrow CFU-c. These findings indicate that murine macrophages, which otherwise had no effect on exogenously stimulated human CFU-c, are induced by increasing concentrations of murine CSF to elaborate a low molecular weight, nonspecies-specific inhibitor of committed granulocyte-macrophage stem cell proliferation (17).

As shown in Figure 6-4(B), the same concentrations of murine CSF that promoted the elaboration of CFU-c inhibitory factor induced the coincident production of PGE by 1×10^5 murine peritoneal macrophages incubated under conditions identical to those for the preparation of the macrophage supernatants used in the previous experiment. Note the inverse relationship between the CSF-stimulated levels of PGE [Figure 6-4(B)] and the CSF-induced human CFU-c inhibitory activity [Figure 6-4(A)], both represented as the constitutive contribution of 1×10^5 macrophages. In the absence of PGE production, the diffusible contribution of 1×10^5 murine macrophages is entirely without CFU-c inhibitory activity. More striking, however, is the manner by which small elevations in the CSF-

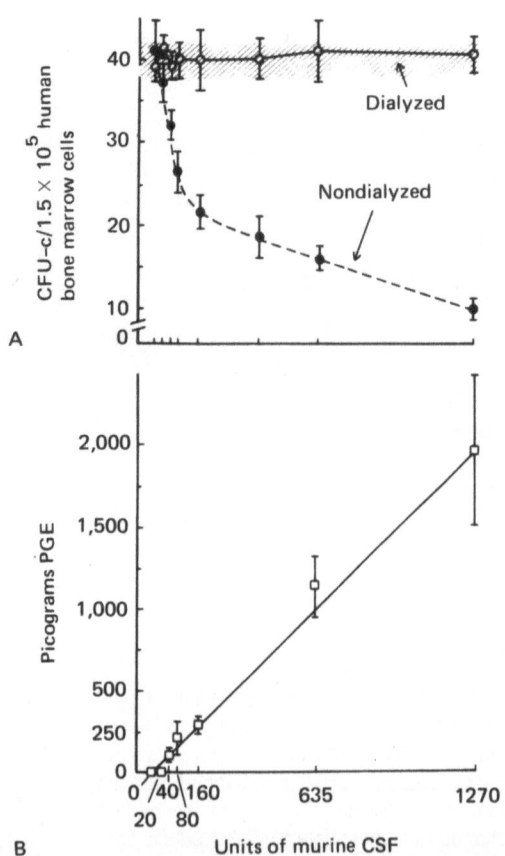

FIGURE 6-4. The coincident generation of a diffusible inhibitor of human bone marrow CFU-c proliferation and prostaglandin E (PGE) by murine peritoneal macrophages in response to increasing concentrations of murine CSF. (A) Dialyzed and nondialyzed supernatants from macrophages cultured in the absence and in the presence of various units of WEHI-3 CSF were assayed against normal human bone marrow CFU-c as described in Figure 6-3, and the number of CFU-c that had undergone clonal proliferation in four replicate cultures to generate colonies was determined after 7 days of incubation. Cross-hatched area, control colony formation by human bone marrow CFU-c ± SE, in the presence of a feeder layer containing 1×10^6 washed human blood leukocytes. The production of PGE by CSF-treated macrophage was determined by radioimmunoassay. The results are expressed as the mean level of PGE ± standard deviation of the mean elaborated by 1×10^5 murine peritoneal macrophages in four replicate cultures. From Kurland, Bockman, Broxmeyer, and Moore (17). Reproduced by permission of *Science*. Copyright 1978 by the American Association for the Advancement of Science.

induced production of PGE by macrophages can account for the marked diminution in the numbers of proliferating CFU-c.

The ability of CSF to stimulate PGE production suggests a feedback mechanism whereby the macrophage can control its synthesis of PGE by sensing

the CSF concentration in its external milieu. In this regard, endotoxin, an agent known to both markedly elevate serum levels of CSF *in vivo* and to stimulate CSF production by monocytes and macrophages *in vitro*, was found to augment dramatically PGE synthesis by murine peritoneal macrophages (18). Since the macrophage constitutively elaborates both CSF and PGE, the possibility that CSF production may be required for continued PGE synthesis by the macrophage was examined. Two methodologies were employed (Figure 6-5), utilizing peritoneal macrophages from LPS nonresponsive (C_3H/HeJ) and LPS responsive (C_3HeB/FeJ) mice and a number of CSF-inducible macrophage cell lines (37). Both peritoneal and cell line macrophages were exposed to lipopolysaccharide (LPS) (*Salmonella typhosa* W0901, Difco Laboratories) and WEHI-3 CSF for 48 hours, and the supernatant culture fluids were assayed for PGE by radioimmunoassay and for CSF against a target population of $B_6D_2F_1$ mouse bone marrow cells. As shown in Table 6-2, both CSF and PGE were present in the supernatant media of both C_3H/HeJ and C_3HeB/FeJ peritoneal macrophages. In contrast to the ability of endotoxin to markedly increase the production of both CSF and PGE by C_3HeB/FeJ macrophages, no potentiation of either factor from C_3H/HeJ macrophages was observed. However, exposure of both macrophage strains to WEHI-3–conditioned media containing CSF resulted in a dramatic stimulation of PGE synthesis. Similarly, C_3H/HeJ macrophages demonstrated a heightened CSF response following exposure to an alternate macrophage-stimulating agent, such as zymosan, and this was accompanied by a marked augmentation of PGE synthesis. Therefore, the genetically determined LPS nonresponsiveness of the C_3H/HeJ macrophage with respect to stimulation of CSF production, extends also to the synthesis of PGE. However, if an exogenous source of CSF is provided, or alternatively, if the macrophages are exposed to another agent that stimulates CSF production, the C_3H/HeJ macrophage is induced to synthesize large quantities of PGE. These findings clearly indicate that endotoxin stimulates PGE synthesis by a mechanism involving heightened CSF production.

Similar findings have been obtained using continuous macrophage cell lines (Table 6-3). The murine myelomonocytic leukemia, WEHI-3 (43), adapted to grow in continuous culture (38) and the macrophage cell lines, SK-2 (31), were found to be constituitive producers of both CSF and PGE, and the potentiation of CSF production by LPS coincidently augmented PGE synthesis. The RAW-264 macrophage tumor induced by Abelson leukemia virus (39) and J774, a Balb/c macrophage cell line

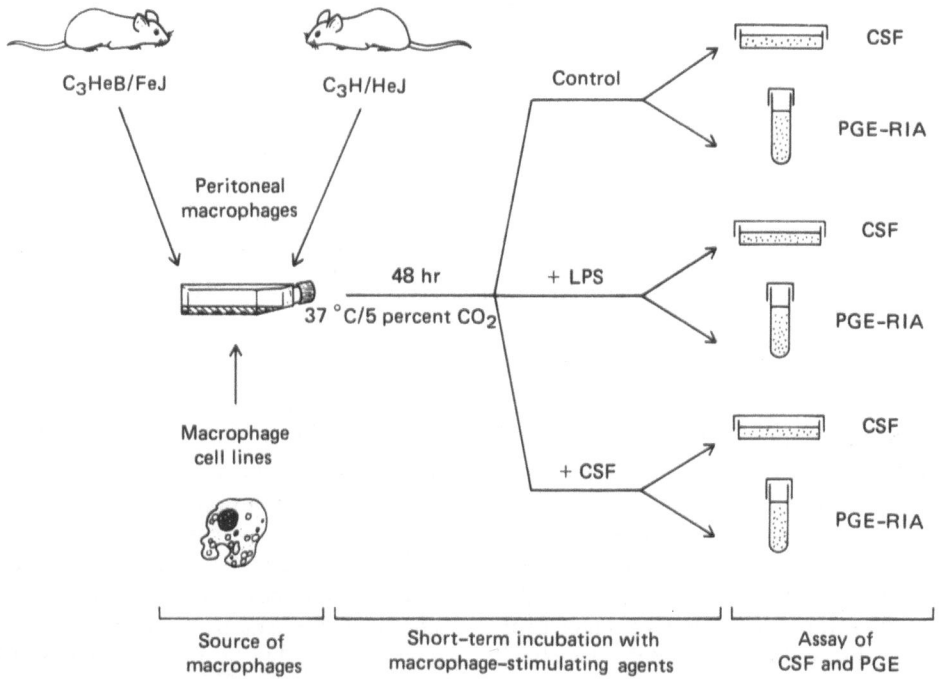

FIGURE 6-5. Experimental methodologies used in determining the relationship between colony-stimulating factor (CSF) production and the synthesis of prostaglandin E (PGE) by normal and neoplastic macrophages. Normal C_3HeB/FeJ and C_3H/HeJ peritoneal macrophages and various macrophage cell lines were cultured in the absence and in the presence of lipopolysaccharide (LPS) and CSF from WEHI-3–conditioned medium. Culture supernatants were harvested after 48 hr and assayed for CSF using a target population of $B_6D_2F_1$ bone marrow cells, and by radioimmunoassay for PGE.

(38), were not constituitive producers of either CSF or PGE, but following exposure to endotoxin, the cells of these lines were induced to synthesize significant amounts of both CSF and PGE. Similarly, incubation of the inducible macrophages in the presence of a source of CSF alone provided the necessary stimulus for an increased PGE synthetic response. In contrast, neither a murine T-cell lymphoma, EL_4, nor the lymphocytic B-cell lymphoma, RBL-3 (41), elaborated any PGE in the ab-

sence or presence of either endotoxin or CSF (18). Thus, the capability to produce PGE by normal monocytes and macrophages extends to populations of neoplastic mononuclear phagocytes, and indeed, the PGE synthetic response by both normal and neoplastic macrophages depends upon an appropriate stimulation of CSF production. This relationship between CSF production and PGE synthesis is shown schematically in Figure 6-6.

Further evidence to suggest that basal CSF pro-

TABLE 6-2 Effect of LPS, Zymosan, and CSF on PGE Production by C_3HeB/FeJ and C_3H/HeJ Peritoneal Macrophages

Peritoneal Macro- phages[c]	CSF PRODUCTION[a]			PROSTAGLANDIN E PRODUCTION[b]			
	Control	LPS	Zymosan	Control	LPS	CSF	Zymosan
C_3HeB/FeJ	26 ± 6	74 ± 9	n.d.[d]	200 ± 30	1905 ± 182	6445 ± 480	n.d.[d]
C_3H/HeJ	49 ± 5	51 ± 4	n.d.[d]	469 ± 74	198 ± 120	1949 ± 130	n.d.[d]
C_3H/HeJ	18 ± 2	n.d.[d]	50 ± 2	369 ± 62	n.d.	1184 ± 154	2084 ± 110

[a]Number of CFU-c $\pm$ S.D./7.5×10^4 $B_6D_2F_1$ bone marrow cells stimulated by 0.1 ml of macrophage-conditioned media.

[b]Picograms PGE $\pm$ S.D. determined by radioimmunoassay.

[c]3×10^5 adherent peritoneal macrophages/ml incubated under conditions described in Figure 6-5.

[d]No data.

TABLE 6-3 Production of CSF and PGE by Murine Macrophage Cell Lines

MACROPHAGE CELL LINES[a]	CSF PRODUCTION[b]		PROSTAGLANDIN E PRODUCTION[c]		
	Control	+LPS	Control	+LPS	+CSF
WEHI-3.A	84 ± 4	161 ± 18	173 ± 91	1040 ± 85	n.d.[d]
SK-2	16 ± 3	91 ± 9	296 ± 13	3462 ± 149	n.d.[d]
RAW-264	0	111 ± 10	0	305 ± 26	393 ± 18
J774	0	128 ± 14	0	1033 ± 61	1670 ± 88

[a] 5×10^5 viable cells/ml cultured, as described in Figure 6-5.
[b] Number of CFU-c ± S.D./7.5 $\times$ 10^4 $B_6D^2F_1$ bone marrow cells stimulated by 0.1 ml of macrophage cell line-conditioned media.
[c] Picograms PGE ± S.D.
[d] No data.

duction is necessary for ongoing PGE synthesis by phagocytic mononuclear cells has been obtained using an extract of mature human granulocytes, which is reported to diminish CSF production by monocytes and macrophages (3,4). When human monocytes were incubated in the presence of the granulocyte extract, a marked suppression of PGE synthesis was observed [Table 6-4(A)]. This effect is more pronounced with lower numbers of monocytes when synthesis of PGE is completely inhibited. With higher monocyte numbers, the effect of the granulocyte extract decreases, but is still sufficient to cause a significant reduction in PGE synthesis. However, the PGE synthesis inhibitory effect of the granulocyte extract is completely reversed in the presence of concentrated, dialyzed human monocyte-conditioned medium used as an

exogenous source of human CSF [Table 6-4(B)]. In this regard, the granulocyte extract does not function as an inhibitor of the prostaglandin synthetase cyclo-oxygenase, as does indomethacin, since the addition of CSF to indomethacin-treated monocytes or macrophages does not reverse the suppression of PGE synthesis. Thus, the granulocyte extract, conceivably by reducing the availability of endogenous CSF within the mononuclear phagocyte, is capable of limiting the basal generation of PGE. Preliminary evidence using a radiorelease assay on murine macrophage monolayers has indicated that semi-purified, serum-free WEHI-3 CSF liberates the essential precursor of prostaglandin, arachidonic acid, from membrane-bound phospholipid stores, possibly by acting on phospholipase. It is therefore conceivable that the granulocyte ex-

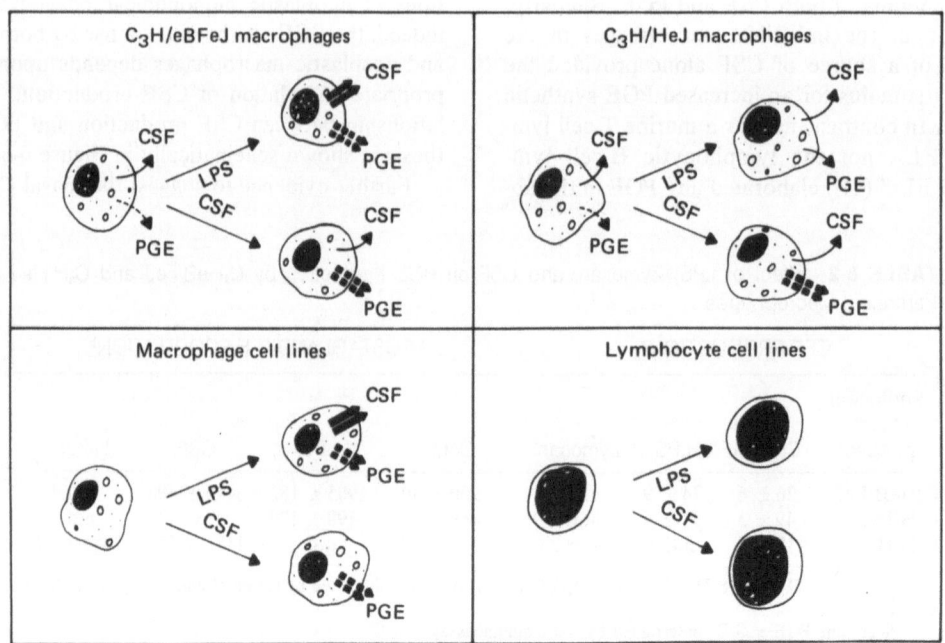

FIGURE 6-6. Role for colony-stimulating factor in macrophage prostaglandin E (PGE) synthesis.

TABLE 6-4 Effect of Human Granulocyte Extract and CSF on PGE Production by Human Monocytes

	A. PICOGRAMS PGE/ADHERENT MONONUCLEAR CELLS ($\times 10^5$)[a]			
	2.3	4.6	6.9	9.2
Control	256 ± 70	711 ± 210	1130 ± 270	1730 ± 340
GE (5 percent v/v)	0	0	375 ± 130	1080 ± 160
IND ($1.4 \times 10^{-7}M$)	0	0	0	0

B. PICOGRAMS PGE/2.4×10^6 ADHERENT MONONUCLEAR CELLS	
Control	6910 ± 260
GE (5 percent, v/v)	4662 ± 100
CSF (1:4)	8558 ± 580
IND ($1.4 \times 10^{-7}M$)	0
GE + CSF	9645 ± 780
IND + CSF	0

[a]Normal adherent human peripheral blood mononuclear cells (monocytes) were incubated in the absence and the presence of human granulocyte extract (GE), indomethacin (IND), colony-stimulating factor (CSF) provided by dialyzed human monocyte-conditioned medium or in various combinations of the different factors. Culture fluids were harvested after 48 hr and assayed for PGE. Results are expressed as picograms PGE $\pm$ SEM.

tract inhibits PGE synthesis by decreasing endogenous CSF, which, in turn, reduces the amount of arachidonic acid available to the prostaglandin synthetase complex (Figure 6-7).

The findings to date suggest that both CSF and PGE have their origins in a common regulatory cell identified functionally and histochemically as a macrophage. Alternatively, CSF and PGE may be produced by two distinct, and as yet unseparable, macrophage subpopulations, as illustrated in Figure 6-8. The sequence shown in Figure 6-8(I) and 6-8(II), depict two such macrophage subsets (A and B), each of which synthesize either CSF or PGE, but not both. Since the synthesis of CSF and PGE are causally related, it would appear that endotoxin, a stimulus for CSF production by macrophage A, would also be an appropriate stimulus for PGE production by macrophage B [Figure 6-8(I)]. In a similar manner, endotoxin may stimulate macrophage A to produce and release CSF, which, in turn, is capable of stimulating PGE synthesis by macrophage B [Figure 6-8(II)]. Though dual origins of CSF and PGE is feasible, the findings obtained using continuous macrophage cell lines, which conceivably represent homogeneous populations of cells, would support the premise that a single macrophage is the source of both CSF and PGE [Figure 6-8(III)].

These observations point to the macrophage as a surveillance cell, which, under steady-state conditions, is elaborating basal levels of CSF and prostaglandin E. The constitutive contribution of CSF to granulopoiesis and monocyte production is rapidly increased in physiologically perturbed circumstances, such as infection. Progressive increases in CSF levels would promote recruitment of additional granulocytes and monocytes, by acting on the marrow CFU-c population, and would also promote local macrophage proliferation. This process would be self-limiting, since a progressive increase in CSF beyond a critical concentration within the milieu of the macrophage is ultimately sensed and serves to stimulate the coincident production of PGE. The lability of the prostaglandin molecule (1), coupled with its rapid and efficient clearance by the lung (9,12) and presumably other tissues, dictates that the inhibitory effects of PGE on the responsiveness of the myeloid progenitor cells to CSF occur via short-range cellular interactions and are transitory in nature. Thus, the continuation of the myelopoietic-modulating effects of PGE would be subject to a persistent elevation in local CSF levels, which, in turns, maintains the mononuclear phagocyte in an accelerated state of PGE synthesis.

THE REGULATORY ROLE OF MACROPHAGES IN LYMPHOPOIESIS

The macrophage has been implicated in the elaboration of a spectrum of molecules that alter or modulate the proliferation and differentiation of lymphoid cells. These include lymphocyte-activating factor (LAF), which potentiates the mitogenic response to T-lymphocytes to lectin and histocompatibility antigens (10), as well as factors that increase the helper function of T-lymphocytes (45) and the promotion of soluble-mediator production by lymphocytes (36). The role of the macrophage in the regulation of humoral immune responses is, however, less well understood, since macrophages are required for humoral immune responses in vitro (13), yet mitogen-induced B-lymphocyte proliferative responses are suppressed in the presence of

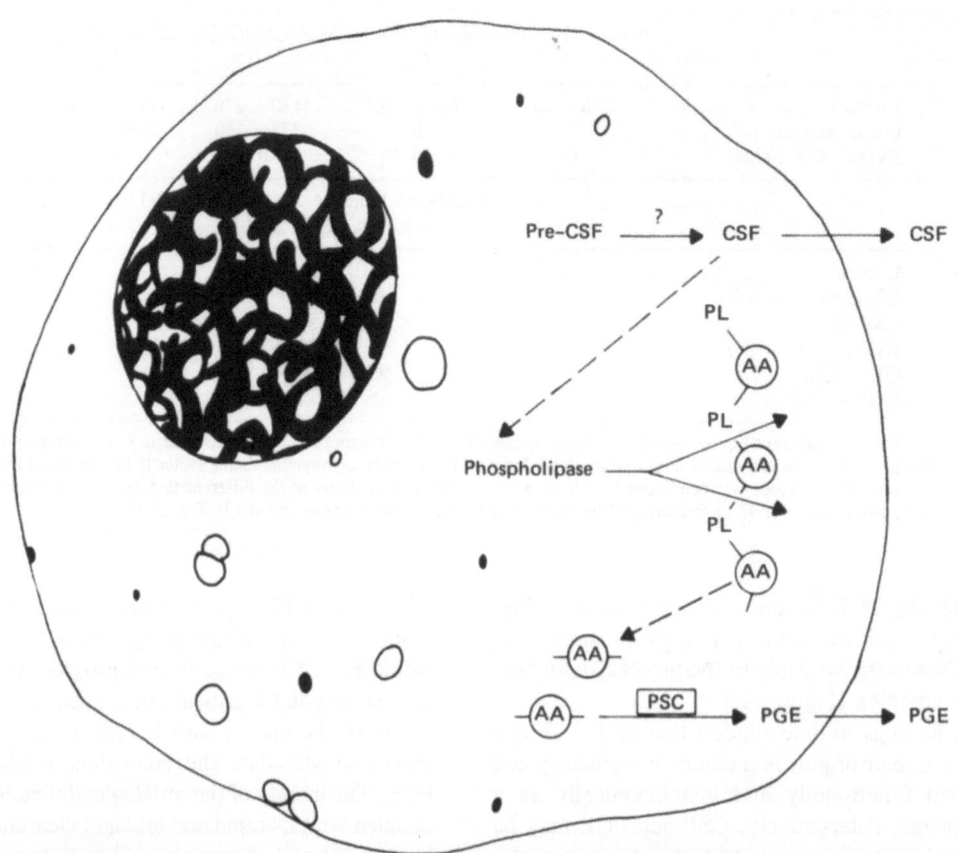

FIGURE 6-7. Schematic diagram illustrating the action of colony-stimulating factor (CSF) on prostaglandin E (PGE) synthesis by the macrophage. The CSF, presumably by acting on phospholipase, liberates arachidonic acid (AA) from membrane-bound phospholipid (PL) and makes it available to the prostaglandin synthetase complex (PSC) for enzymatic conversion to PGE.

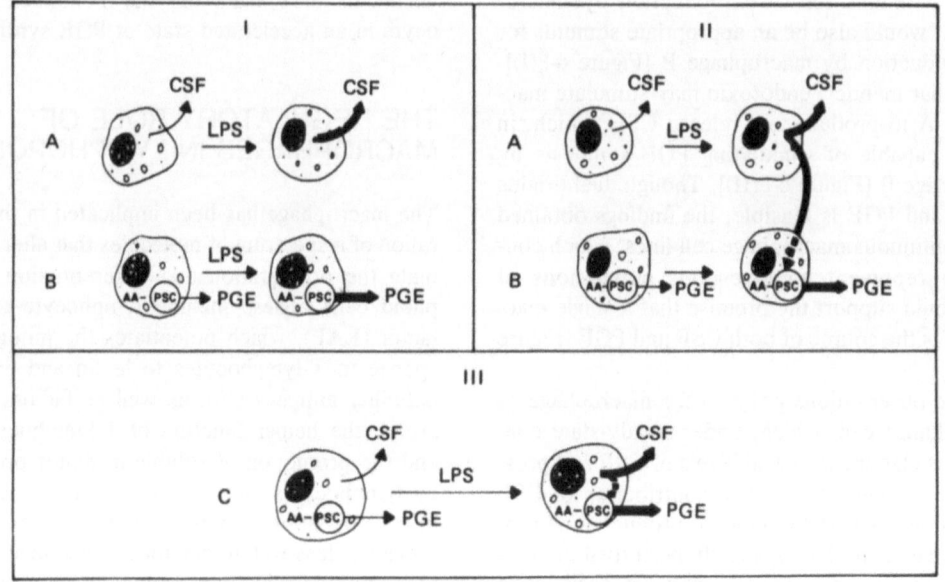

FIGURE 6-8. Single- versus dual-macrophage origin of colony-stimulating factors (CSF) and prostaglandin E (PGE).

macrophages (25). The recently developed system of *in vitro* murine B-lymphocyte cloning has thrown some light on the immunoregulatory role of macrophages. In this system, a functional subpopulation of murine B-lymphocytes proliferate in semisolid agar culture to form colonies (15,30). This process is dependent upon 2-mercaptoethanol (27) and B-cell mitogens native to laboratory grade agar (16). Kurland et al. (20) have investigated the influence of diffusible macrophage-derived factors on this clonal proliferation, using a two-layer culture system that prevented macrophage-lymphocyte contact and permitted B-cell activation to be critically assessed under conditions of extremely low cell density. Adherent peritoneal macrophages potentiated both the number and size of developing B-cell colonies, particularly when low numbers of spleen or lymph node cells or macrophage-depleted lymphoid cell suspensions were used (Table 6-5). Macrophage-depleted lymph node cells or neonatal spleen cells gave virtually no colonies, but colony formation was restored by the presence of an optimal number of macrophages. When the number of macrophages exceeded that required for optimal stimulation, colony formation was suppressed, an effect that was largely prevented by indomethacin. Similar stimulatory and inhibitory activities were also present in media conditioned by varying numbers of peritoneal macrophages (20). The diffusible inhibitor was identified as PGE, which suppresses B-lymphocyte cloning at concentrations as low as 10^{-8} to 10^{-10} M (22). The stimulatory factor may be similar to the 5×10^4 molecular weight factor described by Namba and Hanoaka (35), which is produced by adherent phagocytic cells and stimulates IgM plasmocytoma cell proliferation. In this regard, conditioned medium from a murine myelomonocytic cell line (WEHI-3), which elaborates both CSF (22) and T-lymphocyte activating factor (23), cannot be substituted for macrophages in the initiation of B-lymphocyte colony formation, suggesting the non-identity of this B-lymphocyte stimulating activity with other known factors that alter hemopoietic and lymphoid function (Kurland, submitted for publication).

The role of the antigen, or the mitogen, that stimulates B-lymphocyte proliferation is more complex than a simple interaction with the responsive B-cell, since such agents as LPS and sheep red blood cells (sRBC) may also modulate the elaboration of immunoregulatory factors by the macrophage (20). In the presence of low numbers of macrophages, sRBC facilitates the production and/or release of the B-cell colony-stimulating factor, whereas LPS promotes the elaboration of inhibitory factors from high numbers of macrophages. This heightened suppressor activity of macrophages exposed to LPS is largely ameliorated in the presence of indomethacin (Kurland, submitted for publication).

THE MACROPHAGE AND OTHER HEMOPOIETIC CELLS

An adequate case may be made for considering the phagocytic mononuclear cell population and, specifically, the resident macrophages within the hemopoietic tissues, as of central importance in controlling the proliferation and differentiation of granulocyte-macrophage committed stem cells and B-lymphocytes. A more general regulatory role of macrophages in hemopoietic differentiation is also suggested by studies implicating the production of stimulatory factors influencing other hemopoietic cell lineages.

Regulation of T-lymphopoiesis is also beginning to show a number of parallels with myelopoiesis. Colony formation by murine and human PHA-stimulated T-lymphocytes (CFU-TL) in soft agar me-

TABLE 6-5 Macrophage Requirement for B-Lymphocyte Colony Formation by Adult CBA/CUM Spleen, Lymph Node, and Neonatal Spleen Cells[a]

NUMBER OF PEC USED. ADHERENT MØ. SOURCE[b]	ADULT SPLEEN			ADULT LYMPH NODE		NEONATAL SPLEEN, UNSEPARATED
	Unseparated	G-10 Sephadex separated	Carbonyl iron separated	Unseparated	G-10 Sephadex separated	
—	310 ± 21	120 ± 17	98 ± 4	31 ± 2	4 ± 1	0
5×10^3	288 ± 18	221 ± 20	187 ± 12	68 ± 1	64 ± 4	32 ± 4
1×10^4	316 ± 18	310 ± 18	325 ± 18	164 ± 8	169 ± 6	81 ± 6
5×10^4	295 ± 12	420 ± 18	405 ± 13	258 ± 13	286 ± 6	118 ± 10
1×10^5	240 ± 17	364 ± 21	372 ± 11	213 ± 16	266 ± 14	133 ± 8
5×10^5	148 ± 17	230 ± 17	232 ± 6	147 ± 8	162 ± 12	36 ± 7

From Kurland, Kincade, and Moore (20). Reproduced by permission of the *Journal of Experimental Medicine*.
[a] 5×10^4 nucleated cells/culture.
[b] Peritoneal exudate cells (PEC) harvested 4 days' post-intraperitoneal injection of thioglycollate medium in CBA/CUM mice.

dium is stimulated by conditioned media from adherent spleen cells and peritoneal macrophages (6), as well as from PHA-stimulated human mononuclear cells (47). In contrast, it has been reported that a dialyzable inhibitor of CFU-TL is found in the supernatant fluids of adherent, but not non-adherent spleen cells, and of human monocytes (47). The possibility exists that the mononuclear phagocyte may elaborate both stimulatory and inhibitory factors in T-lymphocyte proliferation. Though not as yet characterized, the stimulatory factor may be similar to lymphocyte activating factor (LAF), whereas the dialyzable inhibitor has recently been found to be a prostaglandin (Rozenszajn, personal communication).

A committed unipotential megakaryocytic stem cell (CFU-m) with a frequency of 10 to $20/10^5$ nucleated bone marrow cells can proliferate in agar culture in the presence of an appropriate stimulating factor to produce colonies of polyploid, platelet-producing megakaryocytes (28; Williams, Moore, and Murphy, unpublished observations). *In vivo* studies have indicated the role of thrombopoietin in the regulation of platelet production, and active material has been obtained from a number of sources (26). The relationship between thrombopoietin and megakaryocyte colony-stimulating factor is unclear; however, the latter has been obtained in media conditioned by mitogen-stimulated mouse spleen cells (28), embryonic kidney cells (33), the murine myelomonocytic cell line WEHI-3 (Moore and Williams, unpublished observation), as well as mouse peritoneal macrophages (34).

The *in vitro* techniques for the detection of erythroid stem cells in the presence of erythropoietin have not, so far, implicated macrophages in any regulatory aspect of normal erythropoiesis. It has, however, recently been reported that erythroid colonies spontaneously developed in soft agar cultures of cells derived from patients with certain proliferative diseases of the hemopoietic system, such as polycythaemia vera (14). In serially sectioned colonies, single or multiple macrophages were frequently observed in the center of the erythroid colony. This observation is reminiscent of the "erythroblast islands" in bone marrow, where one or more central macrophages are surrounded by concentric rings of differentiating erythroblasts, a process that was long thought to subserve a local erythropoietic effect.

The intimate interrelationship of macrophages with proliferating myeloid and lymphoid cell populations in marrow, spleen, and lymph nodes suggest that such cells may be uniquely situated to stimulate cell differentiation and modulate proliferation by elaboration of cell-line specific stimulatory macromolecules and an opposing nonspecific

activity, PGE (Figure 6-9). In contrast to myelopoiesis, in which PGE and colony-stimulating factor have been found to be mutually antagonistic and in which their synthesis is intimately interrelated, the relationship between PGE and the other hemopoietic stimulators of macrophage origin are not clearly understood. However, the generality of the concept of a regulatory role for the macrophage in hemopoiesis is based on the detection of diffusible activities operating in low density clonal assay systems, which would certainly be sufficient to suggest physiological relevance to *in vivo* regulation. A more general action of PGE on hemopoiesis is indicated by its ability to stimulate DNA synthesis in multi-potential stem cells (7), which may be prerequisite for differentiation of the cell population into the more committed progenitor cell compartments. Indeed, the local action of the prostaglandins may account for some of the properties hitherto ascribed to the hemopoietic inductive microenvironments of spleen and marrow (29), in controlling the size of erythroid, myeloid, and lymphoid cell populations.

SUMMARY

The capacity of the mononuclear phagocyte (Mϕ) to elaborate opposing biological regulators of *in vitro* granulopoiesis has been established. An optimal number of Mϕ from both murine peritoneal exudates and human peripheral blood stimulated the clonal proliferation of syngeneic committed granulocyte-macrophage progenitor cells (CFU-c) by the production and release of colony-stimulating factors (CSF). Conversely, high numbers of Mϕ possessed little stimulating activity and were markedly suppressive, due to the accumulation of a dialyzable, nonspecies-specific inhibitor of CFU-c proliferation. The release of this inhibitory principle by Mϕ was markedly potentiated by a soluble source of CSF, since concentrations of CSF that promoted maximal CFU-c proliferation induced a state of heightened Mϕ suppressor activity, which masked the direct stimulatory actions of CSF. Similarly, low numbers of murine Mϕ, which had no effect on exogenously stimulated human CFU-c, became dose-dependently suppressive when incubated with increasing concentrations of murine CSF. This inhibitory principle has been putatively identified as prostaglandin E (PGE) by means of a sensitive radioimmunoassay of medium conditioned by Mϕ, and the levels of PGE have been found to correlate inversely with the net stimulatory activity of Mϕ. Cell density and adherence separation procedures have shown that PGE is released by both murine and human Mϕ, but not by

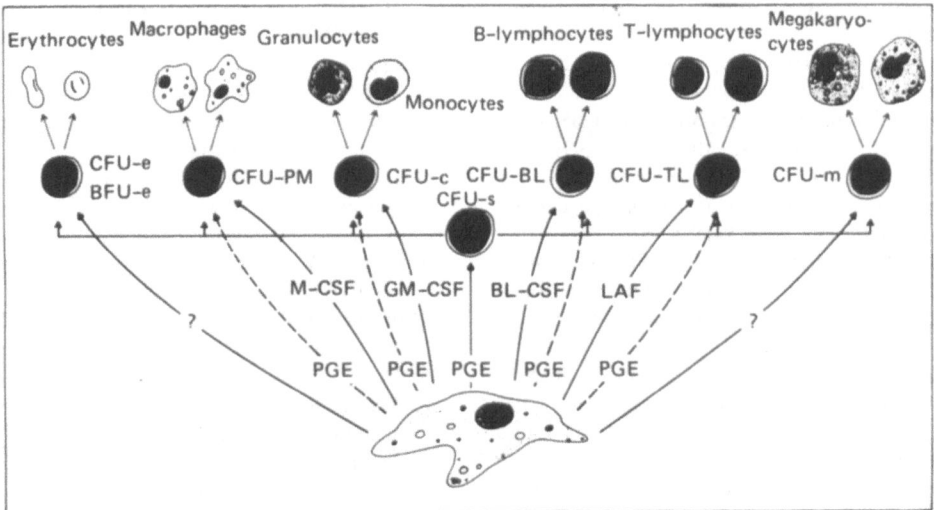

FIGURE 6-9. Regulatory interactions of the macrophage in hemopoiesis.

granulocytes or lymphocytes. The addition of CSF to Mϕ markedly increased the production of PGE, as well as promoting their suppressor activity. Similarly, endotoxin augmented both Mϕ suppressor activity and PGE production, presumably secondary to the stimulation of endogenous CSF by Mϕ. The prostaglandin synthesis inhibitor indomethacin (IND), at concentrations that completely inhibited PGE production by Mϕ, largely prevented the inhibition of CFU-c proliferation by Mϕ, thereby allowing a complete linear stimulation of CFU-c over the range of Mϕ tested. The actions of IND correlated only with its ability to actively inhibit PGE synthesis by Mϕ and was incapable of preventing CFU-c suppression when added to Mϕ supernatants immediately following their active generation. These data indicate a unique regulatory function of the Mϕ in the positive and negative feedback control of committed granulocyte-macrophage progenitor cell proliferation. Accumulation of Mϕ-derived CSF in its own mileu induces the elaboration of PGE, which directly opposes the stimulatory action of CSF. This central surveillance and effector role of Mϕ thereby represents an exquisite self-regulating unit that functions to prevent excessive myelopoiesis associated with elevation of CSF.

REFERENCES

1. Anderson, N.H., and Ramwell, P.W. Biological aspects of prostaglandin. *Arch. Int. Med., 133*:30, 1974.
2. Bradley, T.R., and Metcalf, D. The growth of mouse bone marrow cells *in vitro. Aust. J. Exp. Biol. Med. Sci., 44*:287, 1966.
3. Broxmeyer, H.E., Moore, M.A.S., and Ralph, P. Cell-free granulocyte colony inhibiting activity derived from human polymorphonuclear neutrophils. *Exp. Hemat., 5*:87, 1977.
4. Broxmeyer, H.E., Mendelsohn, N., and Moore, M.A.S. Abnormal granulocyte feedback regulation of colony-forming and colony-stimulating activity-producing cells from patients with chronic myelogenous leukemia. *Leukemia Res., 1*:3–12, 1977.
5. Chervenick, P.A., and LoBuglio, A.F. Human blood monocytes: Stimulators of granulocyte and mononuclear colony formation. *Science, 178*:164, 1972.
6. Claesson, M.H., Whittingham, S., Rodger, B.M., and Burgess, A.W. Colony growth of human T lymphocytes in agar: Effect of a soluble factor from adherent cells. *Eur. J. Immunol.,* in press, 1977.
7. Fehrer, I., and Gidali, J. Prostaglandin E_2 as stimulator of hemopoietic stem cell proliferation. *Nature (Lond.), 247*:550, 1974.
8. Ferreira, S.H., Moncada, S., and Vane, J.R. Indomethacin and aspirin abolish prostaglandin release from the spleen. *Nature New Biol., 231*:237, 1971.
9. Ferreira, S.H., and Vane, J.R. Prostaglandins: Their disappearance from and release into the circulation. *Nature (Lond.), 216*:868, 1967.
10. Gery, I., and Waksman, B.H. Potentiation of the T-lymphocyte response to mitogens: II. The cellular source of the potentiating mediator(s). *J. Exp. Med., 136*:143, 1972.
11. Golde, D.W., Finley, T.M., and Cline, M.J. Production of colony-stimulating factor by human macrophages. *Lancet, ii*:1397, 1972.
12. Golub, M., Zia, P., Matsuno, M., and Horton, R. Metabolism of prostaglandins A_1 and E_1 in man. *J. Clin. Invest., 56*:1404, 1975.
13. Hoffman, M., and Dutton, R.W. Immune response restoration with macrophage culture supernatants. *Science 172*:1047, 1971.
14. Horland, A.A., Wolman, S.R., Murphy, M.J., and Moore, M.A.S. Proliferation of erythroid colonies in semi-solid agar. *Br. J. Hemat., 36*:477, 1977.
15. Kincade, P.W., and Ralph, P. Regulation of clonal B-

lymphocyte proliferation by anti-immunoglobulin or anti-Ia antibodies. *Cold Spring Harbor Symposium., 41*:245, 1977.

16. Kincade, P.W., Ralph, P., and Moore, M.A.S. Growth of B-lymphocyte clones in semi-solid culture is mitogen dependent. *J. Exp. Med., 143*:1265, 1976.

17. Kurland, J.I., Bockman, R., Broxmeyer, H.E., and Moore, M.A.S. Limitation of excessive myelopoiesis by the intrinsic modulation of macrophage derived prostaglandin E. *Science, 199*:552, 1978.

18. Kurland, J.I., and Bockman, R.B. Prostaglandin E production by human blood monocytes and mouse peritoneal macrophages. *J. Exp. Med.*, (in press), 1977.

19. Kurland, J.I., Hadden, J.W., and Moore, M.A.S. Cyclic nucleotides in the humoral regulation of hemopoietic progenitor cell proliferation. *Cancer Res., 37*:4534, 1977.

20. Kurland, J.I., Kincade, P.W., and Moore, M.A.S. Regulation of B-lymphocyte clonal proliferation by stimulatory and inhibitory macrophage-derived factors. *J. Exp. Med., 146*:1420, 1977.

21. Kurland, J.I., and Moore, M.A.S. Regulatory role of the macrophage in normal and neoplastic hemopoiesis. *In* S. Baum and G. D. Ledney, eds., *Experimental Hematology Today.* New York: Springer-Verlag, 1977, pp. 51–62.

22. Kurland, J.I., and Moore, M.A.S. Modulation of hemopoiesis by prostaglandins. *Exp. Hemat., 5*:357, 1977.

23. Lachman, L.B., Hacker, M.P., Blyden, G.T., and Handschumacher, R.E. Preparation of lymphocyte activating factor from continuous murine macrophage cell lines. *Cell Immunol.*, (in press) 1977.

24. Lin, H., and Stewart, C.C. Colony formation by mouse peritoneal exudate cells *in vitro. Nature New Biol., 243*:176, 1973.

25. Lipsky, P.E., and Rosenthal, A.S. The induction and regulation of guinea pig B-lymphocyte proliferation *in vitro. J. Immunol., 117*:1594, 1976.

26. MacDonald, H.R. A comparison of platelet size, platelet count and platelet ^{35}S incorporation as assays for thrombopoietin. *Br. J. Hemat., 34*:257, 1976.

27. Metcalf, D. Role of mercaptoethanol and endotoxin in stimulating B-lymphocytes colony formation *in vitro. J. Immunol., 116*:635, 1976.

28. Metcalf, D., MacDonald, H.R., Odartchenko, N., and Sordet, B. Growth of mouse megakaryocyte colonies *in vitro. Proc. Natl. Acad. Sci. (USA), 72*:1744, 1976.

29. Metcalf, D., and Moore, M.A.S. *Hemopoietic Cells.* Amsterdam: North-Holland, 1971.

30. Metcalf, D., Nossal, G.J.V., Warner, N.L., Miller, J.F.A.P., Mandel, T.E., Layton, J.E., and Gutman, G.A. Growth of B-lymphocyte colonies *in vitro. J. Exp. Med., 142*:1534, 1976.

31. Moore, M.A.S., and Dexter, T.M. Stem cell regulation in continuous hematopoietic cell culture. *Transplant. Rev.*, (in press) 1977.

32. Moore, M.A.S., and Williams, N. Physical separation of colony-stimulating cells from *in vitro* colony-forming cells in hematopoietic tissue. *J. Cell Physiol., 80*:195, 1972.

33. Nakeff, A., and Daniels-McQueen, S. *In vitro* colony assay for a new class of megakaryocyte precursor: Colony-forming unit-megakaryocyte (CFU-m). *Proc. Soc. Exp. Biol. Med. 151*:587, 1976.

34. Nakeff, A., and Bryan, J. Modifiers of CFU-m growth *in vitro. Exp. Hemat., 5* (Suppl. 2):80, 1977.

35. Namba, Y., and Hanoaka, M. Immunucytology of cultured IgM-forming cells in mouse. I. Requirement of a phagocyte cell factor for the growth of IgM-forming tumor cells in tissue culture. *Cell Immunol., 12*:74, 1974.

36. Nelson, R.D., and Leu, R.W. Macrophage requirement for production of guinea pig migration inhibitory factor (MIF) *in vitro. J. Immunol., 114*:606, 1975.

37. Ralph, P., Broxmeyer, H.E., Moore, M.A.S., and Nakoinz, I. Induction of myeloid colony-stimulating activity in murine monocyte tumor cell lines by macrophage activators, and in a T cell line by Con A. *J. Exp. Med., 146*:611, 1977.

38. Ralph, P., Moore, M.A.S., and Nilson, K. Lysozyme synthesis by established human and murine histiocytic lymphoma cell lines. *J. Exp. Med., 143*:1528, 1976.

39. Ralph, P., and Nakoinz, I. Antibody-dependent killing of erythrocyte and tumor targets by monocyte-related cell lines: Enhancement by LPS and PPD. *J. Immunol., 119*:950, 1977.

40. Sanner, J.H. Prostaglandin inhibition with a dibenzoxazephine hydrazide derivative and morphine. *Ann. N.Y. Acad. Sci., 180*:396, 1971.

41. Shevach, E.M., Stobo, J.D., and Green, I. Immunoglobulin and theta-bearing murine leukemias and lymphomas. *J. Immunol., 108*:1146, 1972.

42. Stanley, E.R., Cifone, M., Heard, P.M., and Defendi, V. Factors regulating macrophage production and growth: Identity of colony-stimulating factor and macrophage growth factor. *J. Exp. Med., 143*:631, 1976.

43. Warner, N.L., Moore, M.A.S., and Metcalf, D. A transplantable myelomonocytic leukemia in Balb/c mice: Cytology, karyotype and muramidase content. *J. Natl. Cancer Inst., 43*:953, 1969.

44. Weir, D.H. *Handbook of Experimental Immunology,* Philadelphia, Davies Co., 1967, p. 1019.

45. Wood, D.D., and Gaul, S.L. Enhancement of the humoral immune response of T-cell depleted murine spleens by a factor derived from human monocytes *in vitro. J. Immunol., 113*:925, 1974.

46. Yam, L.T., Li, C.Y., and Crosby, W.H. Cytochemical identification of monocytes and granulocytes. *Amer. J. Clin. Path., 55*:283, 1971.

47. Zeevi, A., Goldman, I., and Rozenszajn, L.A. Stimulation and inhibition of human T-lymphocyte colony cell proliferation by hemopoietic cell factors. *Cell Immunol., 28*:235, 1977.

7

Analysis of the Cells Forming Pure and Mixed Erythroid Colonies in Agar after Stimulation by Pokeweed Mitogen-stimulated Spleen-conditioned Medium

G. R. Johnson and D. Metcalf

INTRODUCTION

After it had been demonstrated that mitogen-stimulated lymphoid populations are able to produce the granulocyte-macrophage colony-stimulating factor (GM-CSF) (7,11,12), it was shown that conditioned medium from pokeweed mitogen-stimulated mouse spleen cells is also able to stimulate the formation of eosinophil (10,11) and megakaryocyte (9) colonies.

Recently, it was found that if pokeweed mitogen-stimulated spleen-conditioned medium (PMSCM) is added to agar cultures of mouse fetal liver cells, not only do granulocyte-macrophage colonies develop, but also large, erythroid colonies with cells containing hemoglobin (5). These colonies occur either as single or multiple (burst) colonies and have essentially the same morphology as those described by others as forming in cultures of marrow or fetal liver cells stimulated by high concentrations of erythropoietin (2,3).

However, the PMSCM system differs in two important respects from the usual type of erythropoietin-stimulated culture: (a) no erythropoietin is detectable in the PMSCM or the culture medium and (b) approximately one-half of the erythroid colonies are mixed colonies containing more than 10 percent of hemopoietic cells of other types, mainly macrophages, neutrophils, megakaryocytes, and eosinophils (5). In studies in which cloned single fetal liver cells were used, both pure and mixed erythroid colonies were shown to be clones arising from single progenitor cells (5).

The present review will briefly discuss the nature of the cells forming erythroid colonies in the PMSCM system and the possible significance of this hemopoietic population. Details of the individual experiments discussed in this paper have been published elsewhere (5,6,8).

MATERIALS AND METHODS
Collection of Cells

The mice were of the inbred strains CBA, C_3H, BALB/c, SJL, and $C_{57}BL$ maintained in this Institute and F_1 hybrids of CBA and $C_{57}BL$ mice. Fetal tissues were obtained from pregnant mice (day of vaginal plug = day 0 of gestation).

Organ pools were converted to single cell suspensions by oral pipetting. Cell suspensions were allowed to stand for 5 min to allow undispersed fragments to sediment; then the supernatant cells were centrifuged through fetal calf serum (1200 g, 7 min) to remove cell debris. Viable cell counts were performed using eosin.

Agar Cultures

Cells were cultured in 35-mm plastic Petri dishes containing 1 ml agar medium. The agar medium was an equal volume mixture of 0.6 percent Bacto-agar and double strength Dulbecco's Modified Eagle's Medium. The composition of the double strength medium was 10 g of Dulbecco's Modified Eagle's Medium HG Instant Tissue Culture Powder (Grand Island Biological Co., New York); 390 ml of double-glass distilled water; 3 ml of L-asparagine (20 μg/ml); 1.5 ml of DEAE Dextran (75 μg/ml) (Pharmacia, Sweden) (MW, 2×10^6); 0.575 ml of penicillin (2×10^5 U/ml); 0.375 ml of streptomycin (200 mg/ml); 4.9 g $NaHCO_3$; 250 ml of human plasma (heated to 56°C for 30 min, centrifuged at 3,000 g for 10 min to remove precipitate).

Not all batches of human plasma were able to support the formation of prominent, red erythroid colonies. Some batches of plasma supported the formation of large numbers of erythroid colonies, but these were so pale that the scoring of unstained cultures was difficult. In general, fetal calf serum did not support the formation of large numbers of colonies, and those colonies that did develop almost always were pale. Batches were selected by prior testing.

Cultures contained from 2,000 to 100,000 nucleated cells/ml. Cell counts for fetal liver excluded yolk sac erythroblasts, but these were included in counts on fetal peripheral blood. One-milliliter volumes of the cell suspensions in agar medium were pipetted into culture dishes containing 0.2 ml of spleen-conditioned medium. The contents of the culture dishes were mixed, allowed to gel, then incubated for 7 days in a fully humidified atmosphere of 10 percent CO_2 in air.

Preparation of Spleen-conditioned Medium

The $C_{57}BL$ spleen cells were incubated in 4 ml of medium in Falcon tubes at a concentration of 2×10^6 cells/ml in RPMI-1640 containing 5 percent heat-inactivated human plasma and 0.05 ml of a 1:15 dilution of pokeweed mitogen per milliliter culture medium (Grand Island Biological Co., New York). After incubation at 37°C in a mixture of 10 percent CO_2 in air for 7 days, the media were centrifuged for 10 min at 3,000 g. The supernatant fluid was then harvested and filtered through a Millipore membrane. Not all batches of human plasma were able to support the production of active conditioned medium, and the batches were pretested before use.

Scoring of Cultures

Unstained cultures were scored for erythroid colonies using an Olympus dissection microscope and semi-indirect lighting at $\times$ 35 magnifications.

Red or pink aggregates containing more than 50 cells were scored as erythroid colonies. It has been shown that scoring by counting unstained cultures leads to an underestimation of the number of colonies containing erythroid cells; no false positive colonies are scored, however.

RESULTS

For convenience, the cells forming pure and mixed erythroid colonies in agar cultures stimulated by pokeweed mitogen-stimulated spleen-conditioned medium will be referred to as E-CFC (erythroid colony-forming cells). All cells forming 7-day, erythropoietin-stimulated erythroid colonies will be referred to as burst-forming units, erythroid (BFU-e) [3], although from published descriptions not all such colonies appear to be multiple (burst) [2].

Although E-CFC were originally detected in early (12-day) fetal liver populations in CBA mice, they are not restricted to the liver. A survey of fetal and adult tissues [6] showed that E-CFC were at their highest frequency (approximately 1/100 cells) in the 10-day yolk sac and 11-day fetal liver. Smaller numbers were also present in the fetal circulation from day 10 onwards. Levels of E-CFC in the yolk sac and liver fell sharply with increasing fetal age, and levels were never high in the developing fetal spleen or bone marrow (1 to $2/10^5$ cells). Postnatally, low levels of E-CFC were demonstrable in the spleen and bone marrow, and in adult CBA mice there were approximately 2 E-CFC/10^5 marrow cells.

A curious feature of the E-CFC is the unusual variation in frequency of these cells in various mouse strains, levels in CBA mice being much higher than in C_3H, BALB/c, SJL or $C_{57}BL$ mice. No published data exist on the frequency of BFU-e in fetal tissues or in mice of various strains, and since, in this laboratory, we were unable to stimulate erythroid colony formation in agar using erythropoietin, no comparison is possible at present of the location and frequency of E-CFC with that of BFU-e. The levels of E-CFC in adult bone marrow ($2/10^5$ cells) appear to be much lower than those reported for BFU-e (approximately $25/10^5$ cells) [2].

Physical Differences between E-CFC and BFU-e

Analysis of E-CFC in CBA fetal liver and peripheral blood populations showed them to be nonadherent, mainly cycling, light density cells (1.056 to 1.067 g/cm³), with a peak sedimentation of 6 to 9 mm/hr [6]. Thus, E-CFC clearly are large cells, whereas the low sedimentation velocity of BFU-e in adult tissues (3.9 mm/hr) (3) suggests that BFU-e are considerably smaller cells.

These obvious differences between E-CFC and BFU-e should be interpreted with caution, since there are significant differences between fetal and adult granulopoietic, monocyte-forming, and erythroid populations (4). Until the properties of BFU-e in fetal mice have been analyzed, no direct comparison is possible between E-CFC and BFU-e.

Formation of Mixed Hemopoietic Colonies by E-CFC

Apart from their apparent lack of dependence on erythropoietin for proliferation (5), the most extraordinary feature of E-CFC is the fact that about one-half of the colonies formed by these cells contain major populations of other hemopoietic cells, the commonest being macrophage, neutrophil, megakaryocytic, and eosinophil cells (5). It is difficult to believe that mixed populations of this type could have been missed for years by those working with colonies derived from BFU-e, and the conclusion is that at least some E-CFC must differ from BFU-e in being multipotential. It is conceivable, of course, that some BFU-e may also be multipotential, but this potential may not be expressed fully in cultures stimulated by erythropoietin only.

In the cell separation experiments referred to above, no obvious segregation was observed of the cells that form pure erythroid colonies from the cells that form mixed colonies. It may be that the E-CFC are a single population and require more time or different culture conditions to express their multipotential for cell differentiation. Alternatively, it may be that two populations of E-CFC exist and although both have similar physical properties, one type may form pure erythroid colonies, while a somewhat more ancestral type of E-CFC is able to form colonies of more than one hemopoietic population. This problem should ultimately be resolved by the use of more precise physical separation methods.

DISCUSSION

Chemical fractionation of pokeweed mitogen-stimulated spleen-conditioned medium has shown that the active factor stimulating erythroid colony formation is a glycoprotein of molecular weight 40,000 (8). The physical properties of this molecule determined so far, show that the active factor is not erythropoietin, and this is confirmed by the failure to detect erythropoietin in such conditioned media by *in vivo* assays (5). But it cannot be said that the proliferation of E-CFC–derived colonies or hemoglobin synthesis in such colony cells does not require some erythropoietin, since there is erythropoietin present in low concentrations in the human plasma used in the culture medium. However, the erythropoietin concentrations in the cultures must be less than one-hundredth of those found to be necessary for BFU-e proliferation (2).

Because of a general lack of published information on BFU-e in fetal tissues and the apparent inability of mouse BFU-e to proliferate in agar cultures, no answer can be given to the obvious question—Are E-CFC and BFU-e the same population? It is not inconceivable that E-CFC are simply fetal-type BFU-e, and previous work by others has, in fact, suggested that fetal erythropoiesis is relatively independent of, or unresponsive to, erythropoietin (1).

However, the presence of mixed hemopoietic populations in one-half of the colonies formed by E-CFC strongly suggests that an unknown type of multipotential hemopoietic cell is present. It may be that this type of multipotential cell predominates in fetal life. One possibility is that E-CFC are the immediate ancestors of the various progenitor cells for the different hemopoietic families—BFU-e, GM-CFC, EO-CFC, and MEG-CFC. This, of course, raises the question whether E-CFC are in fact typical hemopoietic stem cells (CFU-s). By definition, such a stem cell should be capable of forming progeny of more than one hemopoietic class, but should also be capable of self-replication. An analysis of colony populations is now in progress to determine whether E-CFC are present at any stage in the *in vitro* colonies derived from E-CFC. The spleen colony-forming potential of the cells in such colonies is also being studied.

SUMMARY

Certain mouse cells are capable of forming pure and mixed erythroid colonies in agar cultures stimulated by pokeweed mitogen-stimulated spleen-conditioned medium in the absence of detectable erythropoietin. These erythroid colony-forming cells (E-CFC) are present in high numbers in the yolk sac and liver early in fetal life, but only low numbers ($2/10^5$ cells) were found in the adult CBA mouse marrow. The E-CFC from CBA fetal liver were large, light density cells, most of which were nonadherent and in cycle. The relation of E-CFC to burst-forming units, erythroid (BFU-e) or to stem cells is not yet clear, but E-CFC could be a multipotential progenitor cell that is the immediate ancestor of BFU-e.

ACKNOWLEDGMENT

This work was supported by the Carden Fellowship Fund of the Anti-Cancer Council of Victoria, the National Health and Medical Research Council, Canberra, and the National Cancer Institute, Washington, Contract No. NOI-CB-33854.

REFERENCES

1. Cole, R.J., Regan, T., White, S.L., and Cheek, E.M. The relationship between erythropoietin-dependent cellular differentiation and colony-forming ability in prenatal haemopoietic tissues. *J. Embryol. Exp. Morph., 34*:575, 1975.

2. Gregory, C.J. Erythropoietin sensitivity as a differentiation marker in the hemopoietic system: Studies of three erythropoietic colony responses in culture. *J. Cell. Physiol., 89*:289, 1976.

3. Heath, D.S., Axelrad, A.A., McLeod, D.L., and Shreeve, M.M. Separation of the erythropoietin-responsive progenitors BFU-e and CFU-e in mouse bone marrow by unit gravity sedimentation. *Blood, 47*:777, 1976.

4. Johnson, G.R., and Metcalf, D. Characterization of mouse fetal liver granulocyte-macrophage colony-forming cells (GM-CFC) by velocity sedimentation at unit gravity. *Exp. Hematol., 6*:246, 1978.

5. Johnson, G.R., and Metcalf, D. Pure and mixed erythroid colony formation *in vitro* stimulated by spleen conditioned medium with no detectable erythropoietin. *Proc. Natl. Acad. Sci. (USA)* (in press) 1977.

6. Johnson, G.R., Metcalf, D. Nature of cells forming erythroid colonies in agar after stimulation by spleen conditioned medium. *J. Cell. Physiol.* (in press) 1977.

7. McNeill, T.A. Release of bone marrow colony-stimulating activity during immunological reactions *in vitro*. *Nature New Biol., 244*:175, 1973.

8. Metcalf, D., Russell, S., and Burgess, A.W. Production of hemopoietic stimulating factors by pokeweed mitogen-stimulated spleen cells. *Transplant. Proc.* (in press) 1977.

9. Metcalf, D., MacDonald, H.R., Odartchenko, N., and Sordat, B. Growth of mouse megakaryocyte colonies *in vitro. Proc. Nat. Acad. Sci. (USA)* 72:1744, 1975.

10. Metcalf, D., Parker, J., Chester, H.M., and Kincade, P.W. Formation of eosinophilic-like granulocytic colonies by mouse bone marrow cells *in vitro. J. Cell. Physiol., 84*:275, 1974.

11. Parker, J.W., and Metcalf, D. Production of colony-stimulating factor in mitogen-stimulated lymphocyte cultures. *J. Immunol., 112*:502, 1974.

12. Parker, J.W., and Metcalf, D. Production of colony-stimulating factor in mixed leucocyte cultures. *Immunology, 26*:1039, 1974.

8

Current Studies on the Proliferation of Cells of the Mononuclear Phagocyte System

Ralph van Furth,
Theo J.L.M. Goud, and
Dick van Waarde

INTRODUCTION

During the last decade, there have been many studies on the origin and kinetics of mononuclear phagocytes and the regulation of their production. These studies have led to the postulation of the mononuclear phagocyte system (MPS) (Figure 8-1) (7). By definition, a system is made up of cells with the same origin, morphology, and function (1). Since the reticuloendothelial system (1) includes a number of cell types—reticulum cells, endothelial cells, dendritic cells, and macrophages—that do not fulfill these criteria, it cannot be considered a true system and it should not be designated as such in medicobiological terminology. This chapter, in which recent studies on the proliferation and kinetics of mononuclear phagocytes are briefly reviewed, also presents evidence that supports the existence of a separate mononuclear phagocyte cell line.

ORIGIN AND KINETICS OF MONONUCLEAR PHAGOCYTES

The Normal Steady State

It has been shown that macrophages do not derive from lymphocytes or originate from mesenchymal cells. The results of chimera studies have established the bone marrow origin of peritoneal, liver, and lung macrophages (2,10,11,14,17,18,21). On the basis of the *in vitro* incorporation of [³H]-thymidine and 1-hr pulse labeling with this compound, it may be concluded that, during the steady state, only the monoblasts and promonocytes actively divide, whereas monocytes and macrophages are essentially nondividing cells (5,6,12). The low percentage (under 5 percent) of *in vitro*-labeled macrophages suggests a slow proliferation of these cells in tissue. However, during treatment with glucocorticosteroids, which rapidly cause a severe monocytopenia that lasts many days, but do not grossly affect the number of macrophages already present in tissue (19), the *in vitro* labeling index of peritoneal and liver macrophages decreases within 24 hr to almost zero. Since glucocorticosteroids do not inhibit DNA synthesis by promonocytes either *in vitro* or *in vivo* (19,20), but do prevent the influx of mononuclear phagocytes into tissue, it can be concluded that the DNA-synthesizing mononuclear phagocytes in tissue are derived from the circulation and are of bone marrow origin. The rapid decrease in the labeling index of macrophages during hydrocortisone administration seems to indicate that a small percentage of DNA-synthesizing mononuclear phagocytes divide only once, shortly after their arrival in tissue.

In vivo labeling studies support the bone marrow origin of macrophages (4,6,8). Proof that peritoneal

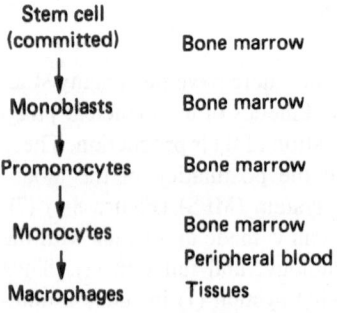

Stem cell (committed) — Bone marrow

Monoblasts — Bone marrow

Promonocytes — Bone marrow

Monocytes — Bone marrow
— Peripheral blood

Macrophages — Tissues

Connective tissue (histiocytes)
Liver (Kupffer cells)
Lung (alveolar macrophages)
Lymph nodes (free and fixed macrophages)
Spleen (free and fixed macrophages)
Bone marrow (macrophages)
Serous cavities (pleural and peritoneal
    macrophages)
Bone tissue (osteoclasts?)
Nervous system (microglial cells)

FIGURE 8-1. The mononuclear phagocyte system [from van Furth et al. (7)].

and liver macrophages derive from the bone marrow was obtained in studies in which the animals received total body irradiation with shielding of both hindlimbs and part of the pelvis, followed 24 hr later by labeling with [³H]-thymidine (3,5). In these animals, in which the peritoneal and liver macrophages are also exposed to the irradiation, the macrophage labeling index is about 10 percent that in normal animals. Since the bone marrow of both hindlimbs constitutes about 10 percent of the total bone marrow mass, the changes in these labeling indices agrees well with what would be expected if tissue macrophages are continuously replaced by circulating monocytes originating from the bone marrow.

The results of the labeling experiments, based on a single injection of [³H]-thymidine, provided quantitative data on the production of monocytes in the bone marrow (Table 8-1) and their transit time in the circulation, which are described in detail elsewhere (4,8). After the first 48-hr period, when there is almost no further production of labeled

monocytes, 50 to 60 percent of the monocytes that have left the circulation become Kupffer cells and 7 to 8 percent of the monocytes migrate to the peritoneal cavity (8).

Acute Inflammation

The main function of mononuclear phagocytes is to remove deleterious material by phagocytosis or pinocytosis, but there are generally too few macrophages in normal tissue to deal effectively with either pathogenic microorganisms or noninfectious substances that may damage the tissue. However, tissue also has available circulating phagocytic cells, which migrate to the affected site and to remove both the inflammatory agents and necrotic tissue, thus facilitating tissue repair.

The production and kinetics of the mononuclear phagocytes during an acute inflammation have been studied in great detail in a mouse model, the inflammatory stimulus consisting of an intraperitoneal injection of newborn calf serum or such particles as polystyrene, latex, or silica (5,6,8,22,23). The findings show that granulocytes first appear in the peritoneal cavity, and after a short time the number of peritoneal macrophages increases. Concomitantly, the number of circulating monocytes also increases.

The problem of whether the increase in the number of peritoneal macrophages during inflammation induced by newborn calf serum is due to local proliferation was studied by *in vitro* and *in vivo* pulse labeling with [³H]-thymidine. The results demonstrate that this kind of inflammatory stimulus does not induce proliferation of resident macrophages (4,8).

To further study the production and kinetics of the mononuclear phagocytes during an inflammatory response, animals were labeled with [³H]-thymidine before administration of the inflammatory stimulus (8). During the first 48 hr following stimulation, monocyte production is markedly increased, the influx of monocytes from the bone marrow into the peripheral blood and the total efflux of monocytes from the circulation to the site

TABLE 8-1 Kinetic Parameters of Mouse Monocytes[a]

	NORMAL STEADY STATE	ACUTE INFLAMMATORY REACTION
Monocyte production rate	0.65×10^5/hr	1.06×10^5/hr at 1–12 hr
		0.78×10^5/hr at 12–24 hr
Monocyte influx from bone marrow into circulation	0.28×10^5/hr	0.57×10^5/hr at 0–48 hr
Monocyte influx into peritoneal cavity	0.01×10^5/hr	0.13×10^5/hr at 0–48 hr

[a]Data from van Furth et al. (8)

of inflammation being doubled. At 48 hr, the inflammatory exudate in the peritoneal cavity contains at least 40 to 50 percent of the monocytes that have left the circulation, whereas in normal mice the percentage of monocytes is 7 to 8 (Table 8-1). It is remarkable that about 70 percent of the newly formed cells arrive at the site of inflammation (8).

It is evident from these findings that, during an acute inflammatory response in the peritoneal cavity, production of monocytes increases, and the majority of the mononuclear phagocytes recruited to the site of the lesion are cells very recently formed in the bone marrow.

REGULATION OF MONOCYTE PRODUCTION

It is possible that the increase in the number of mononuclear phagocytes at the site of an inflammation and the concomitant monocytosis in the peripheral blood are regulated by a humoral control mechanism. It has already been shown that, during certain inflammatory reactions, the serum contains leukocytosis-inducing factors. Most of these factors act to release polymorphonuclear leukocytes in the very early phase of the inflammatory reaction, although granulocyte production has been observed (9). An inhibitor of the granulocytopoietic response (granulocytic chalone) has also been found.

Until recently, detailed studies on the regulation of monocytopoiesis were not available, although a humoral substance has been postulated (15,25). Recently, we found a factor that induces monocytosis during an inflammatory reaction and stimulates the production of monocytes (22,23).

Serum collected during the early phase of an inflammatory reaction induces monocytosis in test animals upon injection, whereas serum from normal mice does not increase the number of monocytes. It is of interest that in mice with an inflammatory reaction, the peak in the activity of the factor in the serum always precedes the monocytosis peak. This active factor increases the rate of production of monocytes by decreasing the cell-cycle time of the promonocytes and increasing the number of these cells, probably acting at the monoblastic level (Figure 8-3) (23); it has been called monocytopoiesis-increasing factor.

The site of production and/or release of FIM seems to be the site of inflammation, in our model, the cells in the peritoneal cavity (23). Extracts of peritoneal cells from normal mice are very active; this activity diminishes rapidly after induction of a sterile peritonitis and subsequently increases to almost normal values. It is striking that the mono-cytosis-inducing activity of peritoneal cell extracts and that of serum are antagonistic, which suggests that FIM is released into the circulation by the macrophages at the site of inflammation.

A protein with a molecular weight between 18,000 and 24,000, FIM does not contain carbohydrate moieties, and it is probably not a glycoprotein nor is it a product of the complement system or a clotting factor; it has no chemotactic or colony-stimulating activity (24).

Recently, evidence for the existence of a monocytopoiesis-inhibitor (MPI) has been obtained. This factor, which has a molecular weight of more than 50,000, affects monoblast proliferation.

On the basis of these studies, a system that controls monocytopoiesis has been postulated (Figure 8-3). According to this view, when an inflammatory reaction is induced in a tissue, the macrophages at the site of the lesion phagocytize the inducing substance and then release FIM, which is transported via the circulation to the bone marrow, where it acts to increase monocyte production. The newly formed monocytes are then transported by the circulation to the site of inflammation. However, since FIM has no chemotactic effect on macrophages, the influx of these macrophages into the lesion must be governed by another (local) mechanism.

When the particulate substance producing the inflammation is removed by the newly arrived mononuclear phagocytes, monocyte production must be returned to normal levels. This could be brought about by a decreased production and/or release of FIM by the macrophages at the site of the lesion. However, there are also indications that a monocytopoiesis inhibitor (MPI), which reduces monoblast proliferation, exists.

IN VITRO PROLIFERATION OF BONE MARROW MONONUCLEAR PHAGOCYTES

Bone marrow cells can be cultured in the presence of a colony-stimulating factor, which makes it possible to investigate the proliferation of immature mononuclear phagocytes. Since the presence of agar or methyl cellulose interferes with the study of the characteristics of these cells, the original method was modified such that leukocyte colonies are grown on a glass surface in a liquid medium (12). The cells adhering to the glass can be directly observed for growth behavior and for morphological, cytochemical, and functional characterization.

Cultures of bone marrow cells in a liquid medium show two kinds of colonies, one of granulocytic cells and the other of mononuclear phagocytes. The granulocyte colonies are characterized by the close proximity of the cells, some even lying on top of other cells. In the mononuclear phagocyte colonies, however, the cells form a monolayer on the glass surface and grow separately; the round cells are more centrally located, the elongated cells are seen toward the periphery. Crowding of cells is only seen in the center of these colonies in the later stages of culture (12).

Three types of cell can be distinguished in the mononuclear phagocyte colonies, the macrophage, the promonocyte, and the monoblast (12). The macrophage is the most mature of these cells and does not synthesize DNA. This cell is extremely elongated, and it is capable of active phagocytosis and pinocytosis (Table 8-2). The promonocyte is a less mature cell with a high [³H]-thymidine labeling index (82 to 88 percent), which indicates that cell division is taking place. The promonocyte is less elongated than the macrophage. Its cytochemical characteristics and functional capacities are intermediate between those of the macrophage and the monoblast (Table 8-2).

The monoblast is the most immature cell seen in the mononuclear phagocyte colony, and its high [³H]-thymidine labeling index (92 to 96 percent) indicates active proliferation. This round cell, which is smaller than the promonocyte, has a nuclear chromatin with a fine dense structure. The almost round nucleus is surrounded by a thin rim of strongly basophilic cytoplasm. Although the cell surface already shows the slight ruffling typical of mononuclear phagocytes, no pseudopods are present. The monoblast contains all the enzymes known to occur in the promonocytes and macrophages of the colony; they are esterase-positive, with α-naphthyl butyrate as a substrate, and contain lysozyme (Table 8-2). A number of monoblasts contain peroxidase-positive granules. In a small number of the monoblasts, C receptors are present; IgG-receptors are present in almost all of them, but the number of receptor sites per cell is lower than in the pro-

monocytes and macrophages (Table 8-2). The functional capacities of the monoblasts are also lower than those of the promonocytes and macrophages. Almost all monoblasts phagocytize, but pinocytosis only occurs in a small percentage of these cells (Table 8-2).

These characteristics, taken together, indicate a cell sequence, monoblast-promonocyte-macrophage, with increasing maturity in that order. Analysis of the increase in the number of each of these three cell types during culture supports the conclusion that in the first few days mononuclear phagocyte colonies consist almost entirely of monoblasts; later, promonocytes and macrophages appear, and still later, macrophages are the predominant cell type (Figure 8-2). With respect to which type of cell initiates the mononuclear phagocyte colony, it has been shown to be the monoblast (12,13).

Various parameters of the proliferating cells have been determined (13). The cell-cycle times of the monoblasts and promonocytes in these colonies were found to be 11.0 to 11.9 hr and 11.4 to 12.8 hr, respectively; the DNA synthesis time was 5.7 hr for the monoblast and 5.5 hr for the promonocyte. The duration of the other phases of the cell cycle of the proliferating mononuclear phagocytes was 0.6 hr for the G_2 phase, 1.8 hr for mitosis, and 3.5 to 3.8 hr for the G_1 phase.

A mathematical analysis of the number of different cell types in culture (Figure 8-2) showed that, *in vitro*, some of the dividing monoblasts and pro-

TABLE 8-2 Characteristics of Mononuclear Phagocytes of Colonies Grown in Liquid Culture[a]

	MONO-BLASTS	PROMONO-CYTES	MACRO-PHAGES
Peroxidase[b]	78	68	19
Esterase[c]	91	90	93
Lysozyme	43	55	98
F_c receptors	94	99	100
C receptors	16	39	75
Phagocytosis	96	100	100
Pinocytosis	16	64	96

[a]From Goud et al. (12).
[b]These data pertain only to colonies with mononuclear phagocytes containing peroxidase-positive granules; in about one-half of the colonies, all the cells are negative.
[c]With α-naphthyl butyrate as substrate.

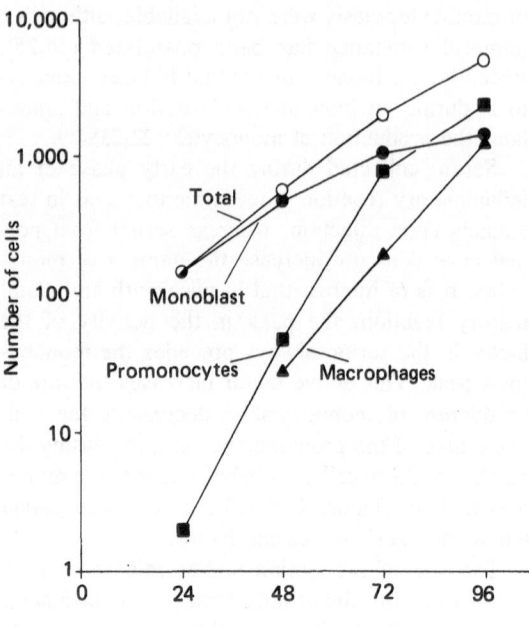

FIGURE 8-2. Total number of monoblasts, promonocytes, and macrophages developing in a culture initiated with 1 × 10⁵ bone marrow cells [from Goud and van Furth (13)].

monocytes replicate themselves, and that after division others differentiate to the next cell stage (13). It is of interest that the proportion of dividing cells that are self-replicating diminishes with the duration of culture. Initially (day 2), about 75 percent of the monoblasts are self-replicating and the remainder give rise to promonocytes, whereas on day 4, about 90 percent of the progeny of the dividing monoblasts are promonocytes (Table 8-3). For the promonocytes, calculation shows that up to day 4 of culture about 60 percent of the cells are self-replicating and about 40 percent give rise to macrophages; on day 4, the percentage of self-replicating promonocytes decreases (Table 8-3). This pattern of proliferation and differentiation of mononuclear phagocytes only occurs *in vitro; in vivo,* the pattern of division is different (Figure 8-3) (13).

Assuming that the monoblast initiates the mononuclear phagocyte colony, we can estimate the number of monoblasts (2.5×10^5) *in vivo* to be one-half the number of promonocytes. In view of this ratio, the most likely pattern for the proliferation of mononuclear phagocytes in the bone marrow is that a monoblast divides once, giving rise to two promonocytes, which, in turn, divide once each to form two non-proliferating monocytes (Figure 8-3) (13).

In addition to the mononuclear phagocyte colonies, colonies of granulocytic cells are also formed in these liquid-media cultures (12). A characteristic feature of the latter colony is their tight structure. In addition to the typical morphology of the granulocytic cells, they are positive for esterase with *N*-acetyl-DL-alanyl-α-naphthylester as substrate and are negative with α-naphthylbutyrate. Cells of granulocytic colonies are unable to pinocytize dex-

TABLE 8-3 Percentage of Dividing Cells Giving Rise to a Different Cell Type in Mononuclear Phagocyte Colonies[a]

DURATION OF INCUBATION (HR)	MONOBLASTS GIVING RISE TO PROMONOCYTES (PERCENT)	PROMONOCYTES GIVING RISE TO MACROPHAGES (PERCENT)
24	0	
48	26.1	41.9
72	58.7	41.2
96	91.2	68.2

[a]From Goud and van Furth (13)

tran sulfate, and only a minority phagocytize bacteria, latex particles, or antibody-coated red cells. This makes it easy to distinguish between granulocyte and mononuclear phagocyte colonies. It was not possible to demonstrate IgG and C receptors on the surface of granulocytic cells, but this may be the result of technical difficulties.

Mixed colonies of both mononuclear phagocytes and granulocytic cells, as described by Metcalf (16), were never observed, nor were colonies of transitional cell forms with the characteristics of both cell lines. This means that, in liquid culture, the colony-forming cell is already committed to form either granulocytic cells or mononuclear phagocytes. No evidence was found to indicate the existence of a common progenitor of granulocytes and mononuclear phagocytes, which Metcalf and Moore (16) consider to be the myeloblast or the (pro)myelocyte. In their studies, however, both kinds of cell were distinguished mainly by morphological criteria.

From the results of the studies discussed above, it may be concluded that the characteristics of the

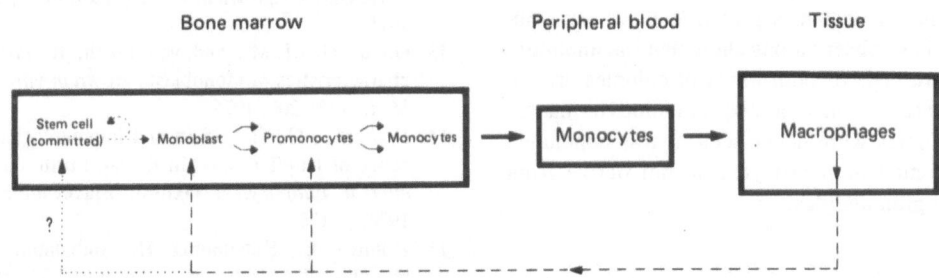

FIGURE 8-3. Schematic representation of the production of kinetics of mononuclear phagocytes and the humoral regulation of hemocytopoiesis (from reference 13). Monoblasts arise from the (committed) stem cell, their two daughter cells being promonocytes. Each of these cells in turn give rise to two monocytes, which leave the bone marrow and migrate via the circulation to the tissues where they become macrophages. In an inflammatory reaction, the macrophages at the site of the inflammation release a humoral factor, the factor increasing monocytopoiesis (FIM), which stimulates monocyte production at the level of the monoblast and promonocyte. Another humoral factor, monocyte production inhibitor (MPI), inhibits monocyte production.

most immature cells of the mononuclear phagocyte and granulocyte cell lines thus far identified, namely the monoblast and the myeloblast, are very different. No evidence indicating that mononuclear phagocytes are derived from granulocytic cells at any stage was found.

SUMMARY

This review deals first with the origin and kinetics of the mononuclear phagocytes during the normal steady state and cites evidence to show that the macrophages in tissue derive from circulating monocytes, which, in turn, arise precursors—promonocytes and monoblasts—in the bone marrow. Next, the production and kinetics of mononuclear phagocytes during an acute inflammatory response were described and compared with the kinetics observed in normal animals.

Since the migration of macrophages to an inflammatory site requires increased monocyte production in the bone marrow, the humoral regulation of monocytopoiesis has been studied. It could be demonstrated that a factor that increases monocytopoiesis (FIM), a protein with a molecular weight of about 20,000, is released from the site of inflammation and affects the mitotic activity of the promonocytes and, conceivably, the monoblasts.

The in vitro proliferation of bone marrow mononuclear phagocytes has been observed in liquid culture, in which the development of colonies attached to a glass surface could be studied. It was shown that mononuclear phagocytes and granulocyte colonies develop independently.

The morphological, cytochemical, functional, and proliferative characteristics of the monoblasts, promonocytes, and macrophages and the cells of the granulocyte colonies grown in vitro were observed. These observations show that the immature and mature cells of both kinds of colonies are totally different. Since mixed granulocyte-macrophage colonies were never seen, it was concluded that mononuclear phagocytes do not derive from immature granulocytes.

ACKNOWLEDGMENT

This investigation was supported in part by the Foundation for Medical Research FUNGO, which is subsidized by the Netherlands Organization for the Advancement of Pure Research (ZWO).

REFERENCES

1. Aschoff, L. Das reticulo-endotheliale System Ergeb. *Inn. Med. Kinderheilk., 26*:1, 1924.
2. Balner, H. Identification of peritoneal macrophages in mouse radiation chimeras. *Transplantation, 1*:217, 1963.
3. Crofton, R.A., Diesselhoff-den Dulk, M.M.C., and van Furth, R. Origin, kinetics, and characteristics of Kupffer cells in the normal steady state. Submitted, 1977.
4. van Furth, R. Modulation of monocyte production. In R. van Furth, ed., *Mononuclear Phagocytes in Immunity, Infection and Pathology*. Oxford: Blackwell Scientific, 1975, p. 161.
5. van Furth, R., and Cohn, Z.A. The origin and kinetics of mononuclear phagocytes. *J. Exp. Med., 128*:415, 1968.
6. van Furth, R., and Diesselhoff-den Dulk, M.M.C. The kinetics of promonocytes and monocytes in the bone marrow. *J. Exp. Med., 132*:813, 1970.
7. van Furth, R., Cohn, Z.A., Hirsch, J.G., Humphry, J.H., Spector, W.G., and Langevoort, H.L. The mononuclear phagocyte system: a new classification of macrophages, monocytes and their precursor. *Bull. Wld. Hlth. Org., 46*:85, 1972.
8. van Furth, R., Disselhoff-den Dulk, M.M.C., and Mattie, H. Quantitative study on the production and kinetics of mononuclear phagocytes during an acute inflammatory reaction. *J. Exp. Med., 138*:1314, 1973.
9. van Furth, R., and van Waarde, D. The humoral regulation of the production of monocytes. In F. Rossi, ed., *Movement, Metabolism and Bactericidal Mechanisms of Phagocytes*. Padova, Italy: Piccin, in press, 1977.
10. Godleski, J.C., and Brain, J.D. The origin of alveolar macrophages in radiation chimeras. *J. Exp. Med., 136*:630, 1972.
11. Goodman, J.W. The origin of peritoneal fluid cells. *Blood, 23*:18, 1964.
12. Goud, Th.J.L.M., Schotte, C., and van Furth, R. The identification of the monoblast in mononuclear phagocyte colonies grown in vitro. *J. Exp. Med., 142*:1180, 1975.
13. Goud, Th.J.L.M., and van Furth, R. Proliferative characteristics of monoblasts grown in vitro. *J. Exp. Med., 142*:1200, 1975.
14. Howard, J.G. The origin and immunological significance of Kupffer cells. In R. van Furth, ed., *Mononuclear Phagocytes*. Oxford: Blackwell Scientific, 1970, p. 178.
15. Komiya, E., Katsunuma, H., Shibamoto, G., Kawakubo, R., Noda, M., Sugimoto, T., Sato, S., Hoshi, K., and Kawashimo, N. Extraktion der neurohumoralen blutregulierenden Wirkstoffe. *Fol. Haemat., Frankfurt, 5*:328, 1961.
16. Metcalf, D., and Moore, M.A.S. Humoral regulation of haemopoiesis. In A. Neuberger and E. L. Tatum, eds., *Haemopoietic Cells*. Amsterdam: North Holland, 1971, p. 362.
17. Pinket, M.O., Cowdrey, C.M., and Nowell, P.C.

Mixed hematopoietic and pulmonary origin of "alveolar macrophages" as demonstrated by chromosome markers. *Amer. J. Path., 48*:859, 1966.

18. Shand, F.L., and Bell, E.B. (1972). Studies on the distribution of macrophages derived from rat bone marrow cells in xenogeneic radiation chimeras. *Immunology, 22*:549, 1972.

19. Thompson, J., and van Furth, R. The effect of glucocorticosteroids on the kinetics of mononuclear phagocytes. *J. Exp. Med., 131*:429, 1970.

20. Thompson, J., and van Furth, R. The effect of glucocorticosteroids on the proliferation and kinetics of promonocytes and monocytes in the bone marrow. *J. Exp. Med., 137*:10, 1973.

21. Virolainen, M. Hematopoietic origin of macrophages as studied by chromosome markers in mice. *J. Exp. Med., 127*:943, 1968.

22. van Waarde, D. Hulsing-Hesselink, E., and van Furth, R. A serum factor inducing monocytosis during an acute inflammatory reaction caused by newborn calf serum. *Cell Tissue Kinet., 9*:51, 1976.

23. van Waarde, D., Hulsing-Hesselink, E., Sandkuyl, L.A., and van Furth, R. Humoral regulation of monocytopoiesis during an inflammatory reaction caused by particulate substances. *Blood, 50*:141, 1977.

24. van Waarde, D., Hulsing-Hesselink, E., and van Furth, R. Properties of a factor increasing monocytopoiesis (FIM) occurring in serum during the early phase of an inflammatory reaction. *Blood, 50*:727, 1977.

25. Willoughby, D.A., Coote, E., and Spector, W.G. A monocytogenic factor released after lymphnode stimulation. *Immunology, 12*:165, 1967.

9

Colony-forming Cells in the Thymus and Mesenteric Lymph Nodes of Mice Engrafted with Lewis Lung Carcinoma Cells

G. D. Ledney, T. J. MacVittie, D. A. Stewart, and G. A. Parker

INTRODUCTION

The *in vitro* culture of granulocyte-macrophage progenitor cells from a number of anatomical sites of the mouse apparently results in the clonal growth of at least two subpopulations of cells. One subpopulation, currently designated colony-forming cells (CFC), has been detected in the antigen-stimulated pleural and peritoneal cavities (4,10,11) and in the blood (9). These cells are also found in the lymph nodes, thymus, spleen, and bone marrow of the mouse (12,13,14). The other subpopulation of cells exhibiting *in vitro* clonal potential is designated the colony-forming unit culture (CFU-c), which are detected in cultures of adult mouse bone marrow, spleen, and peripheral blood. In addition to differing in their tissue origin, CFC differ from CFU-c in a number of other ways. The CFC have a 10- to 15-day lag period prior to the initiation of colony formation, which reaches maximum numbers 25 days after *in vitro* culture. These cells develop along the monocyte-macrophage cell line and have a marked ability to survive in culture in the absence of pregnant mouse uterine extract (PMUE). The CFC obtained from mouse thymus, lymph nodes, bone marrow, and spleen can only be stimulated to grow in the presence of PMUE. Contrary to this, CFU-c reach maximum numbers 10 days after *in vitro* culture, develop along the granulocyte-macrophage cell line, and can be grown in the presence of a number of substances that have colony-stimulating activity (CSA). Disturbances in the CFU-c, as well as in the stem cell compartments of the hemopoietic tissues of mice engrafted with a variety of different tumors, have been reported (1,6,16,17). We have reported earlier that growth of the Lewis lung carcinoma (3LL) in mice results in a profound increase in the number of splenic CFU-c and as well as the total splenic cellularity (7).

Mature macrophages are key elements in the recognition and the subsequent induction of the immune response of the host against neoplasia (8). A depression of macrophage function during neoplasia apparently contributes to the host's increased acceptance of malignant cell growth (5). Regarding this, we observed that the chemotactic responsiveness of peritoneal macrophages of mice injected with 3LL cells is inhibited 1 week after engraftment and enhanced the second week after tumor cell implantation (18). Based on our observations that the development of the 3LL carcinoma in the host produces profound changes in proliferative cell numbers and mature macrophage chemotactic responsiveness, we predicted that this tumor would change the number and the growth characteristics of the monocyte-macrophage progenitor cells found

73

in the thymus and mesenteric lymph nodes. Identification of these changes would be of significance in understanding the host's immune response to the presence of malignancy.

In mice injected with 3LL cells, we observed a decrease in thymic cellularity and little change in mesenteric lymph node cellularity. The number of CFC detected depended upon the tissue of origin, the substances used to stimulate clonal growth, and the time of organ removal for CFC analysis after 3LL cell engraftment of mice.

MATERIALS AND METHODS

Animals

Our C57BL/6-Cum BR (H-2^b) mice were obtained from Cumberland View Farms, Clinton, Tenn. All mice were acclimated to laboratory conditions for 2 weeks prior to use. During that time, they were quarantined until they were certified free of lesions of murine pneumonia complex and of oropharyngeal *Pseudomonas* spp. All animals used in this study were males, 8 to 12 weeks of age. The mice were maintained on a 6 a.m. (light) to 6 p.m. (dark) cycle in filter-covered cages, with Wayne Lab-Blox diet and chlorinated water (15 PPM) *ad libitum*.

Tumor

The Lewis lung (3LL) carcinoma, a transplantable metastatic tumor that arose spontaneously in the pulmonary tissues of a C$_{57}$BL/6 mouse in 1951, was received from the NCI-NIH in its 86th passage in C$_{57}$BL/6 male mice. In the work presented here, tumor cells were derived from the 14th to the 30th subcutaneous passage in C$_{57}$BL/6 male mice at this Institute. The tumor cells were prepared as follows: For the routine maintenence of the tumor line, cells were prepared as a 10 percent concentration of cells in Roswell Park Memorial Institute (RPMI)–1640 (Flow Labs. Rockville, Md.) medium; 0.2 ml was injected into the mice. In the experiments listed in the Results, tumor tissue was taken 12 to 16 days after implantation and cell suspensions were prepared by (a) passage through a tissue press; (b) repetitive expulsion through successively smaller gauge needles, and (c) flotation on a Ficoll-Hypaque solution. Viability, as ascertained by exclusion of Trypan Blue dye, was 90 to 99 percent. No bacterial growth was obtained when tumor cell preparations were cultured in thioglycollate broth.

Hemopoietic Cell Preparations

The mesenteric lymph nodes and thymuses were removed aseptically from mice killed by cervical dislocation and placed in RPMI-1640 medium on ice ($\pm 4°C$). The tissues were minced with scissors in a glass vessel and passed through 6 to 8 layers of nylon mesh. All cell suspensions were washed three times at 250 g and resuspended in RPMI-1640. Viability estimates and nucleated cells counts were performed in a hemocytometer with 0.2 percent Trypan Blue and Turk's solution. Suspensions of cells for growth on agar plates were prepared to the desired concentration in RPMI-1640.

In Vitro Colony-Forming System

The hemopoietic cell double-layer, agar culture technique used was similar to that described by Bradley et al. (2). Extracts prepared from the pooled placentae, membranes, and gravid uteri of pregnant mice (PMUE) were the source of CSA. The PMUE were prepared by a serial procedure, including water extraction, ammonium sulfate fractionation, dialysis, and heat inactivation. Each extract was prepared in a volume such that 1 ml contained the CSA from 1 g of wet tissue used. All cell cultures were performed at the optimum concentration of 3.5 percent (final v/v) of PMUE to culture medium plus agar. The cells from these organs were suspended in a concentration of 10^6 cells/plate. The thymus and lymph node cells were also cultured in the presence of normal human serum (NHS) previously sterilized by passage through a 0.45 μm filter. All these cells were cultured with a 7.5 percent (final v/v) of NHS plus 3.5 percent (final v/v) of PMUE to culture medium plus agar. Colonies of more than 50 cells were counted at 21 to 24 days. The cultured cells were inspected on a daily basis to detect differences in colony growth and cell morphology. The numbers of colonies grown in PMUE and NHS were determined in the following manner. For each sample of cells to be tested, the cells were cultured in with PMUE only and with PMUE and NHS. The values presented in the appropriate figures are based on the differences found between clonal growth in PMUE only and clonal growth in PMUE and NHS.

Histological Preparations

Organ cellularity and CFC activity of the thymus and mesenteric lymph nodes were correlated with the histopathological findings in these organs. Groups of four mice each were killed by cervical dislocation 3, 7, and 14 days after subcutaneous implantation of 2×10^5 cells. Groups of mice injected with RPMI-1640 were used as controls. The organs were fixed in 10 percent buffered formalin, embedded in paraffin blocks, and sections were cut at 6 μm thickness. The sections were stained with Hematoxylin and Eosin and then examined by a comparative pathologist and the principal investigator.

Experimental Design and Statistical Evaluation

To determine if the organ number and concentration of CFC depended upon the number and quality of implanted 3LL cells, mice were injected with either 1×10^4 or 2×10^5 tumor cells or with 2×10^5 irradiated 3LL (2,000 rad ^{60}Co at 600 rad per minute). Mice injected with RPMI-1640 were used as controls. The tumor cell doses are the TD 40 and TD 100, respectively. The 3LL cells irradiated as described did not produce tumors in mice that were observed for 60 days.

The concentration and organ content of CFC were evaluated 3, 7, and 14 days after engraftment of mice with 3LL cells. At each time interval, from three to five mice were killed by cervical dislocation, and the thymuses and mesenteric lymph nodes were removed. The tissues were minced and cell suspensions prepared as previously described.

The data were analyzed by the Friedman's two-way analysis of variance by ranks test (19) because the values were not normally distributed and because of the large biological variability. The analyses were performed by pooling, by replicate, all

of the values obtained from the control animals. A total of six replicate experiments were performed.

RESULTS

Cellularity of the Thymus and Mesenteric Lymph Nodes

The data presented in Figure 9-1 are the thymic and mesenteric lymph node cellularities of mice injected with 3LL cells expressed as a percentage of control animals. The percentage cellularities of all 3LL cell-injected groups of mice were combined and plotted against time, since (a) the cellularities of both organs were not significantly affected by the three different tumor cell doses and (b) the thymic cellularity varied significantly with the length of time the host animal was associated with 3LL cells. The most important observation was the approximate 50 percent reduction in thymic cellularity during the first week after implantation of the mice with 3LL cells. Thymic cellularity returned to above normal levels by day 14. The cellularity of the mesenteric lymph nodes tended to increase during the first week and was reduced by day 14 after

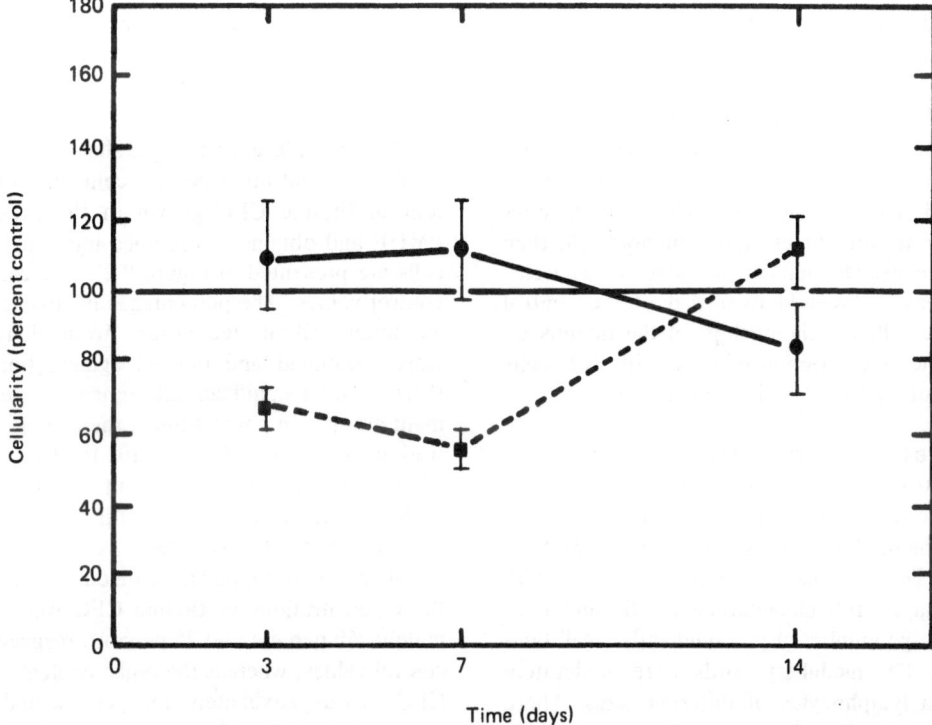

FIGURE 9-1. Percentage cellularity of the thymus and mesenteric lymph nodes of mice injected with 3LL carcinoma cells. There was no statistically significant difference in the percentage cellularities observed for the thymus and mesenteric lymph nodes of mice injected with either 10^4 or 2×10^5 viable 3LL cells or 2×10^5 irradiated 3LL cells. Thus, each point represents the mean value ± standard error of the combined data of all three cell doses from six replicate experiments (n = 18). The only cellularities that differed significantly from controls (—) were that of the thymus (■—■) on days 3 and 7 after the mice were injected with tumor cells. Control cellularity for mesenteric lymph nodes was $18 \pm 3 \times 10^6$, and for thymus $76 \pm 7 \times 10^6$. Mesenteric lymph nodes (●—●).

the injection of the mice with 3LL cells. In control mice, the average cellularity of these nodes was $18\pm 3 \times 10^6$, and $76\pm 7 \times 10^6$ for the thymus.

Histopathology of the Thymus, Mesenteric Lymph Nodes, and Tumor

The Thymus

In control mice, the cortex represented about 70 percent of the thymic cross-sectional area, with the medulla comprising the remaining 30 percent. The cortical area was packed with lymphocytes, and reticular tissue was difficult to identify. In the medulla, there was an approximate 1:1 ratio of lymphoid cells to epithelial cells. About 25 percent of the medulla was open spaces. One to two Hassel's corpuscles were observed, on average, in each high power field (450 X) in the medullary area. No unusual cell elements were detected in any thymic area. In mice injected with 3LL tumor cells, the overall thymic cellularity was reduced, as described elsewhere in this chapter. The cortical cell area was only marginally greater than the medullary area on day 3 after the mice had been injected with 3LL cells. The cortex of these mice was less cellular than that of the controls. There was an approximate 2:1 ratio of lymphocyte to epithelial cells in the medulla. Compared to the controls, there was a threefold increase in the Hassel's corpuscle content of the medulla. On day 7, there was a return to normal of the ratio of cortex:medulla on cross section, but the overall organ size and cellularity was lower than control values. The reduced cortex was filled with lymphocytes. The medulla contained two to three times more lymphoid cells than epithelial cells. The number of Hassel's corpuscles was increased threefold over that of the control mice. The cellular relationships of the thymus on day 14 after the injection of mice with 3LL cells was essentially that described for day 7.

The Mesenteric Lymph Nodes

In control mice, the mesenteric lymph node follicles were few to moderate in number and varied in size. The perifollicular (T-cell) area was well developed. The medullary sinuses were lined with spindle-shaped reticuloendothelial cells and contained a large number of predominantly small lymphocytes. The medullary cords were moderately filled with lymphocytes of different sizes. There were more large lymphocytes in the cords than in the sinuses.

In mice injected with 3LL cells, the follicles and medullary areas did not differ from those of the controls on days 3 and 7. On day 14, however, there were fewer follicles, and the medullary sinuses were nearly empty. The reticuloendothelial cells of the sinus linings were plump and appeared active,

and the sinuses contained a few more histiocytes than did the sinuses of the controls. The medullary cords were full. In the lymph nodes of many of the mice injected with tumor cells, there was an increase in the number of trabeculae and trabeculae-associated cells over that noted in the control mice. These elements were most noticeable on day 3 and decreased in prominence by day 14. The trabeculae-associated cells were large, with a moderate to large amount of indistinctly bounded amorphous acidophilic cytoplasm. These cells had large ovoid nuclei and some nuclei had an irregular margin. The nuclei contained a moderate amount of evenly distributed, dispersed chromatin, with a few superimposed small chromatin clumps. A few nuclei had a small, centrally placed nucleolus.

The Tumor

The 3LL tumor was removed on days 3, 7, and 14 after implantation and examined microscopically. A few neutrophils and an occasional mononuclear cell were seen within the viable tissue of the tumor on each of the three days mentioned. The neutrophilic infiltration tended to be greater during the early stages of tumor development. A large number of neutrophils, many of which were small and hypersegmented, were seen at the points of tumor erosion through the skin. The periphery of the tumor was often necrotic.

Thymic Colony-forming Cells

The concentration per 10^6 cells and organ content of thymic CFC grown in the presence of PMUE and obtained from mice injected with 3LL cells are presented in Figure 9-2 as percentages of control values. The percentages of thymic CFC of all tumor cell-injected groups from all replicates were combined and plotted against time, since there were no significant differences between treatment groups. In control mice, the average concentration of thymic CFC per 10^6 thymus cells was 19 ± 7, while the organ content of this cell was 1540 ± 660. In mice injected with 3LL cells, there were profound disturbances in concentration and organ content of thymic CFC on days 3 and 7, when the concentrations of thymic CFC were approximately 250 percent and 75 percent, respectively, of control values, whereas the organ content of thymic CFC was approximately 150 percent and 50 percent, respectively, of control values. Both thymic CFC values from mice treated with 3LL cells approached control animal values on day 14.

Mesenteric Lymph Node Colony-forming Cells

The concentration per 10^6 cells and organ content of mesenteric lymph node grown in the pres-

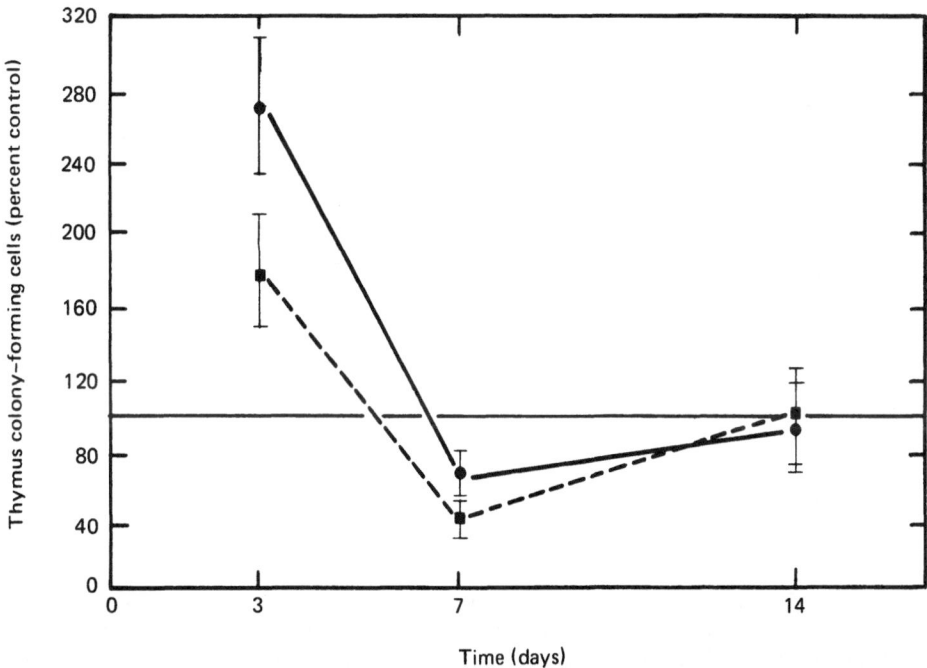

FIGURE 9-2. The percentage CFC values of thymus cells obtained from mice injected with 3LL carcinoma cells and grown in the presence of PMUE. The thymus cells were prepared from groups of mice 3, 7, and 14 days after implantation with 3LL cells and were grown under the conditions described in Materials and Methods. There were no statistically significant differences in the CFC values obtained from mice injected with the three quantities of 3LL cells. Thus, each point represents the mean value ± standard error of the combined data of all three cell doses from six replicate experiments (n = 18). The thymic CFC values from 3LL-injected mice differed significantly from controls on days 3 and 7. The concentration (●—●) and organ content (■---■) of thymic CFC grown in PMUE and obtained from control mice were 19 ± 7 and 1540 ± 660, respectively. Control (—).

ence of PMUE and obtained from mice injected with 3LL cells are presented in Figure 9-3 as percentages of control values. The percentages of mesenteric lymph node CFC values from all tumor cell-injected groups from all replicates were combined, as described. In control mice, the average CFC per 10^6 mesenteric lymph node cells was 23±7, while the organ content of this cell was 383±110. In mice injected with 3LL cells, there was an approximate 200 percent increase in mesenteric lymph node CFC over that measured for control mice on day 3. On days 7 and 14, the mesenteric lymph node CFC values tended to be greater than the control CFC values, but the difference was not significant.

Growth of Colony-forming Cells in PMUE and Normal Human Serum

Clonal CFU-c growth was enhanced by the addition of NHS to a number of substances having CSA (15,22). In initial studies, it was determined that CFC growth could be potentiated by the addition of NHS. Figure 9-4 shows the percentage maximum increase of thymus grown in 3.5 percent (v/v) PMUE and various concentrations of NHS. Bone marrow CFU-c growth is presented in Figure

9-4 for comparison, and to demonstrate the effectiveness of the NHS. The maximum thymic CFC response was obtained when cells were cultured in 3.5 percent PMUE and 7.5 percent NHS (v/v). It has been established that both the size and the number of bone marrow CFU-c is enhanced when the cells are grown in the presence of and substances having CSA (22). Figure 9-5 shows the growth curves for the time of appearance of cultures of thymic CFC, mesenteric lymph node CFC, and bone marrow CFU-c grown only in 3.5 percent PMUE or in 3.5 percent PMUE and 7.5 percent NHS. The lag period before the initiation of colony formation in both thymus and mesenteric lymph node cell cultures was diminished by the presence of NHS in the culture. The lag period for thymic CFC and mesenteric lymph node CFC decreased from 9 to 6 days and from 14 to 10 days, respectively. The maximum number of colonies was obtained at 22 to 25 days regardless of the substances used to promote growth. The number of thymic CFC was increased about twofold; 70 CFC/10^6 when grown in PMUE and NHS as compared to 30 CFC/10^6 when grown only in PMUE. A dramatic sevenfold increase in mesenteric lymph node CFC

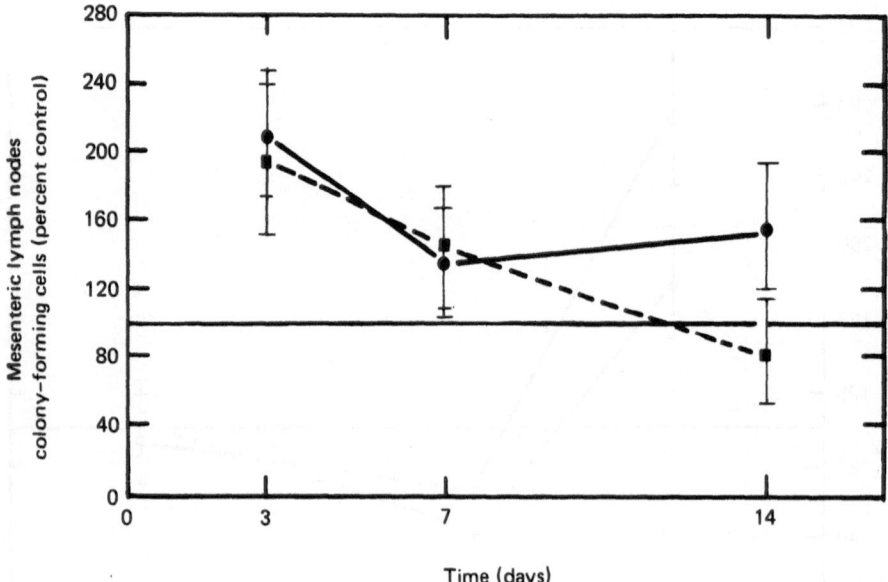

FIGURE 9-3. The percentage CFC values of mesenteric lymph node cells obtained from mice injected with 3LL carcinoma cells and grown in the presence of PMUE. The node cells were prepared from groups of mice 3, 7, and 14 days after implantation with 3LL cells and were grown under the conditions described in Materials and Methods. There were no statistically significant differences in the CFC values obtained from mice injected with the three quantities of 3LL cells ($n = 18$). Thus, each point represents the mean value ± standard error of the combined data of all three cell doses from six replicate experiments. Mesenteric lymph node CFC values from 3LL-injected mice differed significantly from controls on day 3. The concentration (●—●) and organ content (■---■) of node CFC grown in PMUE and obtained from control mice were 23 ± 7 and 383 ± 110, respectively. Control (—).

was seen when node cells were grown in PMUE and NHS. That is, at 22 to 25 days of culture, about 140 colonies grew in PMUE and NHS, whereas only 20 colonies grew in PMUE alone.

Thymic CFC and Mesenteric Lymph Node CFC Grown in PMUE and NHS

The data presented in Figures 9-1 through 9-3 substantiates the observation that CFC obtained from the thymus and mesenteric lymph nodes of mice injected with 3LL cells and grown in PMUE undergo profound population disturbances as compared to that obtained from control mice. Additionally, the data presented in Figures 9-4 and 9-5 support the observation that the CFC obtained from normal mouse thymus and mesenteric lymph nodes undergo substantial population changes in response to culturing in PMUE and NHS as compared to growth only in the presence of PMUE. Based on these data, the hypothesis was advanced that the CFC from the thymus and mesenteric lymph nodes of mice injected with 3LL cells may have a growth response, in the presence of NHS and PMUE, that differs from the response of these cells grown in PMUE only. This hypothesis was tested by culturing mouse thymus and mesenteric lymph node cells in 3.5 percent PMUE and 7.5 per-

cent NHS and comparing the CFC growth obtained to the growth of cells from control mice. The percentage thymic content and concentration/10^6 cells of thymic CFC are presented in Figure 9-6. The percentages from all tumor cell-injected groups from all replicates were combined and plotted against time, since there was no significant differences between treatment groups. The average organ content of thymic CFC in control mice was 1560 ± 310. The concentration per 10^6 cells of thymic CFC in control mice was 25 ± 5. The most important observation made was that the concentration of thymic CFC grown in both PMUE and NHS was significantly increased during the first week after the mice were injected with 3LL cells. In mice injected with tumor cells, the organ content of this thymic CFC did not deviate significantly from that of the control mice. Figure 9-7 shows the percentage mesenteric lymph node content and concentration per 10^6 cells of mesenteric lymph node. Again, the data obtained from tumor-injected mice were combined because there were no significant differences between treatment groups. The average organ content of node CFC in control mice was 1360 ± 390. The concentration per 10^6 cells of node CFC in control mice was 75 ± 20. The major observation was that both the concentration and content

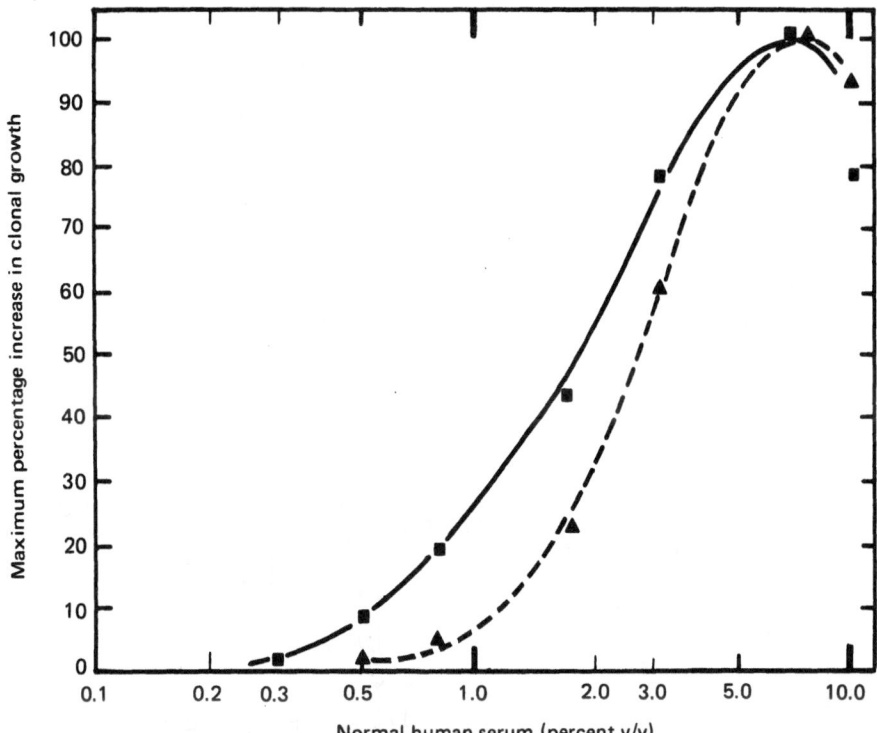

FIGURE 9-4. The maximum percentage increase in clonal growth of cells grown in 3.5 percent pregnant mouse uterine extract (PMUE) and various concentrations of normal human serum (NHS). Data points are mean values of replicate experiments (triplicate cultures) of 2.5×10^4 bone marrow cells (■—■) or 1×10^6 thymus cells (▲---▲). The maximum percentage increases were calculated for each NHS concentration as follows: Number CFC in PMUE + NHS − CFC in PMUE/Number CFC in PMUE.

of node CFC were below control animal CFC numbers on day 3. However, there were no significant differences between control and tumor cell-injected animals on any of the three test days.

DISCUSSION

The thymus and mesenteric lymph nodes of normal mice were recently shown to contain a heretofore undetected population of monocyte-macrophage progenitor cells, designated colony-forming cells, that are solely dependent upon PMUE for growth (12,14). The most important observations documented in this review reveal that (a) the injection of mice with Lewis Lung carcinoma cells results in significant perturbations in the numbers of lymphoid organ-derived CFC and (b) in both normal and tumor cell-injected mice, subpopulations of CFC may exist as based upon the quantities and temporal growth characteristics of the cells cultured in PMUE and NHS versus those for cells grown in PMUE.

A profound reduction in thymic cellularity was noted for a 1-week period after mice were injected with 3LL cells. Thymic hypocellularity in mice in-

jected with 3LL cells has also been reported by other workers (21). The thymic hypocellularity was associated with a reduction in the number of cortical lymphocytes on day 3 and 7 and a lower number of medullary lymphocytes on day 7. Our observation of a thymic cortical hypocellularity in mice injected with 3LL cells supports the observations reported by Snodgrass et al. (20). Except for an increase in the number of Hassel's corpuscles, no unusual thymic-associated cells were detected.

In mice injected with 3LL cells, the mesenteric lymph node cellularity tended to increase the first week and decrease by day 14 after engraftment. The modest hypercellularity was associated with an increase in trabeculae and trabeculae-associated cells. The slight hypocellularity was coincident with a reduction in the trabecular elements mentioned, a reduction in follicle number, and an emptying of the medullary sinuses.

In the thymus, the early host-response to the tumor resulted in a twofold increase in thymic CFC grown in PMUE. This was not due to the selective elimination of lymphocytes, which might account for a concentration enrichment of CFC, but rather, to an actual increase in the thymic *content* of these

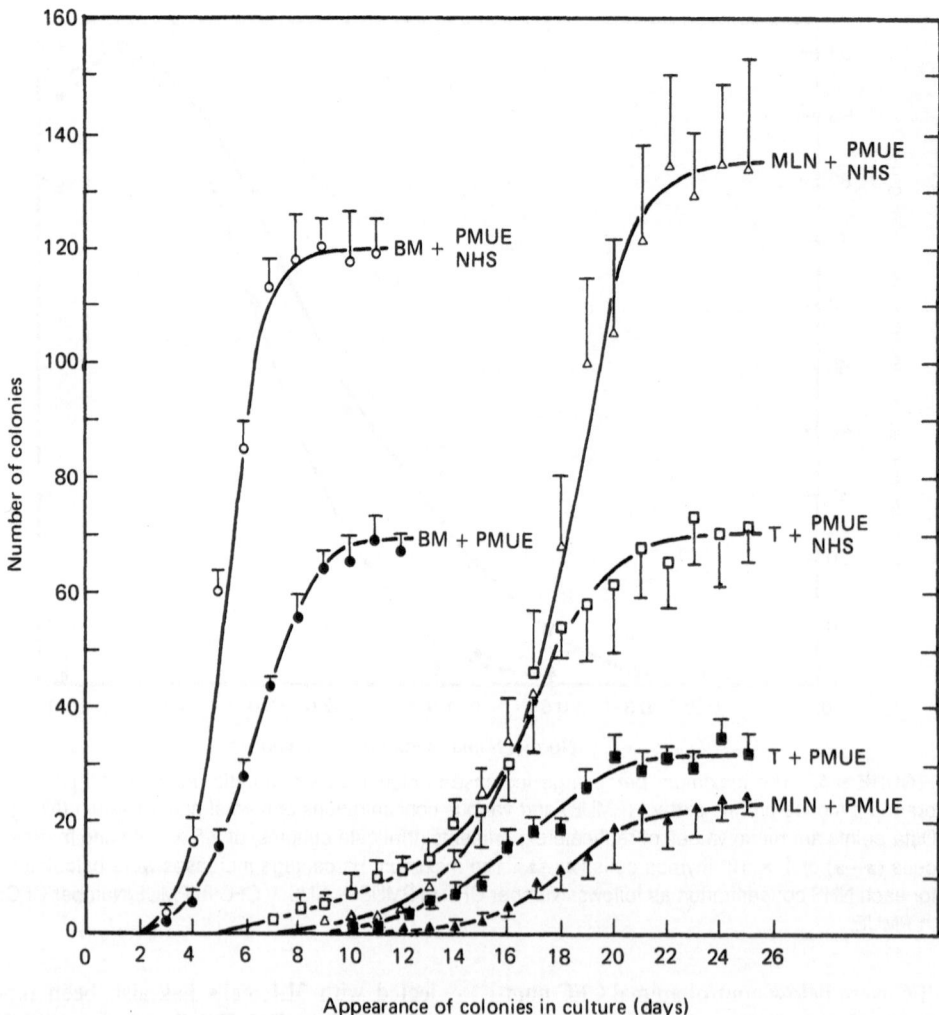

FIGURE 9-5. The growth curves of thymic (T) and mesenteric lymph node (MLN) CFC and bone marrow (BM) CFU-c. All cells were grown either in the presence of 3.5 percent PMUE (solid symbols) or in 3.5 percent PMUE and 7.5 percent NHS (open symbols). Data points are mean values $\pm$ standard errors of six replicate experiments (triplicate cultures) of either 2.5×10^4 bone marrow or 1×10^6 thymus or 1×10^6 lymph node cells. MLN-CFC ($\blacktriangle$, $\triangle$), T-CFC ($\blacksquare$, $\square$), BM-CFU-c ($\bullet$, $\circ$).

elements. These cells may have developed in response to the tumor from proliferative cells already in the thymus or they may have migrated there from the bone marrow. Regarding this, it is known that macrophage and granulocyte precursor cells are increased in the bone marrow, spleen, and blood of mice injected with tumor cells (1,6). Since the thymic cellularity was little changed between days 3 and 7, the threefold reduction in thymic CFC between those days may be the result of a selective elimination or a recruitment to other tissue sites, but our histopathological study of the tumor gave us little support for concluding that these cells migrated into the tumor. Recovery to above control-level cellularity on day 14 resulted in a return to control numbers of thymic CFC.

In the mesenteric lymph nodes, there was a two-fold increase in CFC grown in PMUE on day 3, which dropped to a level 50 percent higher than the controls on day 7 and 14. The slight hypercellularity on days 3 and 7, and the hypocellularity on day 14, could not account for the increased concentrations of node CFC on those days. As in the case of the thymic CFC, the increase in node CFC may have been due to the activity of proliferative cells already in the mesenteric lymph nodes or cells migrating there from the bone marrow. The node hypocellularity seen on day 14 paralleled the reduction of CFC in that organ. It should be stated that the histopathological changes and alterations in numbers of CFC took place in the absence of direct lymphatic drainage of the tumor site. In addition, these changes were not induced by the adherence or penetration of the tumor to the peritoneal wall. The addition of normal human serum to cultures of cells grown in a maximum stimulating amount of

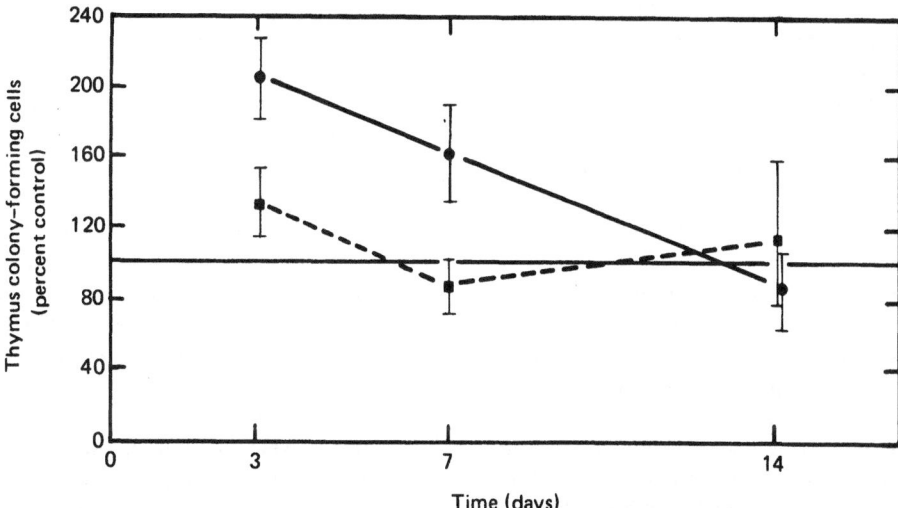

FIGURE 9-6. The percentage CFC values of thymus obtained from mice injected with 3LL carcinoma cells and grown in the presence of PMUE and NHS. The thymus cells were prepared from groups of mice 3, 7, and 14 days after implantation with 3LL cells and were grown utilizing the conditions described in Materials and Methods. There were no statistically significant differences in CFC values obtained from mice injected with the three amounts of 3LL cells. Thus, each point represents the mean value ± standard error of the combined data of all three cell doses (six replicate experiments) (n = 18). Thymic CFC concentration values from 3LL-injected mice differed significantly from controls on days 3 and 7. The concentration (•—•) and organ content (■---■) of thymic CFC grown in PMUE and NHS and obtained from control mice were 25 ± 5 and 1560 ± 310. Control (—).

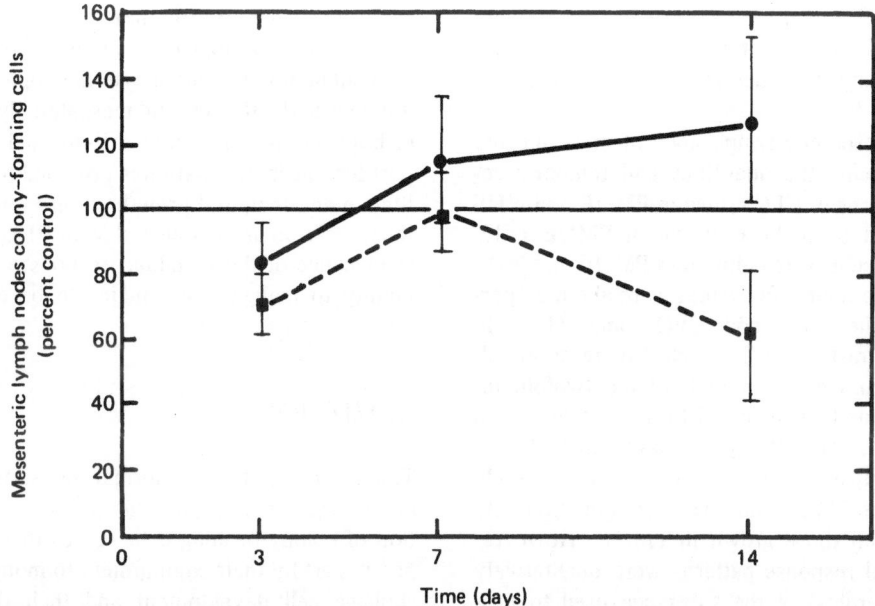

FIGURE 9-7. The percentage CFC values of mesenteric lymph nodes obtained from mice injected with 3LL carcinoma cells and grown in the presence of PMUE and NHS. The node cells were prepared from groups of mice 3, 7, and 14 days after implantation with 3LL cells and were grown utilizing the conditions described in Materials and Methods. There were no statistically significant differences in the CFC values obtained from mice injected with the three amounts of 3LL cells. Thus, each point represents the mean value ± standard error of the combined data of all three cell doses (six replicate experiments) (n = 18). Mesenteric lymph node CFC values from 3LL-injected mice did not differ significantly from controls. The concentration (•—•) and organ content (■---■) of node CFC grown in PMUE and NHS and obtained from control-mice were 75 ± 20 and 1360 ± 390. Control (—).

PMUE enhanced the growth of both granulocyte-macrophage- and monocyte-macrophage-committed cells. Maximal stimulation was accomplished by the addition of 7.5 percent NHS to 3.5 percent PMUE. Clonal growth was not sustained in cell culture containing only NHS. When lymphoid cells were grown in PMUE and NHS, the numbers of thymic CFC and node CFC were increased two- and sevenfold, respectively. Quantities of bone marrow increased twofold when the cells were grown in both PMUE and NHS. The lag periods to the first appearance of both thymic CFC and node CFC were reduced 30 percent, while those of the CFU-c remained the same.

The quantities and temporal response patterns of CFC obtained from the thymus of 3LL cell-injected mice and grown in PMUE and NHS differed from those of cells grown in PMUE only. In the thymus, the concentration of CFC grown in PMUE and NHS was increased 50 to 100 percent on days 3 and 7; it returned to control values on day 14. The organ content of this cell was little changed from that of control mice on all three days of testing. When these data are compared to the changes in thymic cellularity and histopathology, it can be inferred that the increase in this thymic CFC is relative, brought about by the selective loss of thymic lymphocytes. The concentration of thymic CFC grown in PMUE and NHS returned to control values on day 14 at which time the thymic cellularity of 3LL cell-injected mice returned to normal values.

In the mesenteric lymph node of mice injected with 3LL cells, the quantities and temporal response patterns of CFC grown in PMUE and NHS also differed from those grown in PMUE only. When node cells were cultured in PMUE and NHS on day 3, the node CFC values were about 25 percent below the control node CFC values. This is in marked contrast to (a) the modest increase in cellularity discussed earlier and (b) the twofold increase of node CFC detected for node cells grown in PMUE only on that day. On days 7 and 14 after tumor cell implantation, the values of CFC for node cells grown in PMUE and NHS were quantitatively different from those grown in PMUE. However, the temporal response patterns were qualitatively similar, regardless of the substance used to promote CFC growth.

The quantitative differences in responses noted for thymic CFC and mesenteric lymph node CFC obtained from both normal and tumor cell-injected mice and grown in PMUE or in PMUE and NHS may be interpreted two ways. It could be argued that the addition of NHS to the culture media increased the plating efficiency of the cells. Such a phenomenon was reported by Bradley et al. (3) for mouse bone marrow cells grown in mouse embryo extract to which washed fresh mouse red blood cells were added. In our work, changes in plating efficiency do not adequately explain the different temporal growth responses of CFC obtained from tumor cell-injected mice. Rather, the data could be better interpreted as supporting the existence of two subpopulations of CFC, as was proposed for mouse marrow CFU-c grown in PMUE and rat red cell lysate (23). A more definitive answer to whether or not there are two subpopulations of CFC awaits such cell purification procedures as density gradient separation.

The mechanism by which the numbers of thymic CFC and mesenteric lymph node CFC are increased, particularly during early genesis of the tumor, is unknown. A number of explanations have been proposed to explain progenitor cell increases in mice injected with transplantable solid tumors (6). The most likely explanation for CFC increases in our tumor system is that 3LL cells activate progenitor cells through colony-stimulating activity (CSA). In preliminary experiments, we have detected such activity when supernatants from 3LL cells grown in vitro were added to soft agar cultures of normal mouse bone marrow cells. More work must be done to determine if growth of the 3LL carcinoma in vivo results in CSA production.

In conclusion, it is tempting to interpret our data as signifying the existence of two subpopulations of small numbers of monocyte-macrophage progenitor cells in the thymus and mesenteric lymph nodes of both normal and tumor cell-injected mice. Support for this interpretation is provided by the quantitative and temporal growth pattern differences of these cells in mice injected with 3LL cells. The significance of these cellular responses in host immunity to malignancy remains to be determined.

SUMMARY

The in vitro culture of normal mouse thymus and mesenteric lymph node cells results in the formation of colony-forming cells (CFC) that are identified in part by their commitment to monocyte-macrophage cell development and their dependence upon pregnant mouse uterine extract (PMUE) for growth. The behavior of these cells in the lymphoid organs of tumor cell-injected mice and their in vitro growth requirements may be of importance in understanding host-tumor relationships. This idea was tested by engrafting $C_{57}B1/6$ mice with Lewis lung (3LL) carcinoma cells and examining the thymus and the mesenteric lymph nodes for their CFC content 3, 7, and 14 days after tumor-cell implantation.

The cells were grown in the PMUE only or in PMUE and normal human serum (NHS).

In cells obtained from 3LL cell-injected mice and grown in PMUE only, there was an approximate twofold increase in thymic CFC on day 3. A reduction to 50 percent of control thymic values was observed on day 7. The numbers of thymic CFC in control and tumor cell-injected mice were similar on day 14. The numbers of mesenteric lymph node CFC on day 3 were increased twofold over control values and tended to be higher than the controls thereafter.

With normal thymic and mesenteric lymph node cells grown in PMUE, maximum growth enhancement was noted by the addition of 7.5 percent (v/v) of NHS, which reduced the lag period 30 percent for both types of cells, even though the maximum numbers of colonies were counted 25 days after culturing, as is the case for cells grown only in PMUE. The number of thymic CFC and node CFC grown in PMUE and NHS was increased twofold and sevenfold, respectively, over that for cells grown in PMUE only.

Thymus cells obtained from tumor cell-injected mice and grown in PMUE and NHS resulted in a twofold increase over controls of the thymic CFC concentration on days 3 and 7. By day 14, the thymic CFC values returned to control values. Lymph node cells grown in PMUE and NHS resulted in quantities of mesenteric lymph node CFC, which did not differ significantly from control mice. However, the mesenteric lymph node CFC values of 3LL-injected mice tended to be lower than those for control mice on day 3.

In conclusion, the growth kinetics of the thymic CFC and mesenteric lymph node CFC are profoundly altered by the development of the 3LL carcinoma. The different numbers of CFC obtained after culturing in PMUE only or in PMUE and NHS could be interpreted as the growth of two subpopulations of CFC.

ACKNOWLEDGMENT

The authors wish to acknowledge the excellent services of the technical and secretarial staff of the Armed Forces Radiobiology Research Institute without which this work could not have been done.

REFERENCES

1. Baum, M., and Fisher, B. Macrophage production by the bone marrow of tumor-bearing mice. *Cancer Res., 23*:2813, 1972.
2. Bradley, T.R., Stanley, E.R., and Sumner, M.A. Factors from mouse tissue stimulating colony growth of mouse bone marrow cells *in vitro. Aust. J. Exp. Biol. Med. Sci., 49*:595, 1971.
3. Bradley, T.R., Telfer, P.A., and Fry, P. The effect of erythrocytes on mouse bone marrow colony development *in vitro. Blood, 38*:353, 1971.
4. Chu, J-Y., and Lin, H. Induction of macrophage colony-forming cells in the pleural cavity. *J. Reticuloendothel. Soc., 20*:299, 1976.
5. Fauve, R.M., Hevin, B., Jacob, H., Gaillard, J.A., and Jacob, F. Antiinflammatory effects of murine malignant cells. *Proc. Nat. Acad. Sci. (USA) 71*:4052, 1974.
6. Hibberd, A.H., and Metcalf, D. Proliferation of macrophage and granulocyte precursors in response to primary and transplanted tumors. In D.W. Weiss, ed., *Immunological Parameters of Host-Tumor Relationships.* New York: Academic Press, pp. 202-210.
7. Ledney, G.D., Moniot, J.V., Gambrill, M.R., and MacVittie, T.J. Mitogenic and colony forming unit responses of spleen cells from mice engrafted with Lewis lung (3LL) carcinoma cells. *Exp. Hematol., 4* (Suppl.):189, 1976.
8. Levy, M.H., and Wheelock, E.F. The role of macrophages in defense against neoplastic disease. *Adv. Cancer Res., 20*:131, 1974.
9. Lin, H. Colony formation *in vitro* by mouse blood monocytes. *Blood, 49*:593, 1977.
10. Lin, H., and Stewart, C.C. Colony formation by mouse peritoneal exudate cells *in vitro. Nature (Lond.), 243*:176, 1973.
11. Lin, H., and Stewart, C.C. Peritoneal exudate cells, I. Growth requirement of cells capable of forming colonies in soft agar. *J. Cell. Physiol., 83*:369, 1974.
12. MacVittie, T.J., and McCarthy, K.F. The detection of *in vitro* monocyte-macrophage colony-forming cells in mouse thymus and lymph nodes. *J. Cell. Physiol., 92*:203, 1977.
13. MacVittie, T.J., and Provaznik, M. The detection of *in vitro* monocyte macrophage colony-forming cells in mouse marrow, spleen, and peripheral blood. *Exp. Hematol., 5* (Suppl. 2):28, 1977.
14. MacVittie, T.J., and Weatherly, T.L. Characteristics of the *in vitro* monocyte-macrophage colony-forming cells detected within mouse thymus and lymph nodes. In S.J. Baum and G.D. Ledney, eds., *Experimental Hematology Today.* New York: Springer-Verlag: 1977, pp. 147-156.
15. Metcalf, D., MacDonald, H.R., and Chester, H.M. Serum potentiation of granulocyte and macrophage colony formation *in vitro. Exp. Hematol., 3*:261, 1975.
16. Milas, L., and Tomljanovic, M. Spleen colony-forming capacity of bone marrow from mice bearing fibrosarcoma. *Rev. Europ. Etudes Clin. Biol. 16*:462, 1971.
17. Nooter, K., and Bentvelzen, P. Cell density-dependent growth in agar of bone marrow cells from tumor-bearing BALB/c mice in the absence of a colony-stimulating factor. *Cancer Res., 35*:117, 1975.
18. Sheil, J.M., Crawford, R.M., and Ledney, G.D. Peritoneal macrophage chemotactic response and accumulation in mice injected with Lewis lung (3LL) carcinoma cells. *Exp. Hematol., 4* (Suppl.):127, 1976.

19. Siegel, S. *Nonparametric Statistics*. New York: McGraw-Hill, 1956, 312 p.
20. Snodgrass, M.J., Morahan, P.S., and Kaplan, A.M. Histopathology of the host response to Lewis lung carcinoma: Modulation by Pyran. *J. Natl. Cancer Inst., 55*:455, 1975.
21. Treves, A.J., Cohen, I.R., and Feldman, M. A syngeneic metastatic tumor model in mice: The natural immune response of the host and its manipulation. *Israel J. Med. Sci., 12*:369, 1976.
22. Weatherly, T.L. Enhancement of *in vitro* colony formation by human serum. *Exp. Hematol., 4* (Suppl.):35, 1975.
23. Williams, N., and van den Engh, G.J. Separation of subpopulations of *in vitro* colony-forming cells from mouse marrow by equilibrium density centrifugation. *J. Cell. Physiol., 86*:237, 1975.

PART III

Transplantation Immunology

S. Thierfelder

10

Expression of Normal and Leukemia-associated Antigens on Blood Cell Malignancies

S. Thierfelder, H. Rodt,
E. Thiel, B. Netzel, G. Jäger,
G. Hoffmann-Fezer, D. Huhn,
P. Dörmer, R. J. Haas,
G. F. Wündisch,
Ch. Bender Götze, and B. Lau

INTRODUCTION

For several decades, investigators have been trying to define antigens that are expressed on leukemic or other malignant cells but not on normal cells (for a review, see references 10,14). In man, conclusive evidence for tumor-specific antigens is still lacking.

The discovery of antigens characteristic of subsets of lymphocytes, the T- and B-cells in mice (9,21,26 review, reference 26), provided, nevertheless, an antigen-dependent classification principle, which was exploited in the diagnosis of human leukemias (2,3,17,19,26,33). Shortly after the reports of surface Ig on cells of chronic lymphoid leukemia in 1971 (review, reference 26), a report of E rosette formation by lymphoblasts from a cell line derived from a child with acute lymphoid leukemia brought up the possibility that acute lymphoid leukemia can express the T antigen (19). In 1973, several almost simultaneous studies indicated that acute lymphoid leukemia could be T, B or non-T non-B (2,17, 25,26,33). Most non-T non-B acute lymphoid leukemia was recently found to express the common acute lymphoid leukemia antigen (4,11,31). More membrane antigens and receptors have been introduced as markers of subsets of normal and leukemic cells in the hope of refining the diagnosis of leukemia by membrane phenotyping of blood cell malignancies (44).

This chapter deals with the expression and distribution of surface markers in various leukemias. It summarizes the production of our antisera against surface markers. It describes test systems developed in our laboratory, which permit the detection of surface antigens in tissue section as well as their quantitation on single cells or cell suspensions. It demonstrates the absence of these antigens on hemopoietic cell precursors as measured in colony-forming and diffusion chamber assays. It also underlines the therapeutic possibilities of surface marker antisera in autologous and allogeneic bone marrow transplantation.

MATERIALS AND METHODS
Cell Preparation
Normal and leukemic lymphocytes were separated from peripheral blood and bone marrow by density gradient centrifugation on a Ficoll-Isopaque gradient. Granulocytes and cells of myeloid leukemias were prepared by removing red blood cells through dextran-sedimentation followed by a removal of lymphocytes with Ficoll-Isopaque. Thymocytes were prepared from thymi of 1- to 3-year-old children undergoing cardiac surgery. Leukemic cell lines were donated by Dr. Minowada, Roswell Park Memorial Institute and Dr. Schneider, Uni-

versitätskinderklinik, Erlangen. Lymphoblastoid B-cell lines were established by infection of normal lymphocytes with supernatant of the B 95-8 EBV-producing cell line at our Institute.

Antisera

Anti-T-cell globulin (ATCG) and common acute lymphoid leukemia antiserum (anti-cALL) were raised in rabbits, specifically absorbed, and purified as described previously (27,31). The Anti-cALL was produced against cells from untreated patients with common acute lymphoid leukemia. The production of antisera against myeloid cells in the murine bone marrow and against human granulocytes was performed as reported (16).

Test Systems

Lymphocytotoxicity A two-stage microtest under paraffin oil with Eosin as a dye and rabbit complement was used (8).

Immunofluorescence A standard 18 Zeiss microscope with an IV/F epi-fluorescence condenser was used. The indirect test with the FITC-conjugated F(ab′)2 fragment of a sheep anti-rabbit-IgG was applied.

Complement fixation Complement fixation using human lymphocytes as antigens was performed by a micromethod as described previously (15). Veronal buffered saline (VBS) was used for all dilutions. First, 100 μl of a suspension of target cells (10^7 cells/ml) was added to 100 μl of each antibody dilution, the mixture was then allowed to incubate for 30 min, after which 100 μl of diluted guinea pig complement (GPC′) was added. After an additional incubation of 30 min, 100 μl of sensitized sheep red blood cells (SRBC) were added. After 30 min, the reaction mixture was diluted with 0.9 ml of VBS, sedimented in an Eppendorf centrifuge, and examined for hemolysis at 412 nm in a Zeiss PM 6 spectrophotometer equipped for automatic reading. The 50 percent lysis of sheep red cells was defined as the titer of the antiserum. Sensitization of red cells was performed as described (29). The GPC′ was adjusted to a dilution (about 1:100) that produced 90 percent hemolysis if the cell suspension and antibody dilution were replaced by VBS.

Microphotometric quantitative immunoautoradiography Cell labeling with ^{125}I-labeled antibodies, cell spreading by cytocentrifugation, standardization with a radioactive reference source, autoradiography, and microphotometry of immunoautoradiographs have been described in detail elsewhere (36,37,39). The data were statistically analyzed and plotted in histograms, using a data-processing unit (Interdata).

Immunohistochemistry Immunohistochemical identification of surface markers by the unlabeled antibody enzyme method in which the peroxidase–anti-peroxidase complex was used as a marker was reported on previously (15). Frozen sections were cut at 7 μm in a cryostat, air-dried for 30 min, and fixed in acetone for 5 min. Peroxidase activity was demonstrated by diaminobenzidine. The sections were counterstained with hematoxylin.

Colony-forming unit cells (CFU-c) A double-layer agar technique as described by Pike and Robinson (24) was used with minor modifications. After gelation at room temperature, the agar dishes were incubated for 12 to 14 days at 37°C in a fully humidified atmosphere continuously flushed with 5 percent CO_2. Colonies, defined as groups of 50 cells or more, were counted, using a Leitz Diavert microscope.

Diffusion chamber techniques (1,5) Aliquots of 5 × 10^5 nucleated marrow cells were loaded into Millipore chambers and implanted intraperitoneally into male CBA/J mice that had been irradiated with 650 R. Reimplantation was carried out on day 7. On days 6 and 12, 10 chambers were harvested for each antibody incubation and shaken for 1 hr in 0.5 percent pronase-buffered medium to liquefy the clot. The resulting cell suspensions were counted to determine the total number of nucleated cells. The remaining cells were washed and transferred to soft agar cultures for the CFU-c test. Colonies were counted as described above.

RESULTS AND CONCLUSIONS
Production of Antisera against cALL, T-lymphocytes, and Myeloid Cells

Aliquots of 10^8 cALL blasts were injected into rabbits as indicated in Figure 10-1. The antisera—whether or not it was coated with ATG as suggested by Greaves et al. (11)—had to be absorbed with various tissues, including lymphoblastoid cell lines (Figure 10-1), which removed residual cross-reactivity with normal bone marrow cells (31). The antisera were precipitated using saturated ammonium sulfate solution, purified by DEAE cellulose ion exchange chromatography after absorption, and reconcentrated to 10 mg/ml. The ATCG was prepared by absorption of liver and kidney homogenate, chronic lymphoid leukemia cells, and lymphoblastoid B-cell lines (31). A pool of cells from different patients was used, because the cells of any one patient did not always remove all the non-T-cell activity (29). Mouse bone marrow cells were used to produce antimyeloid globulin (16). Absorption was by liver and kidney homogenate, spleen, and thymus cells. Anti-human-granulocyte globulin was absorbed with kidney homogenate, normal peripheral lymphocytes and T- and B-lymphocytes from chronic lympoid leukemia patients (16). The purification procedure was identical for all antisera.

Antibody Specificity

The anti-cALL did not react with normal leukocytes and thymocytes, nor with bone marrow cells from normal persons or of fetal origin. No cross-reaction with other types of acute or chronic leukemias (Figure 10-2) was observed. The ATCG did not cross-react with normal or leukemic B-lymphocytes, with B-type lymphoblastoid cell lines, or with granulocytes or normal bone marrow cells. There was some cross-reaction with the intermediate form of acute lymphoid leukemia (see below) and with some leukemic cell lines (32). Several lymphoid cell lines were tested with anti-cALL and ATCG in the cytotoxic and immunofluorescent test system (Table 10-1).

Anti-myeloid globulin did not cross-react with mouse thymus lymph node, or spleen cells. The specificity of anti-human granulocyte globulin is given in Figure 10-3. There was borderline cross-reactivity with peripheral lymphocytes and the NALM-1 cell line, established by Minowada (20) from a patient with a Ph'positive chronic myeloid leukemia in blastic crisis. Whether the former is caused by cross-reactions due to contamination with monocytes, and the latter to the myeloid origin of the cell line, is not clear.

Quantitation of Surface Marker Antigens

Quantitative complement fixation The test system was adapted for surface marker serology by Rodt et al. (29). It has the advantage over conventional microcytotoxic or immunofluorescent test systems in that it also works with dead cells. It does not depend on microscopic cell evaluation and excludes false positive results by unspecific binding to F_c receptors. With a homogeneous cell suspension, the amount of surface antigens on different cell populations can be compared semiquantitatively. Figure 10-4 shows the complement fixation of ATCG or anti-Ig against suspensions containing T- or B-cells. The absence of T-antigen or surface Ig on the respective cells in suspension leads to complete hemolysis of the indicator system by unconsumed complement, their presence leads to complement fixation by the corresponding antibody, so that the hemolytic test system shows little or no hemolysis. Thus, CLL-S, a B-cell leukemia,

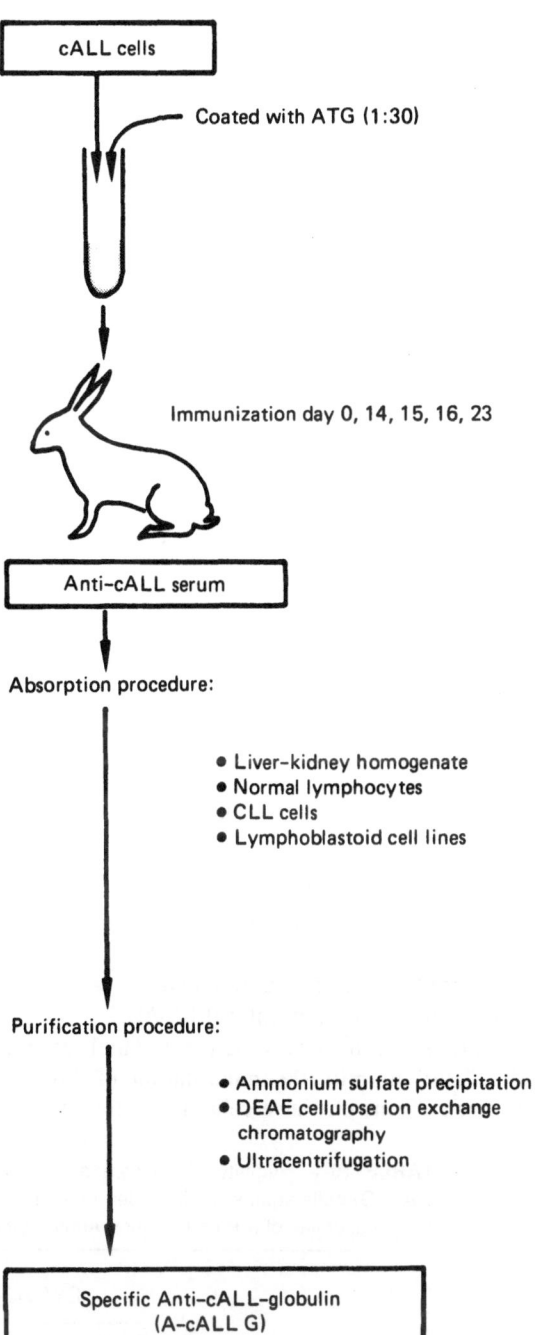

FIGURE 10-1. Production, absorption, and purification of antiserum against lymphoblasts of patients with common acute lymphatic leukemia of the non-T, non-B variety (anti-cALL globulin).

is negative with ATCG and positive with anti-Ig, whereas the opposite holds for T-antigen carrying CLL-P and ALL-F. A chronic T-cell leukemia, CLL-P, fixes the same amount of complement at a clearly lower antibody titer of ATCG than does, for example, ALL-F. This is consistent with the subnormal amounts of T-antigen measured on sin-

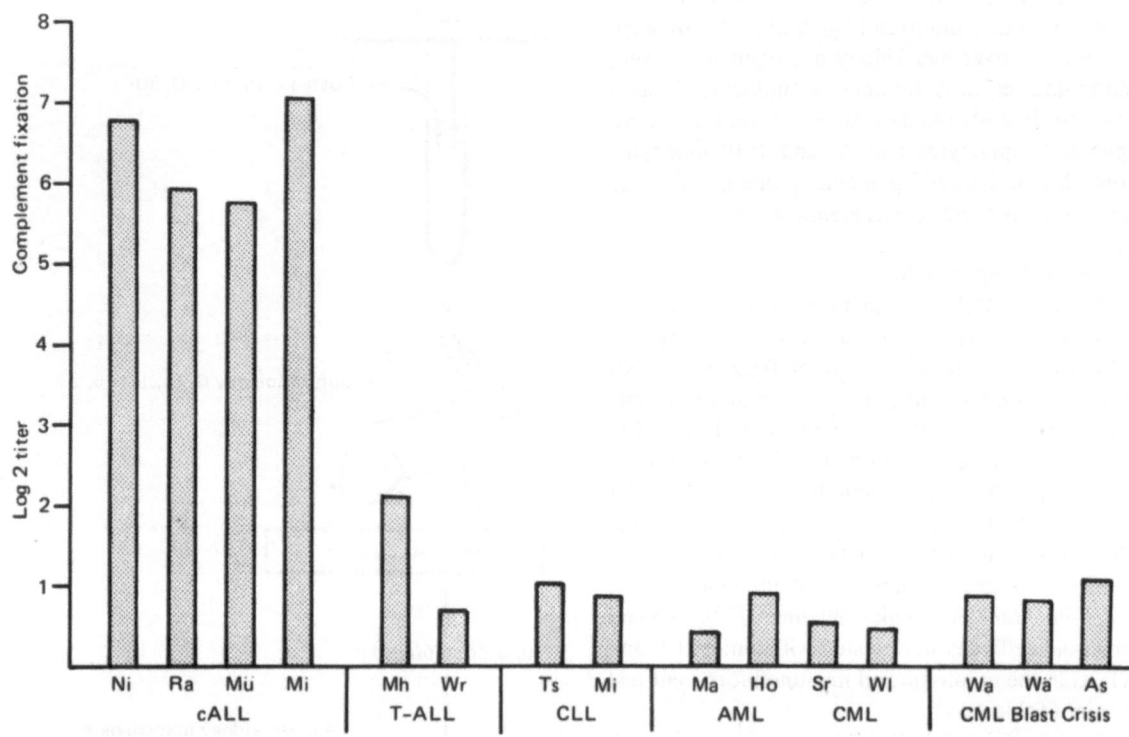

FIGURE 10-2. Activity of anti-cALL against various types of leukemia in the complement-fixation test.

gle cells of this leukemia in quantitative immunoautoradiography (see below).

Microphotometric Quantitative Immunoautoradiography (PhQIA)

This method, developed by Thiel et al. (36,37,39), permits the quantitation of surface marker antigens on single cells. In contrast to immunofluorescent methods, simultaneous morphological and histochemical classification is possible. Here, ^{125}I-labeled surface marker antisera are incubated with cells, which are washed thereafter and processed as autoradiographs. Grain densities on single cells are measured by reflected-light, bright-field microphotometry, using the MPV 2 (Leitz, Germany). By virtue of the linear relationship be-

TABLE 10-1. Activity of Antisera against Common Acute Lymphoid Leukemia (A-cALL-G) or T(ATCG)-cells against Various Cell Lines by the Microcytotoxicity and Immunofluorescence Tests. Two Leukemias of a Pure Common Acute Lymphoid Type are Included as Positive Controls.

CELLS	CYTOTOXIC TITER[a]		INDIRECT IMMUNOFLUORESCENCE	
	A-cALL-G	ATCG	A-cALL-G	ATCG
Null (0) cell lines:				
NALM (CML blast crisis)	7	4	+++	+
REH (cALL)	7	4	+++	+
KM-3 (cALL)	6	4	+	(+)
K-562 (CML blast crisis)	Negative	Negative	–	–
T-cell lines:				
MOLT-4 (T-ALL, E⁺)	Negative	10	–	++++
JM (T-ALL, E⁺)	Negative	9	–	++++
B-cell lines:				
IFH-16 (EBV, NP)	Negative	Negative	–	–
IFH-20 (EBV, NP)	Negative	Negative	–	–
Acute lymphoblastic leukemias:				
cALL Mü	11	1	++++	–
cALL Di	10	1	++++	–

[a] Log base 2 titer.

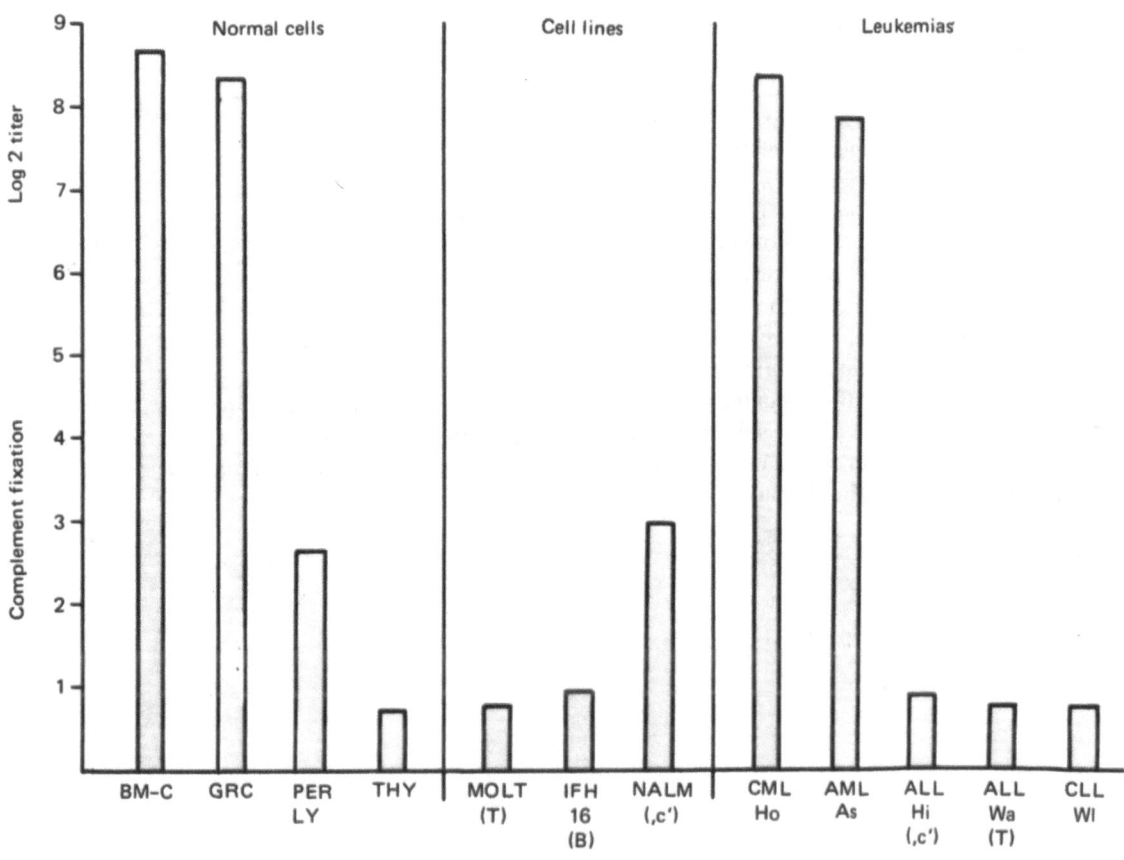

FIGURE 10-3. Activity of anti-granulocyte serum against human bone marrow cells (BM-C), isolated granulocytes (GRC) and peripheral lymphocytes (PER LY), thymocytes, T-, or B-, or cALL (,c')-cell lines, and various leukemias.

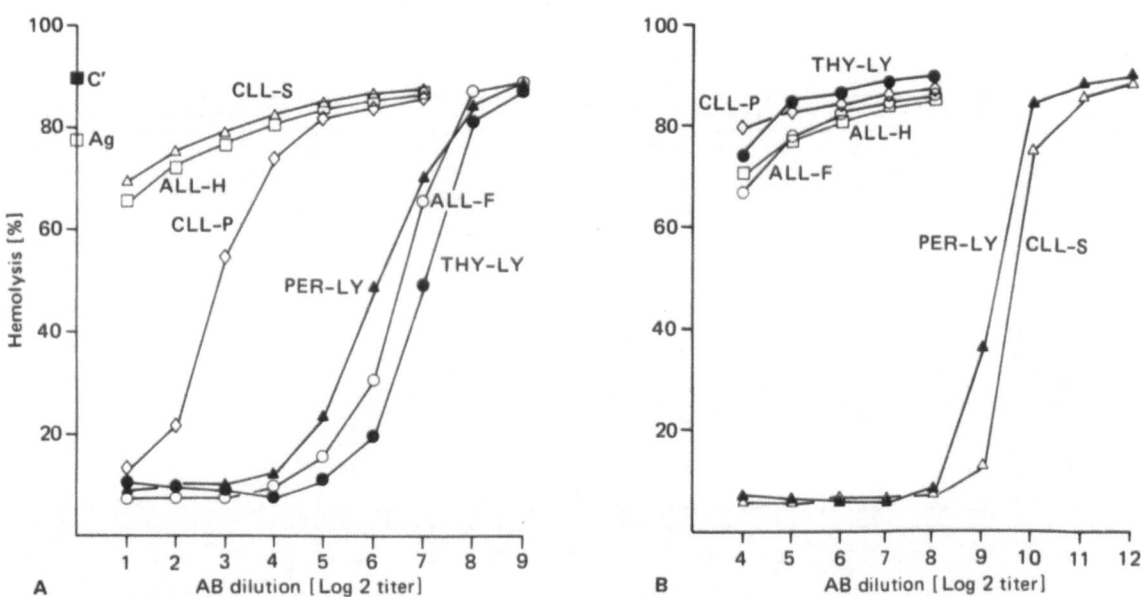

FIGURE 10-4 Complement fixation of ATCG (A) or anti-Ig (B) in the presence of suspensions containing T- and B-cells. Hemolysis, no complement fixation and no activity of the antibody against the cells under investigation. CLL-S, B-cell type; ALL-H, cALL; CLL-P, ALL-F, and thymocytes (THY-LY), T-cell type. Peripheral lymphocytes (PER-LY), T- and B-cell type.

tween the number of silver grains measured and the amount of light reflected, the photometer readings can be taken as equivalent to the number of grains. The data can be plotted as histograms. The introduction of a radioactive standard permits the determination of absolute amounts of ^{125}I-labeled compounds and the calculation of the number of antigenic sites on the surface of a given cell type (39). When mouse bone marrow cells were incubated with labeled antimyeloid globulin and measured by PhQIA, the myelocytes carried fewer labeled antibodies than the mature granulocytes (Figure 10-5), the amount of myeloid antigen increasing with differentiation from myeloblasts to neutrophil granulocytes (16). In contrast, the well-differentiated T-cells in the murine spleen carry less than one-half the T-antigen found on the more immature T-lymphocyte in the thymus (Figure 10-6),

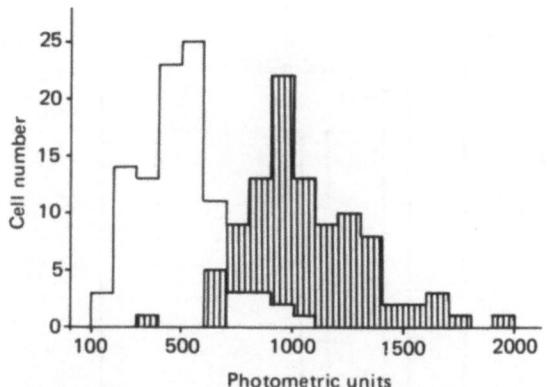

FIGURE 10-6. Histograms of label density on single mouse T-cells from the thymus (shaded (960 ± 271) or the spleen (unshaded) (396 ± 175) after incubation with ^{125}I-labeled globulin reacting specifically with T-cells (PhQIA).

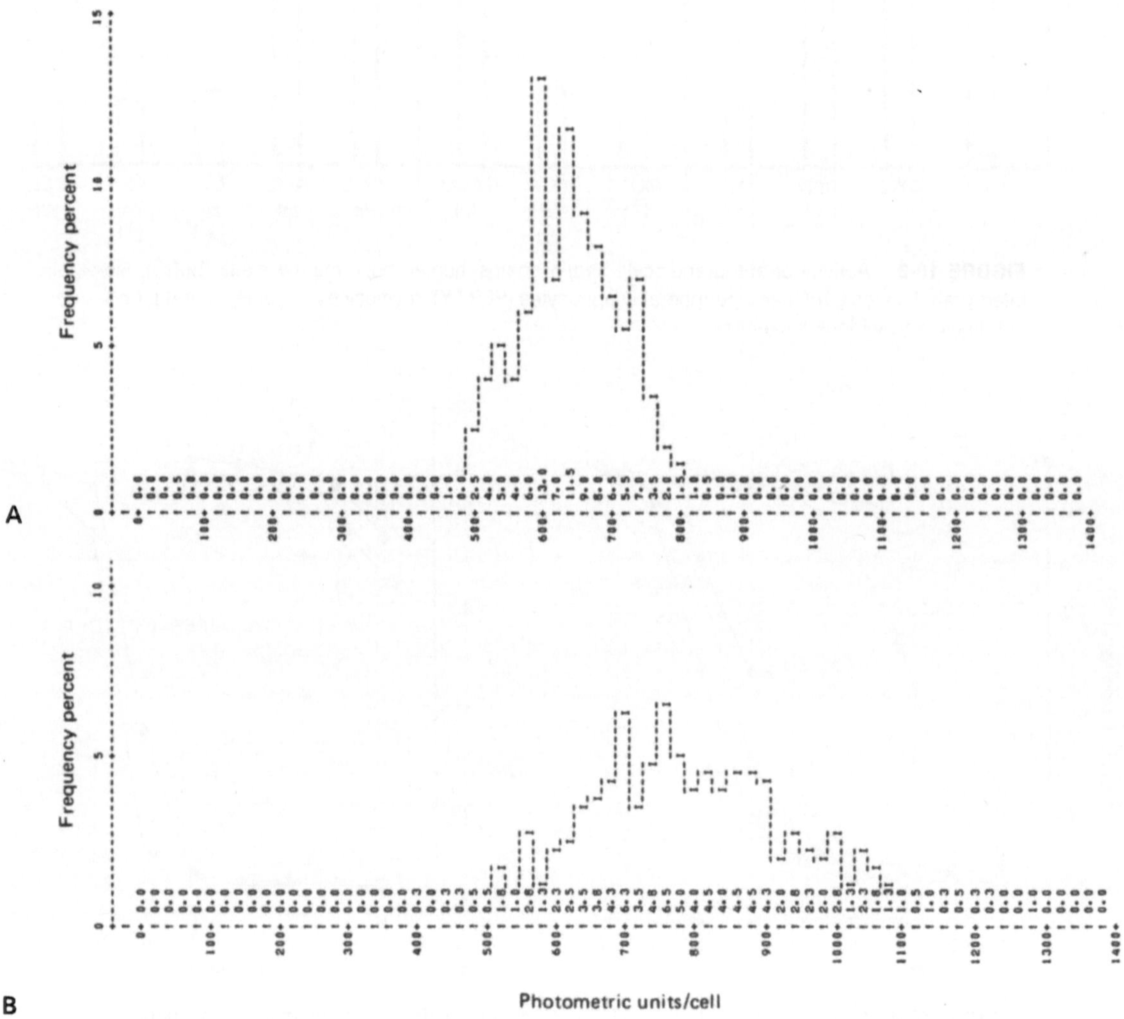

FIGURE 10-5. Histograms of label density (photometric units) on single myelocytes (A) or mature granulocytes (B) after incubation with ^{125}I-labeled anti-mouse-myeloid globulin measured by microphotometric immunoautoradiography (PhQIA).

TABLE 10-2. Membrane Phenotype of Chronic Lymphoid Leukemias; PLL, Prolymphocytic Leukemia.

NUMBER OF CASES	DIAGNOSIS	PHENOTYPE
135 CLL	B-CLL	(T antigen$^-$, SIg$^+$, E$^-$, E$_m^+$, F$_c^+$, CR$^+$)
3 PLL	B-PLL	(T antigen$^-$, SIg$^+$, E$^-$, E$_m^{-/+}$ F$_c^+$, CR$^+$)
3 CLL	T-CLL	(T antigen$^+$, SIg$^-$, E$^+$ at 4°C and E$^-$ at 37°C, E$_m^-$, F$_c^-$, CR$^-$)
1 PLL	T-PLL	(T antigen$^+$, SIg$^-$, E$^+$ at 4°C E$^-$ at 37°C, E$_m^-$, F$_c^-$, CR$^-$)

which confirms earlier measurements of T-antigen by absorption (26). Not only the amount of surface antigen on single cells can be expressed by PhQIA but also the variation of the amount within a given cell population (38,40,42). Figure 10-7 shows the histogram of T-lymphocytes from two normal subjects and a patient with chronic lymphoid leukemia (T-cell) (below). There is a somewhat lower mean concentration of T-antigen on the leukemic cells compared to those from the normal subjects. In addition, a wide variation of normal lymphocytes with different amounts of T-antigen contrasts with the narrow peak of leukemic T cells. This more homogeneous packing of T-antigen on the leukemic cells was taken as an indirect argument in favor of a clonal origin of chronic lymphoid leukemia (38).

Membrane Phenotypes of Chronic Lymphoid Leukemia and Prolymphocytic Leukemia

The great majority (95 percent) of 142 patients with chronic lymphoid leukemia showed the B-cell phenotype (review, reference 34) with surface immunoglobulins and receptors for the F$_c$ portion of IgG and complement components (Table 10-2). The high numbers of mouse rosettes and the prevalence of a single light-chain type in the immunofluorescence test (not shown in Table 10-2) is in agreement with the clonal proliferation of this leukemia.

A small minority of patients carries the T-antigen in accordance with earlier observations (6). In our group, only three cases of chronic lymphoid leukemia (2 percent) had the T-cell phenotype and formed spontaneous E rosettes at 4°C, which dissociated at 37°C. Prolymphocytic leukemia is a rare variant of chronic lymphocytic leukemia (7), with gross splenic enlargement, high lymphocyte counts, drug resistance, and characteristic cell morphology. Three cases of prolymphocytic leukemia with B-cell phenotype and one case with T-cell phenotype were found. In the latter case, a clonal origin of T-cell leukemia could be proven by a chromosome abnormality which was found in all tested karyotypes of this patient's leukemic cells (40,41).

Membrane Phenotypes of Acute Lymphoid Leukemia

The T-antigen is carried by 28 percent of 88 cases of acute lymphoid leukemia observed (Table 10-3). Interestingly, spontaneous E rosette formation with SRBC was not observed in 28 percent of 88 acute lymphoid leukemia patients, and this would have been missed if only the conventional E rosette test had been used; also the receptors for F$_c$ and complement are not diagnostic of T- or B-cells, as can be seen in Table 10-3.

About two-thirds of these cases could not be classified as belonging to the T- or B-cell phenotype, and they lacked Ig- and T-antigens. They did not form spontaneous E rosettes with SRBC. Some of them had receptors for F$_c$ and complement. The discovery of antisera reacting with this leukemia, which is the most common kind in childhood, permits us to define this type of leukemia positively as

TABLE 10-3. Membrane Phenotype of Acute Lymphoid Leukemias: 41 of the 61 0-ALL had been Tested for cALL Antigen.

NUMBER OF POSITIVE CASES[a]	DIAGNOSIS	PHENOTYPE
14 (16)	T-ALL, Ros.$^+$	(T antigen$^+$, cALL antigen$^-$, SIg$^-$, E$^+$ at 4°C or E$^-$ at 37°C, E$_m^-$, F$_c$R$^{-/+}$, CR$^{+/-}$)
(28.5)		
11 (12.5)	T-ALL, Ros.$^-$	(T antigen$^+$, cALL antigen$^-$, SIg$^-$, E$^-$, E$_m^-$, F$_c$R$^{+/-}$, CR$^{+/-}$)
61 (69)	O-ALL	(T antigen$^-$, cALL antigen$^+$, SIg$^-$, E$^-$, E$_m^-$, F$_c$R$^{-/+}$, CR$^{-/+}$)
2 (2.3)	B-ALL	(T antigen$^-$, cALL antigen$^-$, SIg$^+$, E$^-$, F$_c$R$^+$, CR$^+$)

[a]Numbers in parentheses are percent.

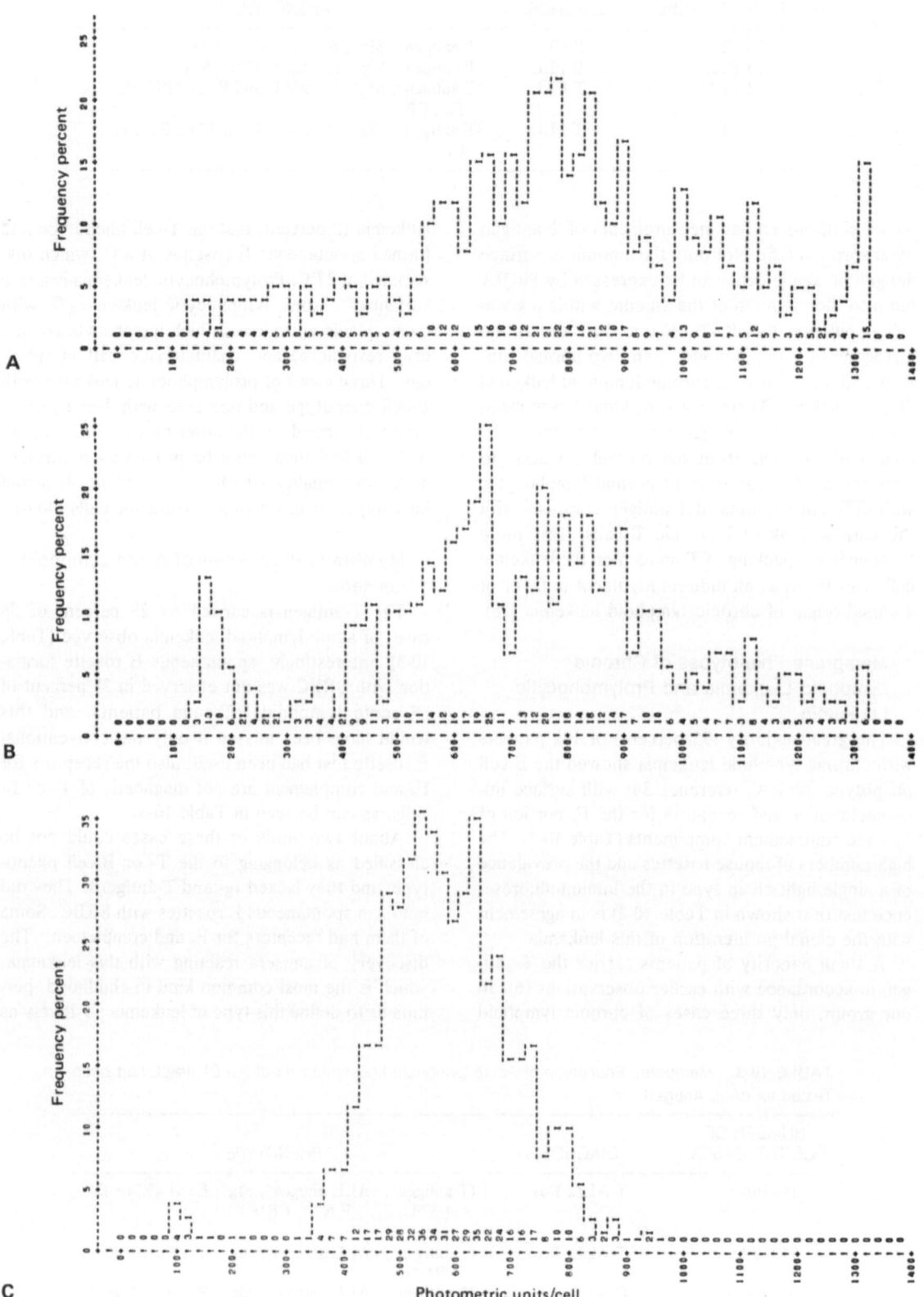

FIGURE 10-7. Histograms of cellular label density on single T-lymphocytes of two normal subjects (B, C) and one patient (A) with T-lymphocyte chronic lymphoid leukemia after incubation with ^{125}I-labeled ATCG (PhQIA). Antibody, anti-T; number of cells measured, 500.

common acute lymphoid leukemia (cALL) (11,12). In 41 cases tested with anti-cALL and ATCG simultaneously, 68 percent were positive with anti-cALL, 24 percent with ATCG, and only 3 percent were negative with both antisera. Several cases also showed a weak reaction with ATCG. Patients with this so-called intermediate type of T antigen-positive cALL did not form spontaneous E rosettes with SRBC. The T antigen is also demonstrable on cell lines of Table 10-1 and was first observed on leukemic cell lines by Minowada et al. (20). The simultaneous occurrence of the two antigens on cell lines excludes the possibility of intermediate cALL being an artifact due to a reaction of ATCG with contaminating nonleukemic T-lymphocytes.

Membrane Phenotypes of Blast Crisis in Patients with Chronic Myeloid Leukemia (CML-BlCr)

Nine CML-BlCr patients were tested with anti-cALL and four of these with anti-granulocyte globulin. Only one of nine cases reacted with anti-cALL. The cells from four cases carried the myeloid antigen (16).

Absence of cALL Antigen on Hemopoietic Precursor Cells

In order to investigate the possible therapeutic applications of anti-cALL in bone marrow transplantation (see below), inhibition by this antiserum of colony formation and proliferation in diffusion chambers of normal bone marrow cells was attempted (23). Table 10-4 shows that incubation with unabsorbed anti-cALL and rabbit complement decreased the number of CFU-c considerably. This inhibitory effect was, however, completely abolished after the antisera had become specific for cALL through absorption. A similar reversal of inhibition of cell proliferation by crude anti-cALL in diffusion chambers was observed with anti-specific cALL (23) (Figure 10-8), and ATCG was also

shown to have no inhibitory effect on CFU-c (22,29) and cell proliferation in diffusion chambers. The absence of cALL antigen on those committed or uncommitted stem cells that were measured by these *in vitro* test systems does not support Greave's position (12) that cALL antigen is a stem cell antigen.

Demonstration of Surface Marker Antigens in Tissue Sections

An immunological classification of malignant lymphomas has been complicated so far by the lack of a method that can localize membrane-labeled cells within lymphoid tissue. Technical difficulties of nonspecific staining could be overcome by adapting the unlabeled antibody enzyme method with peroxidase as a marker, as improved for intracellular antigens by Sternberger and Cuculis (35) and Mason et al. (18) for the demonstration of cell surface antigens in tissue sections. The method involves the use of three layers of antibodies: (a) rabbit antibody against surface antigen, (b) sheep antibody against rabbit IgG in excess to ensure free combining sites for (c) an antibody-antigen complex consisting of rabbit antibodies against peroxidase, and peroxidase. This method permits the localization of T- and B-cells as well as cALL-positive cells in cryostat sections under the light microscope. Figure 10-9 shows the uniform cellular picture of a testicular tumor developed by a child who had undergone HL-A/MLC identical bone marrow transplantation 6 months previously for his common acute lymphoid leukemia at the Kinderpoliklinik, University of Munich. The majority of tumor cells stained with anti-cALL, whereas a few ATCG-positive T-lymphocytes surrounding small venous vessels, were negative with anti-cALL (Figure 10-9). The patient, who is still alive, had had no relapse in the bone marrow at the time of his testicular tumor. The immunohistochemical di-

TABLE 10-4. Effect of Crude (Not Absorbed), Partially (Liver-spleen) Absorbed, and Fully (Including Lymphoblastoid Cell Line, LCL-IFH16) Absorbed Anti-cALL on the Number of CFU-c Following Incubation with Normal Bone Marrow Cells.

	FINAL ANTIBODY DILUTION[a]	CFU-c/2 × 10⁵ NUCL. MARROW CELLS[b]
Anti-cALL		
Not absorbed	1:2 + C	0
	1:4 + C	0
Liver-spleen absorbed	1:2 + C	39 ± 2
	1:4 + C	90 ± 2
Liver-spleen, LCL-IFH 16 absorbed	1:2 + C	78 ± 3
	1:4 + C	121 ± 5
Normal rabbit globulin	1:4 + C	84 ± 4

[a]Final antibody dilution in the incubation volume.
[b]CFU-c from triplicates.

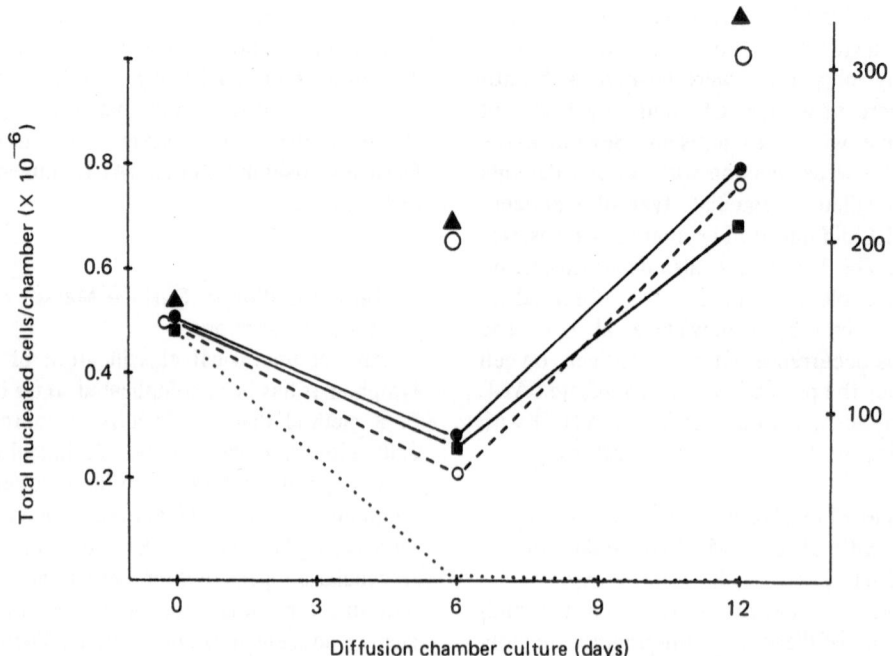

FIGURE 10-8. Effect of crude (■----■) and absorbed (■——■, 1:32 dilution; ●——●, 1:64 dilution) anti-cALL or normal rabbit globulin (○---○, 1:64 dilution) on the number of nucleated cells and of absorbed (▲, 1:32 dilution) anti-cCall, or normal rabbit globulin (○, 1:64 dilution) on the number of CFU-c, following antibody incubation and cultivation of normal bone marrow cells in diffusion chambers.

agnosis of a local relapse of his leukemia in the testicle was therefore of great importance.

Therapeutical Possibilities of Surface Marker Antisera

The testicular tumor of the human bone marrow chimera, mentioned above, was removed surgically. Since it was full of cALL cells, one might consider immunotherapy during operation, with an injection of anti-cALL being given to protect against hematogenic spread. That anti-cALL does indeed remove these leukemic cells from the circulation selectively was observed in another patient with common acute lymphoid leukemia (13). An injection of 3×7 mg of anti-cALL/kg body weight every second day removed the 19 percent circulating cALL cells within hours and for several days. There was no change in the percentage of cALL blasts in the bone marrow.

Our CFU-c and diffusion chamber data, which suggest that ATCG as well as anti-cALL spares hemopoietic cells, led us to propose approaches to allogeneic and autologous bone marrow transplantation whereby the donor's marrow is incubated with surface marker antisera. In allogeneic bone marrow transplantation, a better suppression of T-cell-derived, graft-versus-host reactions can be expected from incubating the donor's marrow with ATCG. In autologous bone marrow transplantation

in leukemic patients, a more complete eradication of residual leukemic cells in the remission marrow may be possible if it is incubated with surface marker antisera.

In experimental bone marrow transplantation, complete suppression of graft-versus-host reactions with rabbit antisera sparing hemopoietic stem cells and inhibiting T-cells has been reported for H-2 incompatible mice from this laboratory, since 1972 (27,28,30).

Autologous bone marrow removed from the leukemic patient, preserved while he is in remission and retransplanted when he is in relapse, is problematical in that remission bone marrow is probably not completely free of leukemic cells. In this context, we showed earlier that irradiated mice injected with syngeneic bone marrow contaminated by high numbers of leukemic T-lymphocytes of the AKR/J strain did not die of leukemia if the leukemic bone marrow had been incubated with stem cell-sparing antiserum against T-cells. It was calculated that the antiserum treatment must have inhibited over 99 percent of the leukemic cells (43). We therefore propose to call "antileukemic autotransplantation" the use of antiserum incubation with autologous bone marrow transplantation in order to inhibit residual leukemic T- or cALL-cells. In the case of ATCG, a simultaneous suppression of normal T-lymphocytes is not critical, since the transferred stem cells will differentiate to produce new T-cells (43). This may also occur with anti-myeloid globulin in acute myeloid leukemia, if the stem cells do not carry the myeloid antigen.

The data presented point up the considerable diagnostic as well as therapeutic implications of surface marker serology. Common acute lymphoid leukemia is probably the best example of the importance of surface markers in early diagnosis and in cytological therapy control. Preliminary data from various clinical centers already suggest that the prognosis of common acute lymphoid leukemia in childhood is much better than that of morphologically indistinguishable T-cell acute lymphoid leukemia (4). Other prognostic features of surface marker serology will evolve with more experience in immunological phenotyping of leukemias and lymphomas.

SUMMARY

Production, specific absorption, and purification of rabbit antisera against membrane antigens on human T-lymphocytes, common acute lymphoid leukemia blasts and myeloid cells are reviewed. Two methods, microphotometric immunoautoradiography (PhQIA) and quantitative complement-fixation (QC'Fix), were developed in order to quantify the amount of surface antigens on single cells and cell suspensions. Microphotometric quantitative immunoautoradiography (PhQIQ) used ^{125}I-labeled rabbit anti-mouse-myeloid globulin to identify a differentiation antigen on cells of the myeloid series; this antigen increased in amount on the more differentiated cells. The ^{125}I-labeled ATCG demonstrated a more homogeneous packing of T antigen on T-lymphocytes from chronic lymphoid leukemia patients when compared to normal T-lymphocytes. Quantitative complement fixation (QC'Fix) had the advantage of not depending on viable cells and microscopic evaluation and permitting the rapid quantitation of mean surface antigen in different homogeneous cell preparations.

Immunological phenotyping was performed on 142 chronic lymphatic leukemic, 88 acute lymphoid leukemic, and nine chronic myeloid leukemic patients. The latter were in blast crisis. The classification of acute lymphatic leukemia revealed a number of T-lymphocytes that did not form E rosettes spontaneously. The unlabeled antibody enzyme method was adapted for the demonstration of surface antigens in tissue sections. Concerning the question as to whether the cALL and the T antigen are stem cell antigens, studies with CFU-c and diffusion chamber assays showed no evidence for the

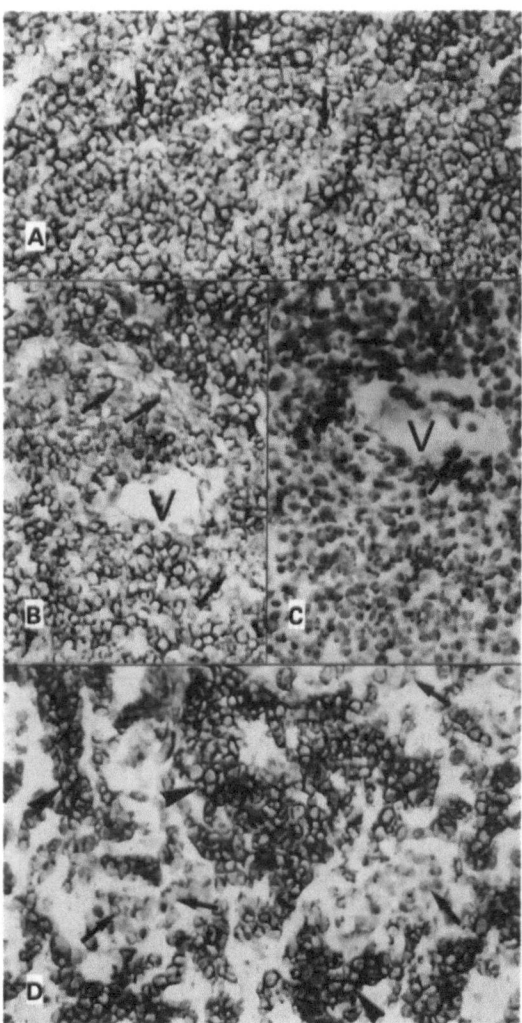

FIGURE 10-9. Immunohistological diagnosis of a local relapse of common acute lymphoid leukemia presenting as a testicular tumor in a child after treatment by bone marrow transplantation (cryostat sections, unlabeled antibody enzyme method). (A) Immunohistological demonstration of cALL-positive tumorous cells (↑):, (B) cALL negative cells (↑) in the immediate neighbourhood of a small venous vessel (V); (C) single ATCG-positive cells (↑) surrounding a small venous vessel (V), the majority of infiltrating tumorous cells are negative with ATCG; (D) cells of testicular tubules (↑) are negative, tumorous cells (▲) are positive with anti-cALL (A-D, X 330).

presence of these antigens on normal hemopoietic progenitors.

Therapeutic approaches to the use of surface marker antisera for a suppression of graft-versus-host disease in allogeneic bone marrow transplantation as well as for an elimination of residual leukemic cells in autologous bone marrow transplantation were proposed and studied in animal models.

ACKNOWLEDGMENTS

The authors are indebted to the technical staff at the Institut für Hämatologie for the skillful assistance in the diagnostical and experimental procedures.

This study was supported in part by SFB 37 and Euratom No. 217-76-1 BIOD.

REFERENCES

1. Benestadt, H.B. Formation of granulocytes and macrophages in diffusion chamber cultures of mouse blood leukocytes. *Scand. J. Haemat.*, 7:279, 1970.
2. Borella, L., and Sen, J. T-cell surface markers on lymphoblasts from acute lymphocytic leukemia. *J. Immunol.*, 111:1251, 1973.
3. Borella, L., Sen, L., Dow, L.W., and Casper, J.T. Cell differentiation antigens versus tumor-related antigens in childhood acute lymphoblastic leukemia (ALL). Clinical significance of leukemia markers. In S. Thierfelder, H. Rodt, and E. Thiel, eds., *Immunological Diagnosis of Leukemias and Lymphomas*. Berlin-Heidelberg: Springer-Verlag, 1977.
4. Borella, L. Differentiation antigens expressed by normal cells of the lymphatic and nervous system. *4th Meeting Intern. Soc. Haematol. Europ. and African Div.*, Istanbul, Sept. 1977.
5. Bøyum, A., and Borgström, R. The concentration of granulocytic stem cells in mouse bone marrow, determined with diffusion chamber technique. *Scand. J. Haemat.*, 7:294, 1970.
6. Brouet, J.C., Flandrin, G., Sasportes, M., Preud' Homme, J.L., and Seligmann, M. Chronic lymphocytic leukemia of T-cell origin. Immunological and clinical evaluation in eleven patients. *Lancet*, 11:890, 1975.
7. Galton, D.A.G., Goldmann, J.M., Wiltshaw, E., Catovsky, D., Henry, K., and Goldenberg, G.J. Prolymphocytic leukemia. *Brit. J. Haemat.*, 27:7, 1974.
8. Götze, D., and Reisfeld, R.A. Soluble murine antigens. I. Immunogenecity of soluble H-2 antigens extracted with hypertonic salt solution. *J. Immunol.*, 112:1643, 1974.
9. Greaves, M.F., and Hogg, N.M. Immunoglobulin determinants on the surface of antigen-binding T- and B-lymphocytes in mice. In B. Amos, ed., *Progress in Immunology*. New York: Academic Press, 1971, p. 110.
10. Greaves, M.F. Clinical applications of cell surface markers. *Progr. Haematol.*, 9:255, 1975.
11. Greaves, M.F., Brown, G., Rapson, N.T., and Lister, T.A. Antisera to acute lymphoblastic leukemia cells. *Clin. Immunol. Immunopath.*, 4:67, 1975.
12. Greaves, M.F., Janossy, G., Roberts, M., Rapson, N.T., Ellis, R.B., Chessels, J., Lister, T.A., and Catovsky, D. Membrane phenotyping: Diagnosis, monitoring and classification of acute (lymphoid) leukemias. In S. Thierfelder, H. Rodt, and E. Thiel, eds., *Immunological Diagnosis of Leukemias and Lymphomas*. Berlin-Heidelberg: Springer-Verlag, 1977.
13. Haas, R., Netzel, B., Rodt, H., Thiel, E., and Thierfelder, S. Use of specific antisera against leukemia-associated antigens in diagnosis and treatment of childhood ALL. *European J. Ped.* (in preparation).
14. Harries, R. Leukemia antigens and immunity in man. *Nature (Lond.)*, 241:95, 1973.
15. Hoffmann-Fezer, G., Rodt, H., Eulitz, M., and Thierfelder, S. Immunohistochemical identification of T- and B-lymphocytes delineated by the unlabeled antibody enzyme method. I. Anatomical distribution of θ-positive and Ig-positive cells in lymphoid organs of mice. *J. Imm. Methods.*, 13:261, 1976.
16. Jäger, G., Hoffmann-Fezer, G., Rodt, H., Huhn, D., Thiel, E., and Thierfelder, S. Myeloid antigens and antigen densities in mice and men. In S. Thierfelder, H. Rodt, and E. Thiel, eds., *Immunological Diagnosis of Leukemias and Lymphomas*. Berlin-Heidelberg: Springer-Verlag, 1977.
17. Kersey, J.H., Sabad, K., Gajl-Peczalski, K., Hallgreen, H.M., Yunis, E.J., and Nesbit, M.E. Acute lymphoblastic leukemia cells with T (thymus-derived) lymphocyte markers. *Science* (Wash.) 182:1355, 1973.
18. Mason, T.E., Phifer, R.F., Spicer, S.S., Swallow, R.A., and Dreskin, R.B. New immunochemical technique for localizing intracellular tissue antigen. *J. Histochem. Cytochem.*, 17:563, 1969.
19. Minowada, J., Ohnuma, T., and Moore, G.E. Rosette-forming human lymphoid cell lines. I. Establishment and evidence for origin of thymus-derived lymphocytes. *J. Nat. Cancer Inst.*, 49:891, 1972.
20. Minowada, J., Tsubota, T., Nakazawa, S., Srivastava, B.I.S., Huang, C.C., Oshimura, M., Sonta, S., Han, T., Sinks, L.F., and Sandberg, A.A. Establishment and characterization of leukemic T-cell lines, B-cell lines and null-cell lines: A progress report on surface antigen study of fresh lymphatic leukemias in man. In S. Thierfelder, H. Rodt, and E. Thiel, eds., *Immunological Diagnosis of Leukemias and Lymphomas*. Berlin-Heidelberg: Springer-Verlag, 1977.
21. Möller, G. Demonstration of mouse isoantigens at the cellular level by the fluorescent antibody technique. *J. Exp. Med.*, 114:415, 1961.
22. Netzel, B., Rodt, H., Hoffmann-Fezer, G., Thiel, E., and Thierfelder, S. The effect of crude and differently absorbed anti-human T-cell globulin on granulocytic and erythrocytic colony formation. *Exp. Hematol.* (in press).
23. Netzel, B., Rodt, H., Lau, B.M., Haas, R.J., Dörmer, P., and Thierfelder, S. Interaction of antisera against human leukemia-associated antigens with hemopoietic stem cells. Exp. Hematol. 5 (Suppl. 2):90, 1977.
24. Pike, B.L., and Robinson, W.A. Human bone marrow colony growth in agar-gel. *J. Cell. Physiol.*, 76:77, 1970.
25. Preud'Homme, J., and Seligmann, M. Surface immunoglobulins on human lymphoid cells. In R. Schwartz, ed., *Progress in Clinical Immunology*, Vol. 2. Grune & Stratton, New York, 1974, p. 121.
26. Reif, A.E., and Allen, J.M. The AKR thymic antigen and its distribution in leukemias and nervous tissues. *J. Exp. Med.*, 120:413, 1964.
27. Rodt, H., Thierfelder, S., and Eulitz, M. Suppression

of acute secondary disease by heterologous anti-brain serum. *Blut, XXV*:385, 1972.

28. Rodt, H., Thierfelder, S., and Eulitz, M. Anti-lymphocytic antibodies and marrow transplantation. III. Effect of heterologous anti-brain antibodies on acute secondary disease in mice. *Europ. J. Immunol., 4*:25, 1974.

29. Rodt, H., Thierfelder, S., Thiel, E., Götze, D., Netzel, B., Huhn, D., and Eulitz, M. Identification and quantitation of human T-cell antigen by antisera purified from antibodies crossreacting with hemopoietic progenitors and other blood cells. *Immunogenetics, 2*:411, 1975.

30. Rodt, H., Netzel, B., Niethammer, D., Körbling, M., Götze, D., Kolb, H.-J., Thiel, E., Haas, R.J., Fliedner, T.M., and Thierfelder, S. Specific absorbed antithymocyte globulin for incubation treatment in human marrow transplantation. *Transpl. Proc., 9*:187, 1977.

31. Rodt, H., Netzel, B., Thiel, E., Jäger, G., Huhn, D., Haas, R., Götze, D., and Thierfelder, S. Classification of leukemic cells with T- and O-ALL-specific antisera. In S. Thierfelder, H. Rodt, and E. Thiel, eds., *Immunological Diagnosis of Leukemias and Lymphomas*. Berlin-Heidelberg: Springer-Verlag, 1977.

32. Rodt, H., Netzel, B., Thiel, E., Jäger, G., Schneider, U., Haas, R.J., Minowada, J., Rosenfeld, C., and Thierfelder, S. Characterization of antigens on leukemic cells and cell lines using specific heterologous antisera. In H. Peters, ed., *Protides of the Biological Fluids*. Oxford 1977 (in press).

33. Seligmann, M., Preud'Homme, J.L., and Brouet, J.C. B- and T-cell markers in human proliferative blood diseases and primary immunodeficiencies, with special reference to membrane bound immunoglobulins. *Transpl. Rev., 16*:83, 1973.

34. Seligmann, M., Brouet, J.C., and Preud'Homme, J.L. The immunological diagnosis of human leukemias and lymphomas: An overview. In S. Thierfelder, H. Rodt, and E. Thiel, eds., *Immunological Diagnosis of Leukemias and Lymphomas*. Berlin-Heidelberg: Springer-Verlag, 1977.

35. Sternberger, L.A., and Cuculis, J. Method for enzymatic intensification of the immunocytochemical reaction without use of labeled antibodies. *J. Histochem. Cytochem., 17*:190, 1969.

36. Thiel, E., Dörmer, P., and Eulitz, M. Quantitative ^{125}I-Autoradiographie einzelner Zellen. *Histochemistry, 43*:33, 1975.

37. Thiel, E., Dörmer, P., Rodt, H., and Thierfelder, S. Quantitative immunoautoradiography at the cellular level. I. Design of a microphotometric method to quantitate membrane antigens on single cells using ^{125}I-labeled antibodies. *J. Immunol. Meth., 6*:317, 1975.

38. Thiel, E., Rodt, H., Huhn, D., and Thierfelder, S. Decrease and altered distribution of human T antigen on chronic lymphatic leukemia cells of T type, suggesting a clonal origin. *Blood, 47*:723, 1976.

39. Thiel, E., Dörmer, P., Ruppelt, W., and Thierfelder, S. Quantitative immunoautoradiography at the cellular level. II. Absolute measurements using labeled standard cells as a source of reference. *J. Immunol. Meth., 12*:237, 1976.

40. Thiel, E., Dörmer, P., Rodt, H., Huhn, D., Bauchinger, M., Kley, H.P., and Thierfelder, S. Quantitation of T-antigenic sites and Ig-determinants on leukemic cells by microphotometric immunoradioautography. Proof of the clonal origin of thymus-derived lymphocytic leukemias. In S. Thierfelder, H. Rodt, and E. Thiel, eds., *Immunological Diagnosis of Leukemias and Lymphomas*. Berlin-Heidelberg: Springer-Verlag, 1977.

41. Thiel, E., Bauchinger, M., Rodt, H., Huhn, D., Theml, H., and Thierfelder, S. Evidence of monoclonal proliferation in prolymphocytic leukemia of T-cell origin. A cytogenetic and quantitative immunoautoradiographic analysis. *Blut* (in press).

42. Thiel, E., Rodt, H., Netzel, B., Huhn, D., and Thierfelder, S. T antigen positive but E-rosette negative acute lymphatic leukemias. (in preparation)

43. Thierfelder, S., Rodt, H., and Netzel, B. Transplantation of syngeneic bone marrow incubated with leukocyte antibodies. *Transplantation, 23*:460, 1977.

44. S. Thierfelder, H. Rodt, and E. Thiel, eds. *Immunological Diagnosis of Leukemias and Lymphomas*. Berlin-Heidelberg: Springer-Verlag, 1977.

11

Leukocyte Transfusions for the Modification of Host-versus-graft Reactions in Dogs

I. Rieder, H. J. Kolb,
E. Schäffer, H. Kolb,
H. Gross-Wilde, S. Scholz,
and S. Thierfelder

INTRODUCTION

Allogeneic bone marrow transplantation has been attempted with promising results in the treatment of severe aplastic anemia and refractory acute leukemia. Failure of engraftment and marrow graft rejection are the most serious problems in the treatment of aplastic anemia with transplantation of marrow from HLA-identical siblings, despite preparation of recipients with otherwise lethal doses of cytostatic drugs or total-body irradiation.

Treatment with high doses of cyclophosphamide (CY) successfully conditioned mice and rats for histoincompatible marrow grafts (12,13). In randomly bred dogs, most grafts of marrow across major histocompatibility barriers failed. The use of even higher doses or a combination of immunosuppressive drugs is limited by their irreversible toxicity to other organ systems. In contrast, combined infusions of marrow and blood leukocytes were favorable for hemopoietic engraftment of DLA-mismatched marrow in irradiated canine recipients (17). We studied the effect of leukocyte transfusions on hemopoietic engraftment in CY-treated dogs. Leukocytes were either transfused the day before each dose of CY or 1 and 2 days after marrow grafting. In addition, the effect of leukocyte transfusions after marrow grafting on mortality from graft-versus-host disease (GvHD) was evaluated in recipients of marrow from DLA-identical littermates following CY and total-body irradiation (TBI) treatment.

MATERIAL AND METHODS

Dogs

Dogs were mainly beagles and in a few instances crosses of beagle and *Bayerischer Gebirgsschweiß-hund*. They were derived from four commercial breeding colonies or bred in our colony. They were dewormed regularly and vaccinated against distemper, hepatitis, and leptospirosis, and they were at least 6 months old before entering the experiment.

Histocompatibility Testing

In all dogs, DLA-antigens were determined with lymphocytotoxic alloantisera (23) and with mixed leukocyte cultures (7).

Hemopoietic Cells

Marrow was aspirated from the forelegs, the hindlegs, and the pelvic crest of anaesthesized living donors. Leukocyte transfusions consisted of buffy coat obtained by repeated bleeding from arteriovenous shunts and subsequent centrifugation (1,500 rpm for 15 min in a PR 6000 International Centrifuge). An average of 7×10^8 marrow cells

101

(range 3×10^8 to 15×10^8)/kg body weight was transfused on day 0. A total of 200 to 400 ml of buffy coat containing an average of 20×10^9 leukocytes was given to the recipients.

Treatment and Evaluation

During the time of CY and TBI treatment, until 5 days after transplantation, dogs received no food or water. They were given fluids parenterally twice daily and antibiotics and blood transfusions as clinically indicated. The blood was irradiated with 2000 R ^{60}Co gamma-rays. Blood counts were obtained before and daily after transplantation within the first 2 weeks and once weekly thereafter. Graft-versus-host disease was diagnosed when an unequivocal skin rash, diarrhea, and jaundice were present. Complete autopsies and histological examinations were carried out on all the dogs that died. The DLA-identical combinations were of the opposite sex, and karyotype analyses of blood and marrow cells were performed 20, 50, and 100 days after grafting and once a year thereafter. Marrow cells were cultured for 3 hr without mitogens and blood lymphocytes were cultured for 3 days in the presence of phytohemagglutinin (PHA).

The criteria for a successful marrow transplant were sustained rises in blood counts and at least 10 percent of normal marrow cellularity of the myeloid elements. Reversion to host type hemopoiesis was assumed when less than 20 percent of cells with donor karyotype were present in marrow and blood. Confirmation of GvHD was by histological examination of liver, skin, and gut.

Experimental Groups

The experimental dogs were given 40 mg of CY/kg on each of three successive days (day -3, -2, and -1), with day 0 being the day of marrow infusion. The TBI-treated dogs received 1,200 R total-body irradiation at a dose rate of 5.5 R/min. The dogs were placed in a wooden cage between two opposing ^{60}Co sources, with a source-to-source distance of 8.5 m.

The DLA-mismatched combinations differed in both haplotypes and were derived from different breeding colonies. Four groups of DLA-mismatched, unrelated combinations were studied as follows:

I. Seven dogs received a 50 ml of donor blood on day -4 and no buffy coat.
II. Seven dogs received 50 ml of donor blood on day -4 and buffy coat on days $+1$ and $+2$.
III. Seven dogs receiving donor buffy coat on days -4, -3, and -2.
IV. Nine dogs received buffy coat from donors DLA-different to the marrow donor on days -4, -3, and -2.

Five groups of DLA-identical littermate combinations were studied:

I. CY-treated recipients received 50 ml of donor blood on day -4 and no buffy coat.
II. CY-treated recipients received donor buffy coat on day -4 or on days -4, -3, and -2.
III. CY-treated recipients received donor buffy coat on days $+1$ and $+2$.
IV. TBI-treated dogs received no buffy coat.
V. TBI-treated dogs received buffy coat on days $+1$ and $+2$.

Statistical Evaluations

The incidence of take, reversion, and GvHD was tested by the Fisher's Exact Test. The degree of chimerism was calculated using the Wilcoxon ranking test.

RESULTS

Cyclophosphamide Toxicity

Table 11-1 gives the mortality of dogs given 40 mg of CY/kg daily for one, two, three, and four days with and without subsequent infusion of marrow. None of the dogs receiving more than 40 mg of CY/kg without marrow infusion and nine of 13 dogs given 120 mg/kg of CY and marrow survived. Five dogs given 160 mg of CY/kg died within 6 days despite marrow infusion. At autopsy, hemorrhagic pneumonitis, enteritis, and cystopyelitis were found in the dogs that died. The bone marrow was devoid of cells in dogs given 80 and 120 mg of CY/kg with-

TABLE 11-1. Cyclophosphamide Toxicity in Dogs

DOSE (40 MG/KG. DAILY)	MARROW INFUSION	NUMBER OF DOGS STUDIED	DOGS SURVIVING TO DAY 30
40	No	4	3
80	No	2	0
120	No	10	0
120	Yes	13	9
160	Yes	5	0

out marrow infusion, with the exception of one dog that died on day 17. In this dog, rare foci of hemopoiesis were found.

DLA-mismatched Marrow Grafts

Engraftment of DLA-mismatched marrow was observed in one of seven dogs given no leukocytes. In three of seven dogs given donor leukocytes after grafting, seven of seven dogs given donor leukocytes before grafting, and four out of nine dogs given leukocytes from a third donor also had suc-

cessful DLA-mismatched marrow transplants (Table 11-2). Dogs without hemopoietic engraftment died with GvHD, often complicated by infections. One dog of the fourth group (No. 181) failed to show a rise in blood cell count, but 20 percent of normal marrow cellularity was present, indicating engraftment. Another dog of the same group (No. 179) showed only a transient rise in blood cell count and died with marrow aplasia. In the remaining dogs, marrow cellularity corresponded well with rises in blood cell count.

TABLE 11-2 Marrow Grafts from DLA-mismatched, Unrelated Donors (Two Haplotype Different) following Cyclophosphamide (120 mg/kg) Effect of Leukocyte Transfusions on Hemopoietic Engraftment

DOGS	MARROW CELLS INFUSED ($\times 10^8$/kg)	LEUKOCYTES	RISE IN BLOOD COUNTS	MARROW CELLULARITY NORMAL (PERCENT)	SURVIVAL TIME (DAYS)	CAUSE OF DEATH
No leukocytes						
L 4	3	—	No	5	5	Hemorrhage
L 27	10	—	No	5	6	Septicemia
N 101	10	—	No	<5	6	Hemorrhagic pneumonia
N 129	8	—	No	5	8	Pneumonia
L 317	5	—	No	5	9	Hemorrhage
N 126	6	—	No	5	10	Hemorrhage
N 128	8	—	Yes	100	11	GvHD
Leukocytes after grafting						
N 146	9	15	No	—	6	Hemorrhage
B 285	5	10	No	<5	9	Septicemia, pneumonia
N 109	3	12	No	5	9	Septicemia
N 98	7	19	Yes	70	11	GvHD
L 313	6	22	Yes	100	12	GvHD
L 394	4	23	No	5	12	Septicemia, pneumonia
N 118	7	16	Yes	70	19	GvHD
Leukocytes before grafting						
N 72	3	18	Yes	60	6	GvHD, pneumonia
N 132	8	22	Yes	30	8	GvHD
N 164	15	41	Yes	60	11	GvHD
L 339	7	15	Yes	30	11	GvHD, septicemia
N 73	3	18	Yes	90	11	GvHD
B 287	8	18	Yes	80	12	GvHD
N 114	4	16	Yes	70	12	GvHD
Leukocytes of third donors before grafting						
N 207	4	20	No	5	5	Hemorrhage
N 181	4	11	No	20	6	Hemorrhage, pulmonary edema
D 3	6	21	Yes	30	7	GvHD, septicemia
N 166	5	21	No	5	7	Hemorrhage
L 351	4	15	Yes	70	7	GvHD
N 179	10	20	(Yes)	5	9	Septicemia
D 45	11	17	No	5	10	Pneumonia
L 346	12	30	No	5	11	Hemorrhage
N 112	2	15	Yes	70	21	GvHD

DLA-identical Littermate Marrow Grafts

A total of 35 CY-treated dogs were grafted with marrow from DLA-identical littermates. All dogs showed evidence of hemopoietic engraftment by rising blood cell counts and restoration of marrow cellularity. Four dogs died with severe infections 7 to 12 days after transplantation and were excluded from evaluation for chimerism and GvHD. Nine of 31 dogs received no leukocyte transfusions, 13 received donor leukocytes before grafting, and nine received donor leukocytes after grafting (Table 11-3).

Reversion to host type hemopoiesis as indicated by a decrease of the proportion of bone marrow and blood cells with donor karyotype to less than 20 percent occurred in the group not given leukocytes and in those dogs given leukocytes before CY and the marrow graft. Reversals did not occur in the group given donor leukocytes after grafting. Two dogs of the latter group and one dog of the first group died with GvHD on days 19, 24, and 68, respectively. The extent of chimerism, as indicated by the proportion of marrow cells and PHA-stimulated blood cells with donor karyotype, is demonstrated in Figure 11-1 (A,B). Median values are indicated by dashes and connected by lines. The majority of the dogs in the three groups were mixed chimeras. At day 50 after transplantation and thereafter, chimerism was lowest in the group of dogs given donor leukocytes before grafting. Complete chimerism in bone marrow occurred in all three groups, but was most frequent in the group given leukocytes after grafting. Lymphoid chimerism in blood increased in dogs not given leukocytes after 50 and 100 days [Figure 11-1(B)]. Lymphoid chimerism was higher at 20 days in dogs given donor leukocytes after grafting ($p < 0.05$).

Donor Leukocyte Transfusions after Grafting and Incidence of Fatal GvHD

The effect of donor leukocyte transfusions after grafting was studied in CY-conditioned and TBI-conditioned recipients. Donor leukocyte transfusions increased the mortality from GvHD in TBI-treated dogs ($p < 0.05$), but not in CY-treated dogs. However, GvHD was not more frequent in TBI-conditioned recipients than in CY-conditioned recipients when leukocytes were not given.

DISCUSSION

Treatment of marrow graft recipients with CY allows engraftment of allogeneic marrow in mice (13), rats (12), dogs (9), monkeys (16), and man (14), but chimerism is frequently incomplete and reversion to host type hemopoiesis has been observed in mice (5) and man (8,15,22). In patients with aplastic anemia conditioned with CY, rejection of the marrow is a serious and frequently fatal complication. Higher doses of immunosuppressive therapy prevented rejection of marrow grafts in dogs that had been sensitized with blood transfusions of the marrow donor (19), but did not improve the results of patients with aplastic anemia (20). Combined infusions of marrow and blood leukocytes appeared to enhance hemopoietic engraftment in TBI-treated, DLA-mismatched dogs (17). In this study, leukocyte transfusions were added to the CY regimen to enhance engraftment and to improve the degree of chimerism. Leukocyte transfusions may be used as a source of large amounts of histocompatibility antigens or as part of a hemopoietic graft. Immunosuppression with CY is particularly strong and selective when the respective antigen is given 1 or 2 days before CY administration (1). The immunosuppressive effect of a given dose of CY in mice tested against sheep red blood cells was greater when a larger dose of stimulating antigens was given before CY (1). Presumably, reactive clones of immunocompetent cells are stimulated to proliferate and can then be eliminated more effectively by CY. Therefore, larger doses of antigens may more effectively stimulate reactive clones. The transfusion of donor leukocytes instead of a single blood transfusion before CY administration significantly improved hemopoietic engraftment of DLA-mismatched marrow (Table 11-2). The more effective immunosuppression was antigen specific, since leukocyte transfusions of third-party dogs before CY administration were less effective. These results indicate that the above-described principles

TABLE 11-3. DLA-Identical, Mixed Leukocyte Cultures, Matched Littermate Grafts

GROUPS	NUMBER OF DOGS STUDIED	DOGS WITH SUCCESSFUL GRAFTS	DOGS WITH FATAL GvHD	REVERSALS/DOGS EVALUATED
No leukocytes	9	9	1	3/9
Donor leukocytes before grafting	13	13	0	6/13
Donor leukocytes after grafting	9	9	2	0/9[a]

[a]Difference $2\alpha = 0.05$ of this group versus other groups.

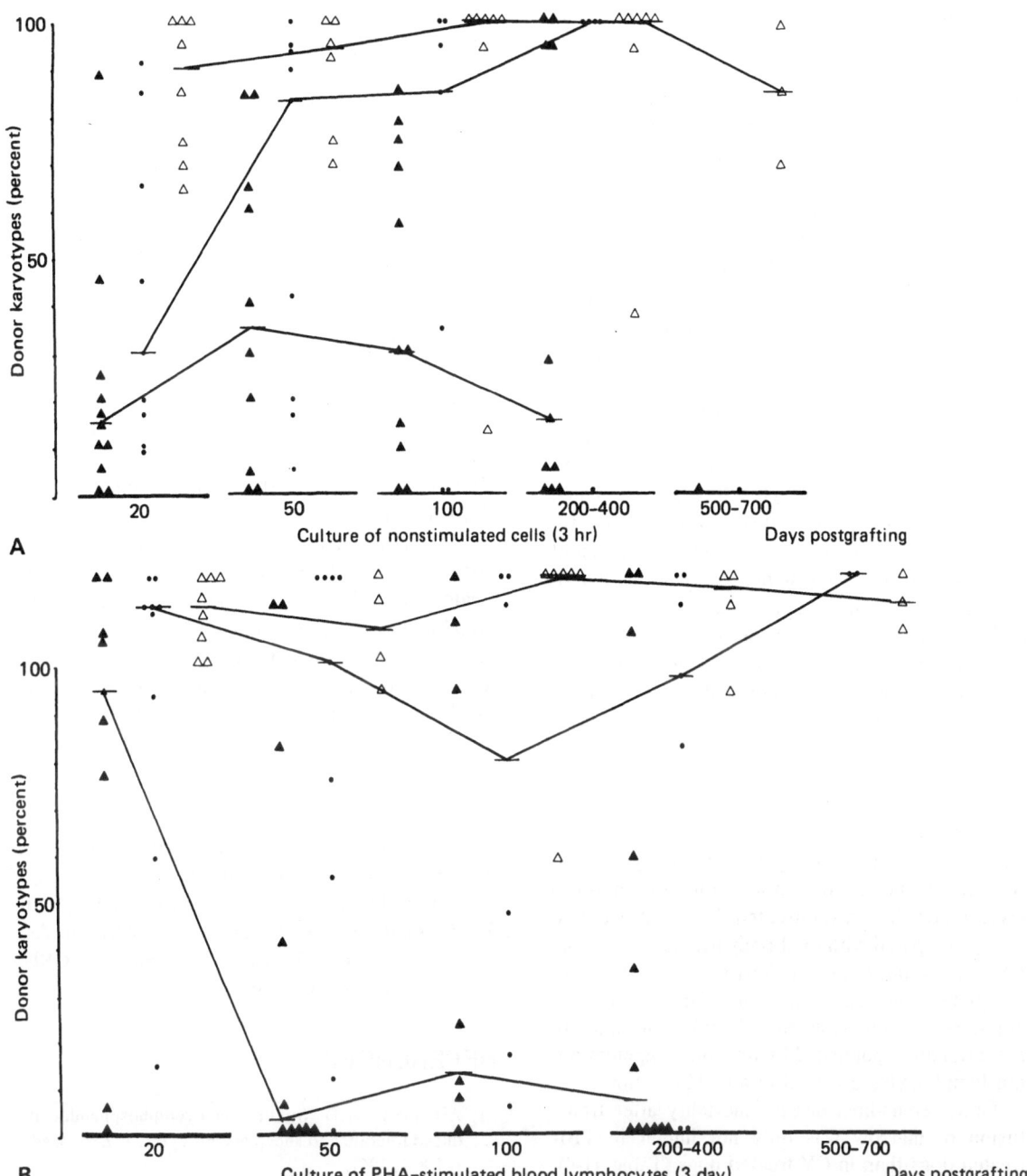

FIGURE 11-1. Chimerism in cy-treated dogs with LD-SD identical marrow grafts. (A) A 3-hr culture of nonstimulated marrow cells; (B) a 3-day culture of PHA-stimulated blood-lymphocytes. Groups given no buffy coat (●), groups given buffy coat before grafting (▲), and groups given buffy coat after grafting (△).

apply also to DLA-mismatched canine marrow grafts. The principles do not apply to DLA-identical littermate grafts, since reversals were more frequent when donor leukocytes were given before CY. These donor–recipient combinations gave negative mixed leukocyte culture tests, indicating that a measurable stimulation by antigens other than DLA-antigens does not occur within the first few days.

The selective effect of CY on proliferating im-

munocompetent cells may be negligible in DLA-identical combinations. In DLA-identical combinations, a higher degree of chimerism and persistent chimerism was obtained when donor leukocyte transfusions were given after grafting. Leukocyte transfusion given after grafting represents an enlargement of the hemopoietic graft by additional hemopoietic precursor cells in the peripheral blood.

Hemopoietic precursor cells capable of repopulating the hemopoietic system have been demonstrated in mice (6) and dogs (3,4,11,18) and may also exist in man (10). Moreover, leukocyte transfusions contain large numbers of immunocompetent lymphocytes, which may also play a role in hemopoietic engraftment. Immunocompetent lymphocytes conferred early lymphoid chimerism as demonstrated by higher proportions of PHA-responsive cells with donor karyotypes in the blood [Figure 11-1(B)].

A lesser degree of early lymphoid chimerism was seen in dogs given marrow only and the least degree in dogs given similarly large amounts of donor leukocytes before CY treatment. Reversals occurred in the latter groups, but did not occur in the group given leukocytes after grafting (Table 11-3). Early lymphoid chimerism as well as permanent chimerism may be related to the number of immunocompetent lymphocytes transferred. Thus, dogs given leukocyte transfusions after grafting retained the largest number of immunocompetent lymphocytes. Dogs given marrow without leukocyte transfusions demonstrated a lower number of cells, and dogs given marrow depleted of immunocompetent leukocytes, by prior leukocyte donations from the marrow donor demonstrated the lowest number of immunocompetent lymphocytes.

Transferred blood lymphocytes may increase chimerism with or without attacking the host by the graft-versus-host (GvH) reaction. Graft-versus-host reactions can suppress hemopoiesis in the host (2) and the host's immune system. Leukocyte transfusions of the marrow donor did not increase GvHD mortality in CY-treated dogs, but it did so in dogs prepared with total-body irradiation. Unlike CY-treated dogs, all the TBI-treated dogs were complete chimeras. Presumably, TBI-treated dogs are more severely depleted of lymphocytes and are more rapidly repopulated by immunocompetent donor lymphocytes involved in a GvH reaction.

Graft-versus-host disease mortality after transfusion of marrow was only not higher in TBI-treated dogs than in CY-treated dogs (Table 11-4).

Probably the number of GvH reactive cells in canine marrow is too small to cause a high GvHD mortality in DLA-identical littermate combinations. This is in contrast to a similar study of Storb et al. (21), who found a high GvHD mortality in DLA-matched littermates, even without additional leukocyte transfusions. However, the group of dogs given marrow only in the reported study received higher radiation doses than the group given additional leukocyte transfusions.

SUMMARY

Leukocyte transfusions of the marrow donor can be used to modify host-versus-graft (HvG) reactions in dogs. In DLA-mismatched hosts, they are effective when given before CY administration, in DLA-identical hosts when given after marrow transplantation. An increased GvHD mortality was observed after leukocyte transfusions in TBI-treated dogs, but not in CY-treated dogs. Additional leukocyte transfusions of the marrow donor may be of value in patients with aplastic anemia treated with CY and HLA-identical marrow grafts for the prevention of marrow graft rejection.

ACKNOWLEDGMENT

The assistance of Mrs. H. Kurr and Mrs. A. v. Arco is gratefully acknowledged.

This study was supported in part by DFG, SFB 37 München, Tp E4, E5 and B1, and by the Association Contract between Euratom and GSF, Contract No. 217-76-1 BIOD.

REFERENCES

1. Aisenberg, A.D. Studies on cyclophosphamide induced tolerance to sheep erythrocytes. *J. Exp. Med.*, *125*:833, 1967.

TABLE 11-4. Effect of Donor Leukocyte Transfusions on GvHD in DLA-identical Littermates

| | | NUMBER OF DOGS | | |
CONDITIONING TREATMENT	LEUKOCYTES	Studied	With sustained engraftment	With fatal GvHD
Cyclophosphamide (120 mg/kg)	No	9	6	1
	Yes	9	9	2[a]
Total-body irradition (1,200 R)	No	6	6	0
	Yes	18	18	10[b]

[a]Not significantly different from group not given leukocytes.
[b]Significantly different from group not given leukocytes (p < 0.05).

2. Boranic, M., Tonković, J., and Blazi, M. Quantitative aspects of destruction of haemopoietic tissue in mouse radiation chimaeras. *Biomedicine, 19*:104, 1973.

3. Cavins, J.A., Scheer, S.C., Thomas, E.D., and Ferrebee, J.W. The recovery of lethally irradiated dogs given infusions of autologous leukocytes preserved at −80°C. *Blood, 23*:38, 1964.

4. Debelak-Fehir, K.M., Catchatourian, R., and Epstein, R.D. Hemopoietic colony-forming units in fresh and cryopreserved peripheral blood cells of canines and men. *Exp. Hematol., 3*:109, 1975.

5. Glynn, J.P., Fefer, A., and Halpern, B.L. Cyclophosphamide-induced chimerism. *Canc. Res., 28*:41, 1968.

6. Goodmann, J.W., and Hodgson, G.S. Evidence for stem cells in the peripheral blood of mice. *Blood, 19*:702, 1962.

7. Grosse-Wilde, H., Vriesendorp, H.M., Netzel, B., Mempel, W., Kolb, H.J., Wank, R., Thierfelder, S., and Albert, E.D. Immunogenetics of seven LD alleles of the DLA-complex in mongrels, beagles and labradors. *Transpl. Proc., 7*:159, 1975.

8. Jeannet, M., Speck, B., Rubinstein, A., Pelet, B., Wyss, M., and Kummer, H. Autologous marrow reconstitution in severe aplastic anemia after ALG pretreatment and HLA-semi-incompatible bone marrow cell transfusion. *Acta Haemat., 55*:129, 1976.

9. Kolb, H.J., Rieder, J., Gross-Wilde, H., Abb, J., Albert, E., Kolb, H., Schäffer, E., and Thierfelder, S. Marrow grafts in LD-SD typed dogs treated with cyclophosphamide. *Transpl. Proc., 8*:555, 1976.

10. McCredie, K.B., Freireich, E.J., Hersch, E.M., Curtis, J.E., and Kaiser, H. Early bone marrow recovery after chemotherapy following the transfusion of peripheral blood leukocytes in identical times. *Proc. Amer. Ass. Canc. Res., 11*:54, 1970.

11. Nelson, B., Calvo, W., Fliedner, T.M., Herbst, E., Bruch, C., Schnappauf, H.P., and Flad, H.D. The repopulation of lymph nodes of dogs after 1200 R whole body X-irradiation and intravenous administration of mononuclear blood lymphocytes. *Amer. J. Path., 84*:259, 1976.

12. Santos, G.W. Effect of syngeneic and allogeneic marrow transfusion on cyclophosphamide-induced lethality in the rat. *Exp. Hematol., 10*:8, 1966.

13. Santos, G.W., and Owens, A.H. Allogeneic marrow transplants in cyclophosphamide-treated mice. *Transpl. Proc., 1*:44, 1969.

14. Santos, G.W., Sensenbrenner, L.L., Burke, P.J., Mullin, G.M., Bias, W.B., Tutschka, P.J., and Slavin, R.E. The use of cyclophosphamide for clinical marrow transplantation. *Transpl. Proc., 4*:559, 1972.

15. Sensenbrenner, L., Steele, A.A., and Santos, G.W. Recovery of hematologic competence without engraftment following attempted bone marrow transplantation for aplastic anemia: Report of a case with diffusion chamber studies. *Exp. Hematol., 5*:51, 1977.

16. Storb, R., Buckner, C.D., Dillingham, L.A., and Thomas, E.D. Cyclophosphamide regimen in rhesus monkeys with and without marrow infusion. *Canc. Res., 30*:2195, 1970.

17. Storb, R., Epstein, R.B., Bryant, J., Radge, H., and Thomas, E.D. Marrow grafts by combined marrow and leukocyte infusions in unrelated dogs selected by histocompatibility typing. *Transplantation, 6*:587, 1968.

18. Storb, R., Epstein, R.B., Radge, H., Bryant, J., and Thomas, E.D. Marrow engraftment by allogeneic leukocytes in lethally irradiated dogs. *Blood, 30*:805, 1967.

19. Storb, R., Floersheim, G.L., Weiden, P., Graham, T.C., Kolb, H.J., Lerner, K.G., Schroeder, M.L., and Thomas, E.D. Effect of prior blood transfusions on marrow grafts: abrogation of sensitization by procarbazine and antithymocyte serum. *J. Immunol., 112*:1508, 1974.

20. Storb, R., Prentice, R.L., and Thomas, E.D. Marrow transplantation for treatment of aplastic anemia. *New Eng. J. Med., 296*:61, 1977.

21. Storb, R., Rudolph, R.H., Kolb, H., Graham, T.C., Michelson, E., Erickson, V., Lerner, K.G., Kolb, H.J., and Thomas, E.D. Marrow grafts between DLA-matched canine littermates. *Transplantation, 15*:92, 1973.

22. Thomas, E.D., Storb, R., Giblett, E.R., Longpre, B., Weiden, P.L., Fefer, A., Witherspoon, R., Clift, R.A., and Buckner, C.D. Recovery from aplastic anemia following attempted marrow transplantation. *Exp. Hematol., 4*:97, 1976.

23. Vriesendorp, H.M., Albert, E.D., Templeton, J.W., Belotsky, S., Taylor, B., Bull, R.W., Cannon, S.D., Epstein, R.B., Gross-Wilde, H., Hammer, C., Krumbacher, K., Léon, S., Meera Cahn, P., Meckey, M.R., Motola, M., Sainson, R., Schnappauf, H., Scholz, S., Schroeder, M.L., Storb, R., Wank, R., Westbruck, D.L., and Zweibaum, A. Joint report of the second international workshop on canine immunogenetics. *Transpl. Proc., 8*:289, 1976.

12

Minor Histocompatibility Systems in Dogs

H. M. Vriesendorp,
A. B. Bijnen,
A. C. M. Van Kessel,
H. Obertop, and
D. L. Westbroek

INTRODUCTION

Over 25 percent of recipients of HLA-identical bone marrow cells die from graft-versus-host disease (GvHD) (6) and, despite maximal immunosuppression, approximately 25 percent of HLA-identical kidneys are rejected within 5 years (1,18). Recently, it has been suggested that GvHD in HLA-identical sibling donor recipient pairs is influenced by minor histocompatibility differences controlled by the sex chromosomes (7,25). However, ethical and practical limitations in the treatment of human patients prevent the proper control of many variables. A better understanding of minor histocompatibility systems is needed to increase survival in humans after transplantation.

Evidently, control of variables can be achieved in a preclinical outbred animal model if the appropriate experimental design is chosen. Studies in inbred mouse strains have shown that an analysis of minor histocompatibility systems can be performed using donor recipient pairs identical for the major histocompatibility complex (MHC), immunization, and no immunosuppression. With a slight modification, these principles might be applied to outbred species. In that case, sibling donor-recipient pairs identical for the MHC should be used. This will exclude the possibility of major histoincompatibility overriding minor histoincompatibility effects. Identity for the MHC in outbred species can currently only be guaranteed in sibling donor-recipient pairs in whom the genotype for the whole MHC region is identical (31). Another complicating factor in the analysis of minor histocompatibility systems is the demonstration of multiple systems on different chromosomes in mice (13,14,23). In all probability, this holds true for other mammalian species as well; a minor histocompatibility analysis will therefore be complicated by the simultaneous influence of several different loci. A prospective identification of the relative contribution of each individual locus is as yet impossible, since minor histocompatibility antigens currently cannot be recognized in man or in outbred experimental animals by serological or other *in vitro* techniques. Retrospectively, a better resolution of the effects of various minor histocompatibility systems can be obtained by exaggerating the influence of a particular system or systems by sensitization of the kidney graft recipient or bone marrow donor (13). Sensitization will have the additional advantage of boosting the usually weak effects of minor histocompatibility systems. The disadvantages of this approach are that the clinical significance of a given system in an unsensitized human patient might be exaggerated and that it will be difficult to achieve sensitization for only one minor histocompatibility system at a time. Another possible procedure that might unravel the effects of multiple minor histocompatibility systems is to analyze separate groups of MHC-identical donor-recipient pairs, subdivided according to other genetic markers (e.g., sex). In this procedure, the working hypothesis is that a minor histocompatibility system is linked to the marker selected for investigation.

In this review, we describe an ongoing analysis on the influence of minor histocompatibility systems in dogs receiving a bone marrow or kidney graft. This species was selected for investigation since the requirements outlined for an analysis of minor histocompatibility systems could be met and earlier experience was gained in this model in the analysis of the major histocompatibility complex.

MATERIALS AND METHODS
Dogs

Donor and recipient were beagles from the Central Breeding Laboratory for experimental animals of TNO, Zeist, The Netherlands. In one of the kidney protocols, the third-party blood transfusion donors were mongrel dogs, in which case the transfusion consisted of blood pooled from seven different dogs (8). Male, as well as female blood, was used. In another kidney protocol, third-party family members were used as transfusion donors. The kidney and blood donor were chosen to be of the same sex.

Transplants

Kidney allografts were performed as previously reported (33). Nephrectomy, both recipient kidneys, were performed on the day of the transplant. Fixed numbers of bone marrow cells per kilogram body weight of the recipient, were injected i.v. 3 to 30 hr after a total-body irradiation (TBI) of 750 rad X-ray (300 KV, 10 mA, HVL 3 mm Cu, dose 16 to 18 rad/min). Antibiotic treatment and hematological support with blood products (irradiated with 1,500 rad) were given whenever appropriate. These procedures will be described in more detail elsewhere (32). A second group of animals received, in addition to bone marrow cells, 0.5×10^8 to 1.0×10^8 popliteal lymph node cells/kg body weight.

Genetics

Typing for lymphocyte-defined (LD) and serologically-defined (SD) determinants of the MHC of the dog (DLA) was performed using previously described techniques (9,12,22,27,30). The sex of the animals was determined by inspection of the external genitalia.

Evaluation of Transplants

Kidney: Animals died from renal failure or were killed by nembutal overdose if the serum creatinine level, which was determined twice a week, rose over 1,000 μmol/liter. Autopsy samples of the kidney graft were studied by light microscopy. Rejection was found to be the most probable cause of death in all cases. The day of death or sacrifice was taken as the rejection day.

Bone Marrow: After a TBI dose of 750 rad and injection of the number of bone marrow cells used in this study, the recipients will become complete hemopoietic chimeras (28,32). A set of four different criteria was used to determine the presence of graft-versus-host disease (GvHD) For diagnosis of GvHD, at least three of the following four criteria were required to be positive.

1. Clinical condition: Clinical evidence of GvHD was observed as (a) erythroderma, crust formation, and hair loss of the skin and paws; (b) mucositis in oropharynx, nose, and eyes; (c) diarrhea; and (d) icterus.

2. Serum biochemistry: Weekly serum protein, urea, alkaline phosphatase, glutamic pyruvate, and glutamic oxalate levels were determined. Levels of SGOT and SGPT over 80 IU (normal lower than 50 IU) and levels of alkaline phosphatase over 16 Bessey-Lowrey units (normal more than 4 U) indicated liver GvHD.

3. Hematology: Hematocrit values, reticulocyte levels, and red blood cell, leukocyte, and differential and thrombocyte counts were determined. Otherwise, unexplained hemolytic anemia, granulocytosis, or thrombopenia were scored as signs indicative of GvHD.

4. Histology: Skin biopsies were taken 25, 50, 75, and 100 days after total-body irradiation. Complete autopsies were performed when the animal died spontaneously or had to be killed 1 to 3 years after transplantation through lack of kennel space. By light microscopy, signs of graft-versus-host (GvH) reactions were found in autopsy material from the digestive tract, skin, lung, liver, and oropharynx. In rarer cases, they were also found in the pancreatic duct, salivary gland, transitional epithelium of the kidney, prostate, and epididymis. A detailed description of these histological observations is in preparation.

RESULTS
Genetic Models for Analysis of Minor Histocompatibility Systems

The kidney and bone marrow recipient pairs were all DLA, SD as well as LD, identical. In this experiment, an attempt was made to look at the effects of one minor histocompatibility system at a time. For this purpose, minor histocompatibility effects were boosted through blood transfusions prior to kidney grafting or through the addition of lymphocytes to the bone marrow cells before transplantation. Moreover, results were analyzed separately for each of the four possible sex combinations between donor and recipient (i.e., M→F; M→M; F→M; F→F). This experimental design might make possible a separation of Y-linked, X-linked, or autosomal minor histocompatibility effects by the application of the criteria outlined in the following paragraphs.

Y-linked Histocompatibility

Incompatibility for this system occurs whenever a male kidney is given to a female DLA-identical sibling or a male recipient is transplanted with bone marrow cells from a female DLA-identical sibling donor. In addition to Y-linked histoincompatibility, X-linked histoincompatibility differences could be present in some of these donor-recipient pairs. The percentage of incompatible pairs would be related to the frequency and number of alleles of the supposed X-linked system. For a simplified model of one locus with n alleles of equal strength and frequency, it can be shown that this percentage follows the function of $n^2\text{-}1/2n^2$ in sibling donor-recipient pairs. The effects of autosomal minor histocompatibility differences might also be seen in a number of cases. Here, the frequency of incompatible pairs would be $3n^3\text{-}n^2\text{-}11n+9/4n^3$ (21). However, neither autosomal nor X-linked effects will be observed if a successful sensitization for Y-linked histoincompatibility is obtained and the analysis is limited to the indicated subgroup of donor-recipient pairs.

X-linked Histocompatibility

The interference of a Y-linked histocompatibility effect with an X-linked histocompatibility effect can be prevented by disregarding the results of male into female MHC-identical kidney grafts or of female into male MHC-identical bone marrow grafts. The percentage of incompatible donor-recipient pairs can be predicted for the other donor-recipient pairs in a hypothetical situation, in which an X-linked minor histocompatibility system (with n alleles of equal strengths and frequency) would be the only system of importance. The predictions for kidney grafts will be, for female donors into male recipients, $2n^2\text{-}n\text{-}1/2n^2$; for female donors into female recipients, $n^2\text{-}2n+1/2n^2$, and for male donors into male recipients, $n\text{-}1/2n$. In an analysis of GvH reactions, the same formula given for kidney donor-recipient pairs can be used if the sex of the donor and the recipient is reversed.

Autosomal Minor Histocompatibility

Differences for autosomal systems should be equally prevalent in the four different sex groups. The incidence of shortened survival or GvH reactions after the kidney graft will depend on the number and the polymorphism of the effective autosomal minor histocompatibility loci. Incompatibility formulas for the probability of incompatibility for one locus with n alleles of equal strengths and frequency have been computed in the past by Newth (15) and Simonson (21). Here, the one of Simonson for sibling donor recipient pairs is used, i.e., $3n^3\text{-}n^2\text{-}11n+9/4n^3$. Evidently, a given incompatibility percentage can also be obtained by combining the histocompatibilities of a number of different autologous systems (21). Figure 12-1 shows the probability curves for incompatibility for one X-linked and one autosomal histocompatibility system as a function of the number of alleles per system. The comparison of the percentage of incompatible grafts found in the different subgroups of MHC-identical donor-recipient pairs to the curves of Figure 12-1 will permit the selection of the best fitting allele number for the X-linked and autosomal histocompatibility system. Subsequently, a X^2 test can be used to indicate which of the two shows the smaller deviation from expectations. No distinction can be made if numerous autosomal systems are effective at the same time. However, if sufficient data are collected, and if only one, or a very few autosomal systems, is effective, the distinction can be made between the effects of a X-linked and an autosomal minor histocompatibility system.

Minor Histocompatibility and Kidney Grafts

All the rejection times are given in Figure 12-2. The survival times of untransfused animals do not show clear differences in the four groups subdi-

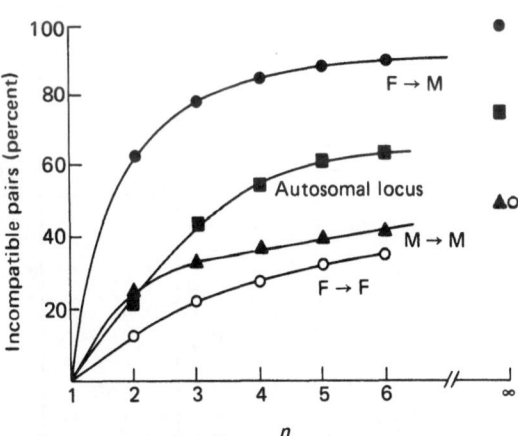

FIGURE 12-1. Probability of incompatibility between sibling donor-recipient pairs for an X-linked or autosomal histocompatibility system, under the assumption of one locus with n alleles of equal strength and frequency.

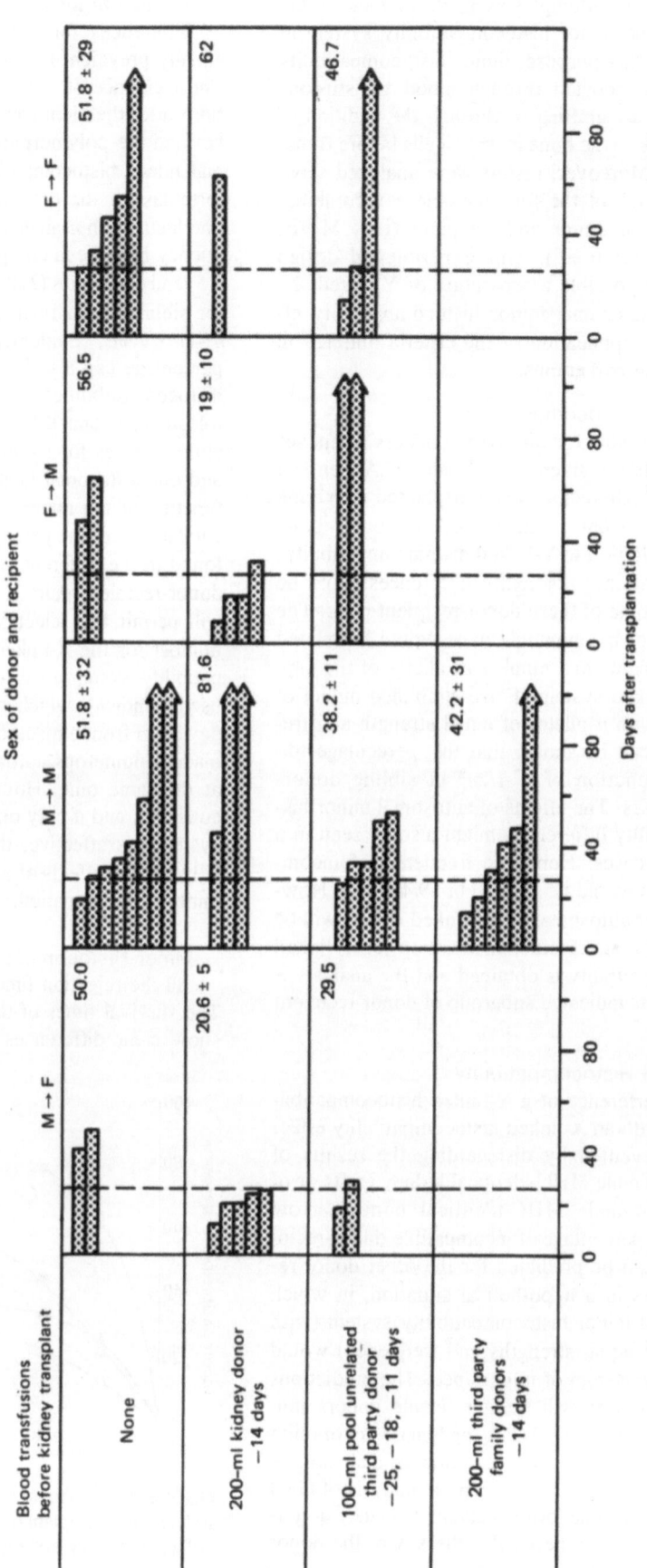

FIGURE 12-2. Rejection of DLA-identical kidney grafts (days). Survival over 100 days is included in the mean survival time computations as 100 days. Marker studies or subsequent kidney graft rejection indicated that these prolonged graft survival times did not reflect the occurrence of monozygotic twin donor-recipient pairs. The mechanisms of this prolonged survival are being investigated. Broken vertical lines, the mean survival time of kidneys in unsensitized recipients (minus one standard deviation). This is the arbitrarily chosen cut-off point for shortened graft survival.

vided according to the sex of the donor and recipient. Results of prior blood transfusions in dog kidney graft recipients have been reported elsewhere (8,16,33). Here, a retrospective analysis for the possible effects of minor histocompatibility systems in such animals is made. Survival times of different experimental groups are pooled for the same sex group because the numbers in each experiment are small and no obvious differences are observed between them.

All grafts obtained from male dogs and transplanted into females appear to have a decreased survival time after prior blood transfusions. Decreased survival time was observed in six out of seven animals compared to the mean survival time (minus one standard deviation) of the group that was not transfused. This is not seen when grafts from male animals are transplanted into other males. Here, four out of 13 animals show decreased survival times, according to the arbitrarily chosen 26-day cut-off point. This difference has a p value of 0.04 in a Fisher Exact Test. These data strongly suggest the presence of a Y-linked minor histocompatibility system that influences the survival of a DLA-identical kidney, if the recipient is sensitized by blood transfusions.

The percentages of shortened graft survival in the M→M, F→M, and F→F groups (cut-off point, 26 days; see Figure 12-2) intersect with the curves of Figure 12-1 around two or three alleles for a X-linked system. The incompatibility percentage for one autosomal locus (i.e., the sums of the previously mentioned three groups, 7/28) is closest to the n = 2 point for that curve. In Table 12-1 the X^2 test values are given for both models. Although the data fit a X-linked histocompatibility model better than an autosomal system, insufficient data have been collected so far. The data do show, however, that the principles outlined above are applicable

and in the future, with the collection of more data, the issue can be decided.

Minor Histocompatibility and Bone Marrow Grafts

In Table 12-2, results for the incidence of GvHD as defined by the criteria outlined in the methods section are summarized. A fourfold increase in the dose of bone marrow cells (from 4×10^8 to 16×10^8/kg body weight) does not appear to affect the incidence of GvHD. The addition of 0.5×10^8 to 1×10^8 lymph node cells/kg body weight does lead to a significantly higher incidence of GvH reactions ($p < 0.02$, X^2 test). After bone marrow cell transplantation, two out of the eight animals with GvHD have died. When lymphocytes are added to the bone marrow cells, six out of the 13 animals with GvHD have died. The difference in lethality is not statistically significant ($p = 0.19$, Fisher Exact Test). In the past, postthymic T-lymphocytes have been implicated as the cells initiating and determining the severity of acute GvH reactions (3). The usually mild and transient GvHD in dogs treated with bone marrow cells only, and the increase in the incidence of GvHD and mortality after the addition of lymphocytes to bone marrow cells indicate that normal dog bone marrow does not contain many postthymic T-cells. This is compatible with the low response of dog bone marrow cells to stimulation by PHA or allogeneic lymphocytes (authors' unpublished observations).

The data, for animals treated with bone marrow cells only, are pooled, since no obvious differences appear to exist in GvHD incidence among the three cell dose levels tested. No evidence is found for an X- or a Y-linked histocompatibility effect in these animals. If the assumption is made that *one* autosomal minor histocompatibility system controls GvHD in these animals, the probability curve in

TABLE 12-1 Search for X-Linked or Autosomal Minor Histocompatibility System Effect in DLA-Identical Kidney Graft Recipients

SEX DONOR → RECIPIENT	PERCENT SHORTENED GRAFT SURVIVAL X-LINKED LOCUS[a]			PERCENT SHORTENED GRAFT SURVIVAL AUTOSOMAL LOCUS[b]		
	Expected	Observed	X^2	Expected	Observed	X^2
M→M N = 14	3.50	3	0.07	3.06	3	0.00
F→M N = 6	3.75	3	0.15	1.34	3	2.06
F→F N = 4	0.50	1	0.50	0.88	1	0.16
		Total X^2 2 D.F.	0.72		Total X^2 2 D.F.	2.22

[a]One X chromosome locus with two alleles of equal strength and frequency (compare Figure 12-1).
[b]One autosomal locus with two alleles of equal strength and frequency (compare Figure 12-1).

TABLE 12-2 Incidence[a] of GvH Reactions in DLA-Identical Sibling Donor-Recipient Pairs

		SEX OF DONOR AND RECIPIENT				
	TRANSPLANT[b]	F→M	M→M	M→F	F→F	Total
Bone marrow cells	4	0/2	3/7	0/2	1/5	4/16
	8	1/2	—	0/2	2/6	3/10
	16	—	0/1	1/4	—	1/5
	Total	1/4	3/8	1/8	3/11	8/31
Bone marrow cells and lymph node cells	2	6/7	1/5	2/3	4/5	13/20
	0.5–1					

[a]Number of animals with GvH/total number of animals tested.
[b]Cell numbers ×10^8 body weight recipient.

Figure 12-1 shows that the incidence found (8/13, 26 percent) would indicate the presence of two to three alleles at such a locus in beagles. Previously, such a system was tentatively identified as being close to the phosphaglucomutase 2 locus (a polymorphic red cell enzyme in dogs) (29).

When lymphocytes are added to bone marrow cells prior to transfusion into dogs, additional minor histocompatibility systems appear to become important in GvHD control, in view of the significantly higher percentage of GvH reactions in such cases. One of the systems is probably Y-linked, since six out of the seven of the female into male grafts do show GvH reactions. The difference seen with bone marrow cells only almost reaches statistical significance in a Fisher Exact Test ($p = 0.09$), especially if one considers that this is a two-tailed test. In addition, the difference in GvHD incidence between the F→M and M→M bone marrow plus lymph node cell groups has a p value of 0.04.

In the other subgroups, the evidence also seems to indicate the presence of additional histocompatibility systems. The increased GvHD incidence in the group treated with bone marrow cells and lymphocytes has a p value of less than 0.05 in a X^2 test. A comparison of the incidence of GvHD, in the curves in Figure 12-1, suggests that at least two alleles are needed to explain the data for one X-linked minor system and four alleles for an autosomal system. In Table 12-3, a X^2 analysis of these different hypotheses is performed. The data indicate that X-linked control is improbable and that autosomal control is more likely. The GvHD frequency observed can also be explained by a model of more autosomal systems, with fewer alleles per system (e.g., two loci, one with two and one with three alleles or three loci with two alleles each) (21).

DISCUSSION

Graft-versus-host reactions in HLA-identical siblings and the rejection of a kidney of a HLA-identical donor do show that minor histocompatibility systems raise problems in human transplantation. Autosomal, X-, and Y-linked histocompatibility systems have been described in the mouse (3,11,23). In view of the postulated homology of mammalian sex chromosomes, the presence of X and Y histocompatibility is also to be expected in other mammalian species (17). The H-Y antigen has been shown to occur in several species, including

TABLE 12-3 Genetic Control of GvHD in Recipients of DLA-Identical Bone Marrow and Lymph Node Cells

SEX (DONOR→ RECIPIENT)	GvHD INCIDENCE UNDER X-LINKED CONTROL[a]			GvHD INCIDENCE UNDER AUTOSOMAL LOCUS[b] CONTROL		
	Expected	Observed	X^2	Expected	Observed	X^2
M→M N = 5	1.25	1	0.05	2.22	1	0.67
M→F N = 3	1.87	2	0.00	1.33	2	0.34
F→F N = 5	0.63	4	18.03	2.22	4	1.43
		Total X^2 2 D.F.	18.09		Total X^2 2 D.F.	2.44

[a]One X-chromosome locus with two alleles of equal strength and frequency (compare Fig. 12-1).
[b]One autosomal locus with four alleles of equal strength and frequency (compare Fig. 12-1).

the dog (20). The preliminary data reported here illustrate that the methodology used in inbred mouse strains to analyze certain aspects of minor histocompatibility systems can be applied successfully to outbred animals. In the dog model, a Y-linked minor histocompatibility system was identified and appeared to be of importance to the survival of DLA-identical kidney in a recipient who had received previous blood transfusions and to the occurrence of GvH reactions in recipients of DLA-identical bone marrow when enough postthymic T-lymphocytes were added to the graft. Suggestive, but as yet inconclusive evidence was obtained for a X-linked minor histocompatibility system with two to three alleles in kidney graft survival. More data are needed before the possibility can be excluded that the same data are explained more efficiently by minor histocompatibility systems on autosomal chromosomes. In bone marrow transplants, there was evidence for autosomal rather than X-linked control of GvH reactions. When the previously postulated PGM$_2$-linked locus are taken into account, at least three loci now appear to be of relevance for a genetic control of GvH reactions in MHC-identical dogs, i.e., a Y-linked locus and at least two autosomal loci.

An extrapolation to human transplantation shows that a search for Y-linked histocompatibility in transfused recipients of a kidney from a HLA-identical sibling donor might be rewarding. The results obtained with bone marrow indicate that dog bone marrow differs from human. Only after the addition of lymphocytes to the dog bone marrow grafts do the GvHD incidence and mortality correspond to the GvHD incidence and mortality in human patients treated with HLA-identical bone marrow for aplastic anemia (26). When higher numbers of lymphocytes than the ones used in the experiments reported here are added, an even higher GvHD incidence might be obtained, similar to the one observed in human leukemic patients treated by bone marrow transplantation (26). A possible explanation for the differences in GvHD percentages found between aplastic and leukemic patients is the more intensive conditioning regimen in the latter group, which permits the "take" of more lymphocytes.

In general, our results illustrate that the incidence and the severity of GvH reactions can be influenced by several "modifying" factors. Some of these factors are genetic, i.e., MHC identity between donor and recipient (24,29), Y-linked histocompatibility, and autosomal minor histocompatibility systems. The already limited availability of MHC-identical donors in human bone marrow transplantation will prohibit the use of further donor selection procedures for minor histocompatibility systems in the majority of cases. Other modifying factors are not genetic, such as the number of postthymic T-lymphocytes in the graft, posttransplantation transfusion care (28), and bacterial decontamination (5). Human bone marrow transplant patients might profit immediately from a change in protocol to provide optimal conditions for these nongenetic graft-versus-host–modifying factors. For example, removal of donor T-lymphocytes by density gradients, anti-lymphocyte serum incubation of bone marrow, or anti-lymphocyte serum pretreatment of the bone marrow recipient have been shown to be effective in the control of GvH reactions in experimental animals (4,10,19). However, the toxicity and the risks of these measures to improve upon the "nongenetic" modifiers of GvHD limit their application. When the genotypes for the relevant minor histocompatibility systems of the bone marrow donor and the recipient can be identified before transplantation, these measures could be utilized in the high risk group only (6). The genetic methodology described here cannot be used to identify minor histoincompatibility (with the exception of Y-linked incompatibility). Attempts to raise serological reagents against the postulated PGM$_2$-linked locus and DLA-identical PGM$_2$-different siblings by repeated skin grafts have failed so far. Graft-versus-host disease or host-versus-graft rejections in DLA-identical donor-recipient pairs and rejections of allografts from DLA-identical donors do not induce antibodies detectable by a microlymphocytotoxicity technique (31; unpublished observations). Recent preliminary experiments have shown that antibodies might be induced by blood transfusions from a DLA-identical donor. The dog model could be exploited further to define conditions under which reagents against minor histocompability antigens can be obtained. Such reagents might be of value in a further analysis of minor histocompatibility problems in dogs as well as in human patients.

ACKNOWLEDGMENTS

The authors gratefully acknowledge the enlightening disagreements with Drs. Löwenberg and van Bekkum, the technical assistance of W. Klapwyk, J. Kennedy, F. Vervat, K. Hillenius, and H. Wiersema, and the secretarial help of Pat Condon.

SUMMARY

The methodology developed for the analysis of minor histocompatibility systems in inbred mice has been modified for application in outbred dogs.

In kidney transplantation, a Y-linked minor histocompatibility effect can be demonstrated after immunization of the recipient prior to transplantation. Evidence for additional X-linked or autosomal minor histocompatibility systems is obtained. In bone marrow transplantation, the incidence of graft-versus-host reactions appear to increase when lymph node cells are added to the graft. Under such conditions, Y-linked as well as autosomal minor histocompatibility systems influence the appearance of graft-versus-host reactions. The possible applications of these results to human transplantation are discussed.

REFERENCES

1. Advisory Committee to the Renal Transplant Registry. Thirteenth Report of the Human Renal Transplant Registry. *Transpl. Proc., 9*:9, 1977.
2. Bailey, D.W. Histocompatibility associated with the X chromosome in mice. *Transplantation, 1*:70, 1963.
3. Bekkum, D.W. van. Use and abuse of hematopoietic cell grafts in immune deficiency diseases. *Transpl. Rev., 9*:3, 1972.
4. Bekkum, D.W., van, Balner, H., Dicke, K.A., Berg, F.G. van der, Prinsen, G.H., and Hollander, C.F. The effect of pretreatment of allogeneic bone marrow graft recipients with antilymphocytic serum on the acute graft-versus-host reaction in monkeys. *Transplantation, 13*:400, 1972.
5. Bekkum, D.W. van, Roodenburg, J., Heidt, P.J., and Waaij, D. van der. Mitagation of secondary disease of allogeneic mouse radiation chimera by modification of the intestinal microflora. *J. Nat. Cancer Inst., 52*:401, 1974.
6. Bekkum, D.W. van. Bone marrow transplantation. *Transpl. Proc., 9*:147, 1977.
7. Bortin, M.M., and Rimm, A.A. Severe combined immunodeficiency disease: Characterization of the disease and results of transplantation. *Trans. Proc., 9*:169, 1977.
8. Bull, R.W., Vriesendorp, H.M., Obertop, H., Bijnen, A.B., Jeekel, J., Gruyl, J. de, and Westbroek, D.L. The effect of prior third party blood transfusion on canine renal allograft survival. Submitted for publication, 1977.
9. Bijnen, A.B., Vriesendorp, H.M., Grosse Wilde, H., and Westbroek, D.L. Polygenic control of mixed lymphocyte culture reactivity in dogs. *Tissue Antigens, 9*:187, 1977.
10. Dicke, K.A. The separation of bone marrow cells on a discontinuous albumin gradient. Thesis, University of Leiden, The Netherlands, 1970.
11. Eichwald, E.J., and Silmser, L.R. Note. *Transpl. Bull., 2*:148, 1955.
12. Gross Wilde, H., Vriesendorp, H.M., Wank, R., Mempel, W., Dechamps, G., Honauer, U., Baumann, P., Netzel, B., Kolb, J.H., and Albert, E.D. Identification of four MLC specificities in the dog. *Tissue Antigens, 4*:229, 1974.

13. Hildemann, W.H. Components and concepts of antigenic strength. *Transpl. Reviews, 3*:5, 1970.
14. Klein, J. The Major H-2 Histocompatibility Complex of the Mouse. Heidelberg: Springer-Verlag, 1975.
15. Newth, D.R. Chance compatibility in homografting. *Transpl. Bull., 27*:452, 1961.
16. Obertop, H., Jeekel, J., Vriesendorp, H.M., MacDicken, I., and Westbroek, D.L. The effect of donor blood on renal allograft survival in DL-A tissue typed beagle littermates. *Transplantation, 20*:49, 1975.
17. Ohno, S. *Monographs on Endocrinology. Sex Chromosomes and Sex-linked Genes.* Heidelberg: Springer-Verlag, 1967.
18. Opelz, G., Mickey, M.R., and Teraski, P.I. Calculations on long term graft and patient survival in human kidney transplantation. *Transpl. Proc., 9*:27, 1977.
19. Rodt, H., Netzel, B., Niethammer, D., Körbling, M., Götze, D., Kolb, H.Y., Thiel, E., Haas, R.Y., Fliedner, T.M., and Thierfelder, S. Specific absorbed antithymocyte globulin for incubation treatment in human marrow transplantation. *Trans. Proc., 9*:187, 1977.
20. Selden, J.R., and Wachtel, S.S. H-Y antigen in the dog. *Transplantation, 24*:298, 1977.
21. Simonson, M. Immunogenetic speculations on human kidney transplants. *Transplantation, 4*:354, 1966.
22. Smid-Mercx, B.M.J., Duyzer-den Hartog, B., Visser, T.P., and Vriesendorp, H.M. Serological studies of canine histocompability antigens. *Transpl. Proc., 7*:361, 1975.
23. Snell, G.D., and Stimpfling, T.H. Genetics of tissue transplantation. In E. L. Green, ed., *Biology of the Laboratory Mouse* (2nd ed.). New York: McGraw-Hill, 1966.
24. Storb, R., Rudolph, R.H., Kolb, H.Y., Graham, T.C., Mickelson, E., Erickson, V., Lerner, K.G., Kolb, H., and Thomas, E.D. Marrow grafts between DL-A matched canine littermates. *Transplantation, 15*:92, 1973.
25. Storb, R., Weiden, P.L., Prentice, K., Buckner, C.D., Clift, R.A., Einstein, A.B., Fefer, A., Johnson, F.L., Lerner, K.G., Neiman, P.E., Sanders, Y.E., and Thomas, E.D. Aplastic anemia (AA) treated by allogeneic marrow transplantation: The Seattle experience. *Transpl. Proc., 9*:101, 1977.
26. Thomas, E.D., Fefer, A., Buckner, C.D., and Storb, R. Critical review: Current status of bone marrow transplantation for aplastic anemia and acute leukemia. *Blood, 49*:671, 1977.
27. Vriesendorp, H.M. Major histocompatibility complex of the dog. Thesis, Erasmus University, Rotterdam, The Netherlands, 1973.
28. Vriesendorp, H.M., Zurcher, C., Bull, R.W., Los, W.R.T., Meera Kahn, Tweel van der, T.G., Zweibaum, A., and Bekkum, D.W. van der. GvH and take of allogeneic bone marrow in tissue typed dogs. *Transpl. Proc., 7* (Suppl. 1):849, 1975.
29. Vriesendorp, H.M. Bÿnen, A.B., Zurcher, C., and Bekkum, D.W. van. Donor selection and bone marrow transplantation in dogs. F. Kissmeyer-Nielsen, ed., *Histocompatibility Testing.* Copenhagen: Munksgaard, 1975, pp. 963–971.

30. Vriesendorp, H.M., Bÿnen, A.B., Westbroek, D.L., and Bekkum, D.W. van. Genetics and transplantation immunology of the DLA complex. *Transpl. Proc.*, 9:293, 1977.

31. Vriesendorp, H.M., Grosse Wilde, H., and Dorf, M.E. The major histocompatibility system of the dog. In D. Götze, ed., *The Major Histocompatibility System in Mammals*. Heidelberg: Springer-Verlag, 1977.

32. Vriesendorp, H.M., Boorman, G., Kessel, A.C.M. van, Hogweg, B., Hendriks, W.D.H., and Zurcher, C. Total-body irradiation as conditioning regimen for bone marrow transplantation in dogs. Submitted for publication, 1977.

33. Westbroek, D.L., Silberbusch, J., Vriesendorp, H.M., Urk, H. van, Roemeling, H.W., Schönherr-Scholtes, Y.H.C.M., and Vries, M.J. de. The influence of DL-A histocompatibility on the function and pathohistological changes in unmodified canine renal allografts. *Transplantation, 14*:582, 1972.

13

The Role of Mononuclear Phagocytes in the Defense against Infections and Malignancies

A. Cruchaud

INTRODUCTION

The participation of mononuclear phagocytes (they will be called "macrophages" for convenience) in the defense mechanisms against infection and malignant growth is an intricate process that requires the interaction of cells as well as the intervention of soluble mediators. The participants in this process are:

1. Macrophages, as accessory cells of the immune response.
2. B-lymphocytes, as precursors of antibody-producing cells.
3. T-lymphocytes, as lymphokine producers.
4. Complement.
5. Macrophages, as effector cells.

MACROPHAGES AND THE IMMUNE RESPONSE

The role of macrophages as accessory cells of the immune response has been recently reviewed (59,73). The uptake of antigen by macrophage involves three phases

First, *antigen binding* or *attachment*, a phenomenon, which is partly dependent on metabolic processes, is usually rapidly followed by *internalization* or *endocytosis*. As will be seen later, these phenomena require receptors for complement and for the Fc fragment of IgG on the plasma membrane of macrophages. Endocytosis occurs after sequential binding of membrane receptors to immunological ligands, which cover the surface of foreign material (antigens), with a subsequent activation of cytoplasmic contractile proteins that resemble a "zipper effect" (41). Antigenic material exposed to and then associated with macrophages may be found in two locations: (a) on macrophage plasma membranes and (b) in phagosomes or phagolysosomes, from which it is slowly released to the extracellular environment (21,22,75).

Second, *antigen processing*, during which phagosomes containing endocytosed antigen combine with primary lysosomes to form secondary lysosomes or phagolysosomes. Lysosomal enzymes are then able to degrade partially or totally before being released to the extracellular environment or to the macrophage membrane (21,22).

Third, *presentation of antigen to immunocompetent lymphocytes*, which may require: (a) *cell-to-cell contact*, which seems essential for lymphocytes to recognize some antigens (59) and occurs only when lymphocytes and macrophages are syngeneic (54); (b) *production of superantigen*, a combination of antigen, RNA, and/or other cellular

119

material that has an enhanced capacity to trigger lymphocytes (6); and (c) *macrophage-secreted soluble factors (monokines)* (57,59,77,81). Among the numerous monokines secreted by macrophages that have been described, the most important are the T-lymphocyte activating factor (LAF), the B-lymphocyte–activating (and maturation) factor (BAF), T-helper cell–inducing factor, a sheep red blood cell solubilizing factor an inhibitor of lymphocyte DNA and protein synthesis, and an inhibitor of the mixed lymphocyte reaction.

The *T-lymphocyte–activating factor* promotes thymocyte and T-lymphocyte viability, a function that can be replaced by 2-mercaptoethanol and vitamin E. Stimulation of T-lymphocytes by LAF allows stimulation of thymocytes by concanavalin A (Con-A) and phytohemagglutinin (PHA); it increases the level of cyclic adenosine monophosphate in thymocytes. The production of LAF is augmented by PHA, Con-A, lipopolysaccharide, and X-irradiation. T-lymphocyte–activating factor is not dialyzable, shows no genetic restriction, does not contain components of the major histocompatibility complex, and is sensitive to some proteolytic enzymes (7,12,13,34,35,60,76).

The *B-lymphocyte activating and maturation factor(s)* potentiates B-cell response in the presence of small numbers of T-lymphocytes (65,77,86).

The *T-helper-cell–inducing factors* include a genetically related factor (GRF) and a nonspecific factor (NMF). These two factors have been recently described (26,27). Genetically related factor is obtained from mouse macrophages exposed to soluble antigens. It induces T-helper cells in the absence of macrophages and antigens, provided the GRF-producing macrophages and the target lymphocytes are identical at the I-A subregion of the H-2 locus. Nonspecific macrophage factor is released by cells incubated in the absence of antigen. It can replace macrophages to stimulate lymphocytes in the presence of particulate antigens. It is not restricted allogenically.

The *sheep red blood cell solubilizing factor* transforms sheep red blood cells *in vitro* into soluble immunogenic material (46), perhaps changing a thymus-dependent antigen into a thymus-independent one.

The *inhibitor of lymphocyte DNA and protein synthesis* is mostly thymidine, with a molecular weight less than 1,000 (12,58).

The *inhibitor of the mixed lymphocyte reaction* is also a low molecular weight factor, which might be identical with the preceding one (30). In addition to factors mediating the immune response, macrophages also secrete products that may directly participate in inflammation as well as in nonspecific defense mechanisms (for review, see reference 74).

These are complement components, (25,69,87); interferons (1,68); plasminogen activators (78); various lysosomal enzymes (15,37,83); collagenase and elastase (84,85); lysozyme (37); cyclic adenosine monophosphate (83); and alpha$_2$-macroglobulin (47), and inhibitor of hydrolytic enzymes.

EFFECTS OF THE HUMORAL IMMUNE RESPONSE ON MACROPHAGE FUNCTIONS

The humoral immune response influences the function of macrophages through antibodies and complement factors. Macrophages have on their plasma membrane, receptors for the Fc fragment of IgG as well as for fragments of C3, i.e., C3b and C3d (10,15,55), though the role of the latter has been questioned (40,71). Receptors for C3b, or fragments of it, and to a lesser extent, receptors for C3d mediate the attachment of opsonized antigens to the macrophage membrane, whereas Fc receptors mediate the ingestion of bound material (10,24,55). It must be stressed that these receptors are not specific. By contrast, cytophilic antibodies, which are bound to the Fc receptors, play the role of temporary specific ligands for the corresponding antigens.

Once internalized, opsonized microorganisms will be inactivated by processes that will be analyzed in a later section. However, the role of antibodies may not be limited to the promotion of foreign material ingestion. It has been shown that for such bacteria as *Mycobacterium tuberculosis* and *Toxoplasma gondii*, antibody coating reverses *in vitro* the inhibition of the phagosome-lysosome fusion that is usually observed with uncoated bacteria (5). Whether this phenomenon is responsible for the killing of microorganisms *in vivo* is still an open question.

EFFECTS OF THE CELL-MEDIATED IMMUNE RESPONSE ON MACROPHAGE FUNCTIONS

Activated T-lymphocytes produce a large number of soluble factors called *lymphokines* (for review, see reference 81). Some of these factors react with macrophages (79,80) either by redirecting their migration or by activating them.

The most important of these lymphokines are the *macrophage migration inhibitory factor* (MIF), which immobilizes macrophages in the vicinity of lymphocytes and thus increases their concentration where they are needed; the *monocyte chemotactic factor* (MCF); the *macrophage growth factor* (MGF) and the *macrophage mitogenic factor* (MMF), which both promote cell viability and proliferation; the *macrophage-activating factor* (MAF), for which the following properties have been reported:

1. It stimulates the secretion of monokines, whose effects on lymphocytes have been discussed earlier.
2. It increases the release of lysosomal enzymes (61).
3. It promotes the secretion of cytotoxins capable of killing allogeneic tumor cells in the absence of macrophages (52,64).
4. It induces the synthesis and release of bactericidal substances (9,66).
5. The MAF from specifically or nonspecifically activated lymphocytes may confer on macrophages the capacity to suppress replication of facultative or obligate intracellular pathogens, such as *Listeria monocytogenes* (31), *T. Gondii* (4,11), and *Mycobacterium leprae* (36).
6. The *specific macrophage arming factor* (SMAF), which confers on normal macrophages the capacity to destroy tumor cells (28,29).

MICROBIOCIDAL MECHANISMS OF MACROPHAGES

Exposure of phagocytic cells to suitable stimuli, such as opsonized microorganisms, provokes the *respiratory burst*, a sequence of events characterized by increased oxygen consumption; NADPH production via the activation of the hexose monophosphate (HMP) shunt; production of superoxide ($\cdot O_2^-$); and production of hydrogen peroxide (H_2O_2) (for review, see references 19,70). These phenomena have been particularly well studied in polymorphonuclear cells, but they seem to occur a certain extent in mononuclear phagocytes as well (50,70); they therefore will be briefly described here.

Under certain conditions, O_2 is reduced to $\cdot O_2^-$, an unstable molecule that is bactericidal, and cytotoxic. This reduction occurs through the oxidation of NADPH and $NADP^+$ in the presence of NADPH-oxidase. The $NADP^+$ is, in turn, reduced to NADPH due to an activation of the HMP shunt in the presence of glucose-6-phosphate dehydrogenase. The $\cdot O_2^-$ ions recombine to produce oxygen and H_2O_2; this reaction may occur spontaneously, but it is also catalyzed by superoxide dismutase. Hydrogen peroxide is bactericidal and cytotoxic and may be detoxified in the presence of catalase to produce H_2O and O_2. Alternatively, it may be halogenated when reduced glutathione is oxidized ($2GSH + H_2O_2 \rightarrow GSSG + 2H_2O$) via the GSH-peroxidase-halide system (49); the halogen ion is mainly Cl^-. Chloride ion is transformed into OCL^-, which is incorporated into the amino acids of the bacterium wall (chloramination) and destroys it.

Both $\cdot O_2^-$ and H_2O_2 can generate singlet molecular oxygen ($^1O_2$), a very unstable molecule, that can oxidize certain substrates within target microorganisms and kill them. This provokes a measurable chemiluminescent effect in leukocytes (3).

Oxidation of $\cdot O_2^-$ to O_2 or to H_2O_2 via the $NADP^+$/NADPH system induces *in vitro* the reduction of colorless nitroblue tetrazolium (NBT) to black formazan, a reaction that can be used to test the integrity of this system.

It is not clear whether this sequence of events in polymorphs occurs in mononuclear phagocytes as well. To be sure, H_2O_2 formation has been demonstrated in human monocytes (8), mouse peritoneal macrophages (62), and rabbit alveolar macrophages (33), but it is not known whether the intermediates in the formation or breakdown of H_2O_2 (i.e. $\cdot O^{2-}$ and $^1O_2$) also have microbiocidal properties in these cells (50). The H_2O_2-myeloperoxidase-halide system probably contributes to the microbiocidal activity of monocytes; however, some mature macrophages do not contain myeloperoxidase (32,67). In this case, catalase may substitute for myeloperoxidase, since this enzyme has microbiocidal activity when combined with a H_2O_2-generating system and either I^- or thyroxine at acid pH (50).

The capacity of macrophages to kill microorganisms is not limited to the oxidative processes resulting from the respiratory burst; several other mechanisms have been described. For instance, the acid pH of phagolysosomes, lysozyme and lysosomal enzymes, and free radicals, as well as interferon, all can destroy endocytosed microorganisms. In addition, it has been shown that macrophages can destroy virus-infected cells that express viral antigens on their membrane in the presence of specific antibodies (51). Also, immune complexes may induce macrophages to kill target cells, as has been shown when *Schistosoma mansoni* antigens combined with anti-*Schistosoma* antibody of the IgE type provoke the adherence of macrophages to schistosomulae, with the subsequent lysis of these larvae (14). Finally, upon activation of lymphokines, macrophages may release bactericidal products that are then capable of killing microorganisms in the absence of macrophages (9,66).

ROLE OF MACROPHAGES IN THE DEFENSE AGAINST NEOPLASTIC DISEASE

Recent reviews indicate that macrophages play a crucial role in the defense against neoplastic disease, although the exact mechanisms of their anti-tumor activity have not been entirely defined (42,53). This role is suggested by observations that suppression of the reticuloendothelial system leads to an increased incidence of, and susceptibility to, neoplastic disease, whereas stimulation of the reticuloendothelial system leads to increased resistance, whether this stimulation is obtained by Bacillus Calmetta-Guerin (BCG), methanol-extraction residue of BCG (MER), *Corynebacterium parvum, Bordetella pertussis,* endotoxin, levamisole, or other nonspecific stimulants.

Macrophages may function in several ways in the prevention of neoplastic disease:

1. As accessory cells of the immune response (cf. Macrophages and the immune response), they contribute to the development of cytotoxic T-lymphocytes as well as the synthesis of antibodies; the antibodies participate in tumor cell destruction either directly or through the mechanism of antibody-dependent cell-mediated cytotoxicity.

2. Macrophages that have been activated by the *in vivo* injection of *Listeria monocytogenes*, peptone, or endotoxin or by *in vitro* exposure to endotoxin or double-stranded RNA have been shown to destroy tumor cells *in vitro* or to inhibit their proliferation (2,45,48).

3. Macrophages armed with lymphocyte-released factors are toxic to target cells. A specific arming factor (SMAF), released by T-lymphocytes sensitized to tumor antigens upon reexposure *in vitro* to sensitizing antigen (28,29), has been found to differ from classical immunoglobulins and, in particular, from cytophilic antibodies. Armed monocytes have been isolated from solid tumors (43). In contrast to the specificity of SMAF, a nonspecific factor has been obtained following incubation of sensitized lymphocytes with antigens that are not from the target tumor cell (20).

4. Macrophages carrying cytophilic antibodies are capable of destroying tumor cells having the corresponding surface antigens (38,39). This capacity is abolished following treatment with trypsin (63) and may be considerably decreased when free circulating tumor antigen "blocks" the antigen-binding site of cytophilic antibodies (53). Under the conditions described in 2, and 3, classical phagocytosis is not responsible for target cell destruction (29,45,48). It is known, how-

ever, that cell-to-cell contact is a prerequisite for macrophage-mediated cytotoxicity in many systems (39,45,48,56). This suggests the possibility that macrophages produce localized lesions in the membrane of target cells; "piecemeal cytophagocytosis" has indeed been demonstrated in electronmicroscopic studies of macrophages and tumor cells in intimate contact (17,18). Also, the hypothesis has been advanced that lysosomal enzymes are involved in target cell destruction (72). Recently, intracellular material was shown to be transferred from macrophages to target cells (44).

5. Finally, macrophages may, following appropriate stimulation, release factors capable of interfering with tumor cell development. For instance, a growth inhibitory factor (GIF) was obtained from cultures of cytophilic antibody-coated macrophages exposed to their target tumor cells (82). Along the same line, a cytotoxin has been described that nonspecifically kills different types of allogeneic tumor cells (52,64). Macrophages are also considered to be a possible source of the recently described tumor-necrosis factor (TNF), which appears in the serum of mice primed with BCG or *C. parvum* and subsequently injected with bacterial endotoxin (16). It is true that these different modes of action are not mutually exclusive and may, on the contrary, combine in the defense against neoplastic disease. They contribute significantly to the mechanisms involved in cancer immunotherapy.

SUMMARY

In this brief review, we have shown that macrophages play various roles in the destruction of microorganisms and tumor cells. It appears, however, that a deficiency in one of the several systems participating in these defense mechanisms may result in the failure of the whole system. For instance, a deficient humoral immune response, or abnormal complement, may prevent bactericidal processes from taking place simply by affecting the membrane attachment and endocytosis of microorganisms. Similarly, cell-mediated immunodeficiencies that cause a decrease in the production of lymphokines will affect the activation and migration of macrophages. Also, macrophages may themselves be defective, which will prevent them from participating normally in defense mechanisms. For example, deficiencies of glucose-6-phosphate dehydrogenase, NADPH-oxidase, or myeloperoxidase will disorganize, to various extents, the respiratory burst. Impaired fusion between phagosomes and lysosomes may also prevent the destruction of otherwise normally endocytosed microorganisms.

ACKNOWLEDGMENT

Supported by the Swiss National Science Foundation (Grant No 3.470.0.75).

REFERENCES

1. Acton, J.D., and Myrvik, Q.N. Production of interferon by alveolar macrophages. *J. Bacteriol., 91*:2300, 1966.
2. Alexander, P., and Evans, R. Endotoxin and double-stranded RNA render macrophages cytotoxic. *Nature New Biol., 232*:76, 1971.
3. Allen, R.C., Yevich, S.J., Orth, R.W., and Steele, R.H. The superoxide anion and singlet molecular oxygen: Their role in the microbiocidal activity of the polymorphonuclear leukocyte. *Biochem. Biophys. Res. Comm., 60*:909, 1974.
4. Anderson, S.E., Bautista, S., and Remington, J.S. Induction of resistance to *Toxoplasma gondii* in human macrophages by soluble lymphocyte products. *J. Immunol., 117*:381, 1976.
5. Amstrong, J.A., and d'Arcy Hart, P. Phagosome-lysosome interactions in cultures macrophages infested with virulent tubercle bacilli. Reversal of the usual nonfusion pattern and observations on bacterial survival. *J. Exp. Med., 142*:1, 1975.
6. Askonas, B.A., and Rhodes, J.M. Immunogenicity of antigens-containing ribonucleic acid preparations from macrophages. *Nature (Lond.) 205*:470, 1965.
7. Bach, F.H., Alter, B.J., Solliday, S., Zoschke, D.C., and Janis, M. Lymphocyte reactivity *in vitro*. II. Soluble reconstituting factor permitting response of purified lymphocytes. *Cell. Immunol., 1*:219, 1970.
8. Baehner, R.L., and Johnston, R.B. Jr. Monocyte function in children with neutropenia and chronic infection. *Blood, 40*:31, 1972.
9. Bast, R.C. Jr., Cleveland, R.P., Littman, B.H., Zbar, B., and Rapp, H.J. Acquired cellular immunity: Extracellular killing of *Listeria monocytogenes* by a product of immunologically activated macrophages. *Cell Immunol., 10*:248, 1974.
10. Bianco, C., Griffin, F.M. Jr., and Silverstein, S.C. Studies on the macrophage complement receptor. *J. Exp. Med., 141*:1278, 1975.
11. Borges, J.S., and Johnson, W.D. Jr. Inhibition of multiplication of *Toxoplasma gondii* by human monocytes exposed to T-lymphocyte products. *J. Exp. Med., 141*:483, 1975.
12. Calderon, J., Williams, R.T., and Unanue, E.R. An inhibitor of cell proliferation released by cultures of macrophages. *Proc. Nat. Acad. Sci. (USA), 71*:4273, 1974.
13. Calderon, J., Kiely, J.M., Lefko, J.L., and Unanue, E.R. The modulation of lymphocyte functions by molecules secreted by macrophages. I. Description and partial biochemical analysis. *J. Exp. Med., 142*:151, 1975.
14. Capron, A., Dessaint, J.P., Joseph, M., Rousseaux, J., Capron, M., and Bazin, H. Interaction between IgE complexes and macrophages in the rat: A new mechanism of macrophage activation. *Eur. J. Immunol., 7*:315, 1977.
15. Cardella, C.J., Davies, P., and Allison, A.C. Immune complexes induce selective release of lysosomal hydrolases from macrophages. *Nature, (Lond.) 247*:46, 1974.
16. Carswell, E.A., Old, L.J., Kassel, R.L., Green, S., Fiore, N., and Williamson, B. An endotoxin-induced serum factor that causes necrosis of tumors (activated macrophage). *Proc. Nat. Acad. Sci. (USA) 72*:3666, 1975.
17. Chambers, V.C., and Weiser, R.S. The ultrastructure of target cells and immune macrophages during their interaction *in vitro*. *Cancer Res., 29*:301, 1969.
18. Chambers, V.C., and Weisers, R.S. The ultrastructure of target L-cells and immune macrophages during their interaction *in vivo*. *Cancer Res., 31*:2059, 1971.
19. Cheson, B.D., Curnutte, J.T., and Babior, B.M. The oxidative killing mechanisms of the neutrophil. *Progr. Clin. Immunobiol., 3*:1, 1977.
20. Churchill, W.H. Jr., Piessens, W.F., Sulis, C.A., and David, J.R. Macrophages activated as suspension cultures with lymphocyte mediators devoid of antigen become cytotoxic for tumor cells. *J. Immunol., 115*:781, 1975.
21. Cruchaud, A., and Unanue, E.R. Fate and immunogenicity of antigens endocytosed by macrophages: A study using foreign red cells and immunoglobulin G. *J. Immunol., 107*:1329, 1971.
22. Cruchaud, A., Berney, M., and Balant, L. Catabolism, physical, and immunologic properties of endocytosed isologous and heterologous gamma-globulins by mouse macrophages. *J. Immunol., 114*:102, 1975.
23. Cruchaud, A., Girard, J.-P., and Hitoglou, S. The functions of human monocytes in normal subjects and in disorders associated with immune deficiency. *Int. Arch. Allerg. Appl. Immunol., 54*:529, 1977.
24. Ehlenberger, A.G., and Nussenzweig, V. The role of membrane receptors for C3b and C3d in phagocytosis. *J. Exp. Med., 145*:357, 1977.
25. Eisenstein, L.P., Schneeberger, E.E., and Colten, H.R. Synthesis of the second component of complement by long-term primary cultures of human monocytes. *J. Exp. Med., 143*:114, 1976.
26. Erb, P., and Feldmann, M. The role of macrophages in the generation of T helper cells. III. Influence of macrophage-derived factors in helper cell induction. *Eur. J. Immunol., 5*:759, 1975.
27. Erb, P., Feldmann, M., and Hogg, N. Role of macrophages in the generation of T helper cells IV. Nature of genetically related factor derived from macrophages incubated with soluble antigens. *Eur. J. Immunol., 6*:365, 1976.
28. Evans, R., and Alexander, P. Role of macrophages in tumor immunity. II. Involvement of a macrophage cytophilic factor during syngeneic tumor growth inhibition. *Immunology, 23*:627, 1972.

29. Evans, R., and Alexander, P. Mechanism of immunologically specific killing of tumor cells by macrophages. *Nature (Lond.), 236*:168, 1972.

30. Fernbach, B.R., Kirchner, H., and Herberman, R.B. Inhibition of the mixed lymphocyte culture by peritoneal exudate cells. *Cell. Immunol., 22*:399, 1976.

31. Fowles, R.E., Fajardo, I.M., Leibowitch, J.L., and David, J.R. The enhancement of macrophage bacteriostasis by products of activated lymphocytes. *J. Exp. Med., 138*:952, 1973.

32. van Furth, R., Hirsch, J.C., and Fedorko, M.E. Morphology and peroxidase cytochemistry of mouse promonocytes, monocytes and macrophages. *J. Exp. Med., 132*:794, 1970.

33. Gee, J.B.L., Vassallo, C.L., Bell, K., Kaslin, J., Basford, R.E., and Field, J.B. Catalase-dependent peroxidative metabolism in the alveolar macrophage during phagocytosis. *J. Clin. Invest., 49*:1280, 1970.

34. Gery, I., and Waksman, B.H. Potentiation of the T-lymphocyte response to mitogens. II. The cellular source of potentiating mediator(s). *J. Exp. Med., 136*:143, 1972.

35. Gery, I., Gershon, R.K., and Waksman, B.H. Potentiating of the T-lymphocyte response to mitogens. I. The responding cell. *J. Exp. Med. 136*:128, 1972.

36. Godal, T., Rees, R.J.W., and Lamvick, J.O. Lymphocyte-mediated modification of blood-derived macrophage function *in vitro*; inhibition of growth of intracellular mycobacteria with lymphocytes. *Clin. Exp. Immunol., 8*:625, 1971.

37. Gordon, S., Todd, J., and Cohen, Z.A. *In vitro* synthesis and secretion of lysozyme by mononuclear phagocytes. *J. Exp. Med., 139*:1228, 1974.

38. Granger, G.A., and Weiser, R.S. Homograft target cells: Specific destruction *in vitro* by contact interaction with immune macrophages. *Science, 145*:1427, 1964.

39. Granger, G.A., and Weiser, R.S. Homograft target cells: Contact destruction *in vitro* by immune macrophages. *Science, 151*:97, 1966.

40. Griffin, F.N. Jr., Bianco, C., and Silverstein, S.C. Characterization of the macrophage receptor for complement and demonstration of its functional independence from the receptor for the F_c portion of immunoglobulin G. *J. Exp. Med., 141*:1269, 1975.

41. Griffin, F.M. Jr., Griffin, J.A., and Silverstein, S.C. Studies on the mechanism of phagocytosis. II. The interaction of macrophages with anti-immunoglobulin IgG-coated bone marrow derived lymphocytes. *J. Exp. Med., 144*:788, 1976.

42. Harris, J.E., and Sinkovics, J.C. *The Immunology of Malignant Diseases* St. Louis: Mosby, 1976.

43. Haskill, J.S., Radov, L.A., Yamamura, Y., Parthenais, E., Korn, J.H., and Ritter, F.L. Experimental solid tumors: The role of macrophage and lymphocytes as effector cells. *J. Reticuloendothel. Soc., 20*:233, 1976.

44. Hibbs, J.B. Jr. Role of activated macrophages in nonspecific resistance to neoplasia. *J. Reticuloendothel. Soc., 20*:233, 1976.

45. Hibbs, J.B. Jr., Lambert, L.H. Jr., and Remington, J.S. Possible role of macrophage mediated nonspecific cytotoxicity in tumor resistance. *Nature New Biol., 235*:48, 1972.

46. Hoffman, M., and Dutton, R.W. Immune response restoration with macrophage culture supernatants. *Science, 172*:1047, 1971.

47. Hovi, T., Mosher, D., and Vaheri, A. Cultured human monocytes synthesize and secrete α_2-macroglobulin. *J. Exp. Med., 145*:1580, 1977.

48. Keller, R. Cytostatic elimination of syngeneic rat tumor cells *in vitro* by nonspecifically activated macrophages. *J. Exp. Med., 138*:625, 1973.

49. Klebanoff, S.J. In R.C. Williams and H.H. Fudenberg, *Phagocytic Mechanisms in Health and Disease*. New York: Intercontinental Medical Book Corp., 1972, pp. 3–21.

50. Klebanoff, S.J., and Hamon, C.B. Antimicrobial systems of mononuclear phagocytes. In R. van Furth, ed., *Mononuclear Phagocytes in Immunity, Infection, and Pathology*. Oxford: Blackwell Scientific, 1975, pp. 507–529.

51. Kohl, S., Starr, S.E., Oleske, J.M., Shore, S.L., Ashman, R.B., and Nahmias, A.J. Human monocyte-macrophage-mediated antibody-dependent cytotoxicity to herpes simplex virus-infected cells. *J. Immunol., 118*:729, 1977.

52. Kramer, J.J., and Granger, G.A. The *in vitro* induction and release of a cell toxin by immune $C_{57}Bl/6$ peritoneal macrophages. *Cell. Immunol., 3*:88, 1972.

53. Levy, M.H., and Wheelock, E.F. The role of macrophage in defense against neoplastic disease. *Adv. Cancer Res., 20*:131, 1974.

54. Lipsky, P.E., and Rosenthal, A.S. Macrophage-lymphocyte interaction. II. Antigen-mediated physical interactions between immune guinea pig lymph node lymphocytes and syngeneic macrophages. *J. Exp. Med., 141*:138, 1975.

55. Mantovani, B., Rabinovitch, M., and Nussenzweig, V. Phagocytosis of immune complexes by macrophages. Different roles of the macrophage receptor sites for complement (C_3) and for immunoglobulin (IgG). *J. Exp. Med., 135*:780, 1972.

56. Melsom, H., and Seljelid, R. The cytotoxic effect of mouse macrophages on syngeneic and allogeneic erythrocytes. *J. Exp. Med., 137*:807, 1973.

57. Nelson, D.S., and Gatti, R.A. Humoral factors influencing lymphocyte transformation. *Progr. Aller., 21*:261, 1976.

58. Opitz, H.G., Niethammer, D., Jackson, R.E., Lemke, H., Huguet, R., and Flad, H.D. Biochemical characterization of a factor released by macrophages. *Cell. Immunol., 18*:70, 1975.

59. Oppenheim, J.J., and Rosenstreich, D.L. Signals regulating *in vitro* activation of lymphocytes. *Progr. Aller., 20*:65, 1975.

60. Oppenheim, J.J., Shneyour, A., and Kook, A.E. Enhancement of DNA synthesis and cAMP content of mouse thymocytes by mediator(s) derived from adherent cells. *J. Immunol., 116*:1466, 1976.

61. Pantalone, R., and Page, R.C. Enzyme production and secretion of lymphokine-activated macrophages. *J. Reticuloendothel. Soc., 21*:343, 1977.

62. Paul, B.B., Strauss, R.R., Jacobs, A.A., and Sharra,

A.J. Function of H_2O_2, myeloperoxidase, and hexose monophosphate shunt enzymes in phagocytizing cells from different species. *Infect. Immunit., 1*:338, 1970.

63. Pearsall, N.N., and Weiser, R.S. Macrophage in allograft immunity. II. Passive transfer with immune macrophages. *J. Reticuloendothel. Soc., 5*:121, 1968.

64. Piper, C.E., and McIvor, K.L. Alloimmune peritoneal macrophages as specific effector cells: Characterization of specific macrophage cytotoxin. *Cell. Immunol., 17*:423, 1975.

65. Schrader, J.W. Mechanism of activation of the bone marrow-derived lymphocyte. III. A distinction between a macrophage-produced triggering signal and the amplifying effect on triggered B lymphocytes of allogeneic interactions. *J. Exp. Med., 138*:1466, 1973.

66. Sharma, S.D., and Middlebrook, G. Antibacterial product of peritoneal exudate cell cultures from guinea pigs infected with *Mycobacterium, Listeria,* and *Rickettsiae. Infect. Immunit., 15*:745, 1977.

67. Simmons, S., and Karnovsky, M.L. Iodinating ability of various leukocytes and their bactericidal activity. *J. Exp. Med., 138*:44, 1973.

68. Smith, T.J., and Wagner, R.R. Rabbit macrophage interferons. I. Conditions for biosynthesis by virus infected and uninfected cells. *J. Exp. Med., 125*:559, 1967.

69. Stecker, V.J., and Thorbecke, G.J. Sites of synthesis of serum proteins. I. Serum proteins produced by macrophages *in vitro. J. Immunol. 99*:643, 1967.

70. Stossel, T.P. Phagocytosis. *New Eng. J. Med., 290*:774, 1974.

71. Stossel, T.P., Field, R.J., Githin, J.D., Alper, C.A., and Rosen, F.S. The opsonic fragment of the third component of human complement (C_3). *J. Exp. Med., 141*:1329, 1975.

72. Temple, A., Loewi, G., Davies, P., and Howard, A. Cytotoxicity of immune guinea-pig cells. II. The mechanism of macrophage cytotoxicity. *Immunology, 24*:655, 1973.

73. Unanue, E.R. The regulatory role of macrophages in antigenic stimulation. *Adv. Immunol., 15*:95, 1972.

74. Unanue, E.R. Secretory function of mononuclear phagocytes. *Amer. J. Pathol., 83*:395, 1976.

75. Unanue, E.R., and Cerottini, J.C. The immunogenicity of antigen bound to the plasma membrane of macrophages. *J. Exp. Med., 131*:1329, 1970.

76. Unanue, E.R., Kiely, J.M., and Calderon, J. The modulation of lymphocyte functions by molecules secreted by macrophages. II. Conditions leading to increased secretion. *J. Exp. Med., 144*:155, 1976.

77. Unanue, E.R., Beller, D.I., Calderon, J., Kiely, J.M., and Stadecker, M.J. Regulation of immunity and inflammation by mediators from macrophages. *Amer. J. Pathol., 85*:465, 1976.

78. Unkeless, J.C., Gordon, S., and Reich, E. Secretion of plasminogen activator by stimulated macrophages. *J. Exp. Med., 139*:834, 1974.

79. Waksman, B.H. Immunoglobulins and lymphokines as mediators of inflammatory cell mobilization and target cell killing. *Cell. Immunol., 27*:309, 1976.

80. Waksman, B.H. Cell-antibody cooperation in host-parasite relations. *Pathol. Biol., 24*:587, 1976.

81. Waksman, B.H., and Namba, Y. On soluble mediators of immunologic regulation. *Cell. Immunol., 21*:161, 1976.

82. Weiser, R.S., Heise, E., McIvor, K., Han, S., and Granger, G.A. R.T. Smith and R.A. Good, eds., In *Cellular Recognition.* New York: Appleton, 1969, pp. 215–220.

83. Welscher, H.D., and Cruchaud, A. The influence of various particles and 3',5'-cyclic adenosine monophosphate on release of lysosomal enzymes by mouse macrophages. *J. Reticuloendothel. Soc., 20*:405, 1976.

84. Werb, Z., and Gordon, S. Secretion of a specific collagenase by stimulated macrophages. *J. Exp. Med., 142*:346, 1975.

85. Werb, Z., and Gordon, S. Elastase secretion by stimulated macrophages. Characterization and regulation. *J. Exp. Med., 142*:361, 1975.

86. Wood, D.D., and Gaul, S.L. Enhancement of the humoral response of T cell-depleted murine spleens by a factor derived from human monocytes *in vitro. J. Immunol., 113*:925, 1974.

87. Wyatt, H.V., Colten, H.R., and Borsos, T. Production of the second (C_2) and fourth (C_4) components of guinea-pig complement by single peritoneal cells: Evidence that one cell may produce both components. *J. Immunol., 108*:1609, 1972.

14

Enhancement of Immunological Responses by Methotrexate Pretreatment as a Result of an Eventual Elimination of Suppressor Cells

S. Orbach-Arbouys, M. Castes, and B. M. Berardet

INTRODUCTION

Several chemotherapeutic agents, by destroying normal lymphoid cells together with tumor cells, exhibit immunosuppressive properties. This phenomenon is particularly well documented for cyclophosphamide (CY), which has been used to condition patients for bone marrow grafting and kidney transplantation.

In contrast, several observations have been reported that show that treatment of experimental animals with CY prior to immunization can lead to augmented T-cell responses as measured by increased levels (7,8,11,12,13) and/or prolonged production of T-cells (9) and of delayed-type hypersensitivity to sheep red blood cells (SRBC). Since, in these studies, antibody responses were concomitantly diminished, it was concluded that the decrease in antibody level was responsible for the increased delayed-type hypersensitivity response. However, Askenase (1) has demonstrated a CY-mediated augmentation of a delayed-type hypersensitivity at drug doses that do not influence antibody response. This suggested to him that there exists a population of CY-sensitive suppressor T-cells.

The results reported here show that a variety of immune responses may be enhanced by the administration of another chemotherapeutic drug, methotrexate (MTX). We chose to study MTX because its mode of action differs from that of CY. This is based on the findings of Bruce (2), who compared the sensitivity of normal hemopoietic and transplanted lymphoma colony-forming cells to chemotherapeutic agents administered *in vivo*. He defined two types of drugs—those whose activity was phase specific (e.g., vinblastine, MTX, and cytosine arabinoside) and those whose activity was not phase specific (e.g, nitrosourea and 5-fluorouracil). Since MTX, like CY, selectively destroys dividing cells, we anticipated a selective effect on some, but not all, T-cell responses. We observed that the *in vivo* plaque-forming cell response to SRBC, *in vitro* phytohemagglutinin (PHA) responsiveness, and the *in vivo* graft-versus-host (GvH) reactivity of spleen cells from MTX-treated mice were greater than those found in controls.

MATERIALS AND METHODS
Animals

Specific-pathogen-free $C_{57}Bl/6$ and $(C_{57}Bl/6 \times DBA/2)$ F_1 6- to 8-week old mice were obtained from the breeding center of the Centre National de la Recherche Scientifique in Orleans, France.

Enumeration of Antibody-forming Cells

Antibody-forming cells were assayed by the Cunningham technique (3) on day 6 after the intraperitoneal injection of mice with 0.2 ml of 10 percent SRBC suspension. Results are expressed as number of direct plaque-forming cells per spleen or, when total numbers of spleen cells were enumerated for each individual mouse, as plaque-forming cells per 10^6 spleen cells.

In Vitro Lymphocyte Culture

Pooled spleens from groups of three mice were aseptically removed and gently squeezed between two sterile glass slides. Cell suspensions thus obtained were filtered through gauze and washed in cold sterile tissue culture medium 199.

For subsequent tests, the suspensions were prepared in RPMI-1640 (Eurobio, France), supplemented with 2 mM glutamine (Gibco, Grand Island, N.Y., USA) containing 5 percent fresh human AB serum (decomplemented by heating at 56°C for 30 min), 100 IU/ml penicillin, and 100 mg of streptomycin/ml.

The responsiveness of spleen cell populations to phytohemagglutinin (PHA) (Wellcome Laboratories, England) was tested by placing 5×10^5 cells in 0.25 ml into each well of a falcon 3040 microplate. Optimal stimulation of spleen cells was obtained with the addition of 0.4 mg PHA.

All cultures were carried out in triplicate. The microplates were covered and incubated at 37°C in an atmosphere of 5 percent CO_2 plus 95 percent air for 48 hr, after which 1 μCi of tritiated thymidine ([^{3}H-]TdR; TMM 48, Commissariat a l'Energie Atomique, Saclay, France; specific activity 27 Ci/mmole) was added to each well 5 hr prior to cell harvesting. Cultures were harvested with a multiple automated sample harvester ("MASH"; Microbiological Associates, Bethesda, Md., USA) on glass fiber filters (Reeve Angels, Clifton, N.J., USA). Glass fiber disks containing the radioactive cells were placed in toluene plus omnifluor (NEN, Dreieichanhaim, West Germany), and counted in a Packard Counter. Results are expressed as mean counts per minute (CPM) of triplicate samples ± standard deviation.

Assay of DNA Synthesis of Lymphoid Cells in Response to Alloantigens

This technique has been described in detail elsewhere (4,5). Mice (C_{57}Bl/6 × DBA/2)F_1 were irradiated at 950 rad and inoculated intravenously with C_{57}Bl/6 lymphoid cells on day 0. On day 4, at the peak of the reaction, the mice were injected intraperitoneally with 2 μCi of 5-^{125}iodo-2-deoxyuridine (IUDR; Radiochemical Centre, Amersham, England). The DNA incorporation of ^{125}IUDR was determined 24 hr later when the spleens of these animals were removed and their radioactivity counted for 2 min in a Packard Autogamma Scintillation Spectrometer. Also counted at this time were standard 0.2 ml samples of IUDR from the same solution used for injecting test animals. Results are expressed as mean uptake of IUDR (in percent of injected dose) per spleen ± standard deviation.

BCG

When necessary, mice were inoculated intravenously with 3 mg of BCG, kindly provided by the Pasteur Institute (Paris, France) in the form called "Immuno BCG," which retains 95 percent viability after 3 months' storage at 4°C.

Obtainment of Purified T-Cells from Spleen

We employed the technique described by Julius (6) using nylon wool columns.

RESULTS

Influence of MTX Administration on the Plaque-forming-cell Response to SRBC

Mice were injected intraperitonally with 0.5 mg of MTX 12, 5, or 2 days before intraperitoneal injection with 0.2 ml of a 10 percent SRBC suspension. The direct plaque-forming cells per spleen were enumerated 6 days after antigen administration. As seen in Table 14-1, MTX treatment enhances the splenic plaque-forming-cell response to SRBC when given 5 days before antigen administration ($p = 0.02$). Since this also represents the maximum effect, all subsequent experiments were carried out using this time schedule.

We have previously reported (10) that the spleen cell reactivity of mice given high doses of BCG is lower than normal, and that this observation may be attributed to the development of suppressor cells. Thus, it was interesting to determine whether MTX could modify the response of the BCG-treated animals.

Mice injected with 3 mg of BCG have a greater number of plaque-forming cells per spleen than do normal mice, but when their spleen reactivity is

TABLE 14-1 Enhancement of the Splenic Plaque-forming-cell Response to SRBC by the Injection of 0.5 mg of Methotrexate 5 Days before SRBC

PROTOCOL	DIRECT SPLENIC PLAQUE-FORMING-CELL RESPONSE[a]
MTX day-12	$57,600 \pm 9,172$
MTX day- 5	$87,767 \pm 29,914$
MTX day- 2	$42,867 \pm 18,642$
Controls	$45,767 \pm 19,380$

[a]Mean ± standard deviation.

TABLE 14-2 Enhancement of the Splenic Plaque-forming-cell Response to SRBC of Normal or BCG-treated Mice by the Injection of 0.5 mg of methotrexate 5 days before SRBC

PROTOCOL	PLAQUE-FORMING CELLS/ SPLEEN[a]	NUMBER OF CELLS/SPLEEN	PLAQUE-FORMING CELLS/ 10^6 CELLS
Nil	$38,000 \pm 8,560$	3×10^7	1,266
MTX, day 5	$30,300 \pm 5,900$	6×10^7	505
BCG, day 14 (3 mg)	$83,520 \pm 7,300$	14×10^7	596
BCG, day 14 (3 mg) MTX, day 5	$197,320 \pm 34,450$	20×10^7	986

[a]Mean ± standard deviation.

evaluated per 10^6 cells, their response is one-half that of normal. The MTX treatment dramatically enhances the response in BCG-treated animals, be it evaluated as plaque-forming cells per spleen or per 10^6 cells (Table 14-2).

It must be emphasized that there is a certain amount of variability in the response of normal mice after MTX injection. In our hands it seems that the older the mice, the greater the enhancement by MTX of their response to SRBC.

Enhancement of the PHA Responsiveness of Spleen Cells after MTX Injection

As can be seen in Table 14-3, the PHA responsiveness of spleen cells from mice given 0.5 mg of MTX 5 days before testing is greater than normal. Similarly, MTX treatment greatly enhanced the PHA responsiveness of spleen cells from mice previously injected with BCG. The response of spleen cells to PHA from mice given BCG only was 50 percent of that found for spleen cells from normal mice.

Enhancement of Graft-versus-host Reactivity by Administration of MTX to the Cell Donor

Graft-versus-host (GvH) reactions were induced in lethally irradiated $(C_{57}Bl/6 \times DBA/2)F_1$ mice by the intravenous injection of 10^7 spleen or 5×10^6 nylon-purified T-cells from normal or MTX-treated $C_{57}Bl/6$ mice. In both MTX and control groups, the cell yield per spleen was fairly comparable, so equal numbers of cells were injected into the mice.

TABLE 14-3 Phytohemagglutinin Responsiveness of Lymphoid Cells after Methotrexate Injection

PROTOCOL	5×10^5 CELLS CULTIVATED	
	Alone	With PHA
(Controls)	549 ± 196	$13,408 \pm 710$
MTX, day 5	$2,799 \pm 531$	$27,246 \pm 2,559$
BCG, day 14 (3 mg)	$5,588 \pm 722$	$7,831 \pm 987$
BCG, day 14 (3 mg) MTX day 5	$3,855 \pm 45$	$20,052 \pm 1,672$

As seen in Table 14-4, ^{125}IUDR incorporation in the spleens of mice undergoing a GvH reaction is much greater when the effector cells (spleen or purified T-cells) are taken from MTX-injected donors. Similar results (Table 14-5) were obtained when BCG-treated donors were used. When the GvH reaction was induced using adult thymectomized donors, the same enhancement of the response by MTX treatment was observed (Table 14-6).

DISCUSSION

All the results reported in this chapter indicate that immune reactivity may be enhanced by MTX administration. Methotrexate was selected for this study because of its phase-specific cytotoxicity,

TABLE 14-4 Enhancement of the GvH Reactivity of Spleen Cells 5 Days after the Injection of 0.5 mg of Methotrexate

INJECTED CELLS	^{125}IUDR SPLEEN INCORPORATION (PERCENT)
10^7 spleen cells	0.117 ± 0.009
10^7 MTX spleen cells	0.240 ± 0.091
5×10^6 T-cells	0.052 ± 0.014
5×10^6 MTX T-cells	0.151 ± 0.043

TABLE 14-5 Enhancement of the GvH Reactivity of Spleen Cells from Normal and BCG-treated Mice 5 Days after the Injection of 0.5 mg of Methotrexate

PROTOCOL	PERCENT SPLEEN ^{125}IUDR INCORPORATION ON DAY 4 OF GvH
Normal spleen	0.099 ± 0.032
MTX, day 5	0.214 ± 0.031
BCG, day 7 (3 mg)	0.079 ± 0.024
BCG, day 7 (3 mg) MTX, day 5	0.159 ± 0.033

TABLE 14-6 Enhancement of the GvH Reactivity of Spleen Cells from Normal and Thymectomized Mice 5 Days after the Injection of 0.5 mg of Methotrexate

DONOR	TREATMENT	PERCENT SPLEEN ^{125}IUDR INCORPORATION ON DAY 4 OF GvH
Normal	Nil Spleen cells	0.257 ± 0.075
	MTX Spleen cells	0.376 ± 0.081
	Nil T-cells, spleen	0.197 ± 0.060
	MTX T-cells, spleen	0.323 ± 0.086
Thymectomized	Nil Spleen cells	0.188 ± 0.037
	MTX Spleen cells	0.430 ± 0.160
	Nil T-cells, spleen	0.221 ± 0.043
	MTX T-cells, spleen	0.600 ± 0.061

which allowed us to expect some selective killing efficiency. We thus eliminated a population defined by its MTX sensitivity in relation to its division potential, and we observed that, on a spleen-equivalent basis, the reactivity of the remaining cells was greater than that of the initial suspension. This finding was also confirmed using spleen cells from BCG-treated animals, which very regularly have a depressed reactivity (on a per cell basis) 2 weeks after injection of 3 mg of BCG: after MTX injection, their reactivity returned to normal values.

Several hypotheses may be advanced to account for this observation. The explanation we favor is that the cells that are killed by MTX exert, both in normal and in BCG-treated mice, a suppressive action on the responding cells. However, it is also conceivable that the restoration that takes place after cell destruction favors production of the reactive cell type.

In an attempt to obtain further information about the nature of the MTX-sensitive cells, we induced a GvH reaction with cells from adult thymectomized animals. Spleen cells taken from MTX-injected thymectomized donors were more reactive than cells from noninjected thymectomized donors, indicating that the putative MTX-sensitive suppressor cell is not found in the pool of short-lived T-cells. These results are consistent with those of Feldmann (unpublished), who found that when cells from thymectomized and from Anti-Lymphocyte Serum (ALS)-treated donors were cultured with antigen in double chambers, the suppressive factors were secreted by the cells from thymectomized donors.

The fact that an appropriate combination of BCG and MTX led to a considerable increase in the anti-SRBC response of the intact animal tends to indicate, if our explanation for the mechanism

of the MTX effect is correct, that despite enhancement of BCG of total splenic plaque-forming cells, development of a high degree of immunosuppression has occurred concomitantly. This is in agreement with our previously published findings (10). It may be possible to quantify the magnitude of the suppressor cells response by evaluating the increase induced by MTX injection.

It is difficult with most protocols to enhance GvH reactivity—the use of adjuvants or immunization do not always yield the expected results. Consequently, in spite of the fact that the exact mechanism responsible for the MTX-induced increase in immune reactivity is not known, this empirical observation should, nevertheless, prove useful for those who would combine MTX treatment with immunotherapy.

SUMMARY

A single intraperitoneal injection of 0.5 mg of methotrexate (MTX) has been found to increase the immune reactivity of $(C_{57}Bl/6 \times DBA/2)F_1$ mice. Five days after injection, spleen cells from MTX-treated mice exhibited greater phytohemagglutinin (PHA) responsiveness and graft-versus-host (GvH) reactivity. Mice given SRBC had at this time greater than normal numbers of direct plaque-forming cells. A similar pattern of response to MTX was observed in mice given 3 mg of BCG intravenously 14 days before testing. These results suggest that MTX treatment may lead to a selective destruction of suppressor cells.

ACKNOWLEDGMENT

We should like to thank Dr. Linda Pritchard for her help in the preparation of the manuscript.

This work was supported by INSERM (Institut National de la Santé et de la Recherche Médicale) CRL No 75-5 0961 and Faculte de Medicine No 77, 26.

REFERENCES

1. Askenase, P.W., Hayden, B.J., and Gershon, R.K. Augmentation of delayed-typed hypersensitivity by doses of cyclophosphamide which do not affect antibody responses. *J. Exp. Med.*, 141:697, 1975.
2. Bruce, W.R., Meeker, B.E., and Valeriote, F.A. Comparison of the sensitivity of normal hematopoietic and transplanted lymphoma colony-forming cells to chemotherapeutic agents administered *in vivo*. *J. Nat. Canc. Inst.*, 37:233, 1966.

3. Cunningham, A.J., and Szenberg, A. Further improvement in the plaque technique for detecting single antibody forming cells. *Immunology, 14*:599, 1968.

4. Gershon, R.K., and Hencin, R.S. The DNA synthetic response of adoptively transferred thymocytes in the spleens of lethally irradiated mice. I. Effect of varying antigen and thymocyte doses. *J. Immunol., 107*:1723, 1971.

5. Gershon, R.K., and Liebhaber, S.A. The response of T cells to histocompatibility 2 antigens—dose-response kinetics. *J. Exp. Med., 136*:112, 1972.

6. Julius, M.J., Simpson, E., and Herzenberg, L.A. A rapid method for the isolation of functional thymus-derived murine lymphocytes. *Europ. J. Immunol. 3*:645, 1973.

7. Katz, S.I., Parker, D., and Turk, J.L. B-cell suppression of delayed hypersensitivity reactions. *Nature (Lond), 251*:550, 1974.

8. Lagrange, P.H., MacKaness, G.B., and Miller, T.E. Potentiation of T cell mediated immunity by selective suppression of antibody formation with cyclophosphamide. *J. Exp. Med., 139*:1529, 1974.

9. Neta, R., and Salvin, S.B. Specific suppression of delayed hypersensitivity: The possible presence of a suppressor B cell in the regulation of delayed hypersensitivity; *J. Immunol., 113*:1716, 1974.

10. Orbach-Arbouys, S., and Poupon, M.F. Active suppression of *in vitro* reactivity of spleen cells after BCG treatment. *Immunology* 1977.

11. Poulter, L.W., and Turk, J.L. Proportional increase in the O carrying lymphocytes in peripheral lymphoid tissue following treatment with cyclophosphamide. *Nature New Biol., 283*:17, 1972.

12. Turk, J.L., Parker, D., and Poulter, L.W. Functional aspects of the selective depletion of lymphoid cells by cyclophosphamide. *Immunology, 23*:493, 1972.

13. Turk, J.L., and Poulter, L.W. Selective depletion of lymphoid tissue by cyclophosphamide. *Clin. Exp. Immunotherp., 10*:285, 1972.

15

Syngeneic and Autologous Bone Marrow Transplantation after Noncurative Chemotherapy with AAFC, DMM, and Cyclophosphamide in Tumor-bearing Mice

Gisela Stinner, J. Torhorst, and G. L. Floersheim

INTRODUCTION

Anticancer drugs are toxic to malignant as well as to normal cells. One of the limiting factors in the use of cytotoxic agents is their toxicity to hemopoietic cells. Therefore, autologous bone marrow transplantation is usually carried out to support hemopoiesis in tumor patients after intensive chemotherapy. There is no risk of immunological reactions (graft versus host) and no immunological barrier to engraftment in autologous bone marrow transplantation.

In an earlier experiment, we performed autologous bone marrow transplantation in tumor patients (especially those with Hodgkin's disease) with therapeutic drug-induced prolonged marrow insufficiency (25,26). Surprisingly, however, after bone marrow regeneration, which allowed us to continue our conservative therapeutic regimens, we could observe a marked increase in relapse-free intervals and an increase of the tumor regression rate. These results could not be due to chemotherapy alone, since the same regimen without bone marrow cells had previously proved to be less effective. The effect of grafting autologous bone marrow cells could be reproduced after repeated treatment in the same patients and also in a clinical study.

At first we thought of an effector mechanism related to tumor-specific antigens. In several murine tumor models, curative effects of sublethal chemotherapy and the adoptive transfer of syngeneic lymphoid cells were reported. But lymphoid cells immune to the tumor-associated antigens were required (2,3,4,14,21). In leukemia, thymus-derived lymphocytes were found to be responsible for rejection or destruction of syngeneic tumors (1,8). Other reports describing the rejection of allogeneic as well as syngeneic tumors indicated not only participation of T-cells but also macrophages and B-lymphocytes (5,6,7,15,16,19,20).

In earlier studies, it was demonstrated that disseminated Moloney lymphoma was eradicated by combining a lethal dose of a chemotherapeutic agent with syngeneic bone marrow and spleen cell injections (9). The present investigation was designed to reproduce similar synergistic effects between syngeneic bone marrow grafting and noncurative chemotherapy in an animal model and to determine the type of cell responsible for tumor cell destruction under these conditions.

To find some evidence for the immunospecificity of the therapy, we infused animals bearing progressively growing syngeneic antigenic tumors with syngeneic nonimmune cells and cells immune to tumor antigens, after chemotherapy and without chemotherapy.

MATERIALS AND METHODS

Mice

Female, 8- to 10-week-old BDF_1 mice ($C_{57}Bl/6 \times DBA/2)F_1$ and DBA/2 mice, male, 2- to 3-month-old BALB/c mice, and female, 3- to 5-month-old CBA mice were used in experiments with leukemia L-1210, fibrosarcoma METH-A, and Moloney lymphoma, respectively. All mice were purchased from Bomholtgard, RY, Denmark, and maintained on commercial feed. For 2 weeks after the administration of cytotoxic agents, polymycin B (10 mg/liter) and neomycin sulfate (100 mg/liter) were added to the drinking water.

Drugs

Anhydro-arabinofuranosyl-fluorocytosine hydrochloride (AAFC), an antimetabolite, was produced by J. J. Fox (Sloan Kettering Institute, N.Y.) (11) and generously supplied by Dr. Bollag (Hoffmann-La Roche Ltd., Basle, Switzerland). A dimethyl homologue of Myleran, DDM, was obtained from Dr. H. B. Wood, Jr. (National Cancer Institute, Bethesda, Md.). Cyclophosphamide (CY) was purchased from ASTA, Brackwede, FGR.

The AAFC and CY were dissolved in distilled water and the DMM in warm ethanol as an aqueous solution (12.5 percent ethanol in isotonic saline). All drug solutions were made up shortly before use and injected by the intraperitoneal (i.p.) route in a volume of 0.2 ml/10 g of body weight in noncurative and nonlethal dose.

Tumors

The following transplantable murine tumors were used: leukemia L-1210, an antigenic tumor growing progressively only in DBA/2 mice and F_1 hybrids. The L-1210 used in this study was maintained by transplantation (i.p.) over approximately 18 years in adult BDF_1 and DBA/2 mice, respectively. To obtain fragments of solid leukemic tumors, we inoculated 10^6 ascites cells (day 8 after i.p. tumor inoculation) subcutaneously (s.c.) into healthy mice, which were sacrificed on day 8. The tumors were removed, pooled, and cut in fragments containing approximately 3×10^6 leukemic cells. Two fragments were then transplanted (s.c.) into new recipient BDF_1 and DBA/2 mice.

The Moloney lymphoma was maintained in CBA mice. It was derived from lymphoma line YBA maintained at the Department of Tumor Biology, Karolinka Institutet, Stockholm (Prof. G. Klein). Tumor-specific antigens have been demonstrated in this lymphoma (18). A methylcholanthrene-induced fibrosarcoma (METH-A) in BALB/c mice was obtained from the Max Planck Institute of Immunobiology, Freiburg, FGR (Prof. Westphal). This tumor is highly immunogenic, with a 100 percent incidence of takes. On day 7 after tumor implantation, 3×10^5 ascites cells were injected into BALB/c mice (s.c.) to obtain solid measurable tumors.

All tumor fragments were implanted subcutaneously by trocar to a ventrolateral site at the lower margin of the thoracic cage. Local tumor growth was observed by measuring the two greatest diameters with calipers.

PREPARATION OF CELL INOCULA

Bone marrow cells for syngeneic grafts were prepared by flushing out the marrow from both femurs and tibia. The spleen, thymus, lymph node, and fetal liver (day 12 to 17 of pregnancy) were removed and each organ pooled and minced. Cells were prepared by gently pressing fragments through a stainless steel mesh. The cell suspension was then filtered through cotton wool. To obtain peritoneal exudate cells, mice were sacrificed and injected (i.p.) with 6 ml of medium, the abdomen was massaged and the cell fluid was withdrawn into a tube, spun, resuspended, and counted. All cell preparations were suspended in TC 199 tissue culture medium (DIFCO, Lab.), washed in the medium, and centrifuged at $250 \times g$ for 10 min. The number of viable cells was determined by Trypan Blue exclusion (95 to 97 percent) and concentrated in TC 199.

TREATMENT OF BONE MARROW CELLS WITH ANTI-THY 1.2 SERUM PLUS COMPLEMENT

Anti-Thy 1.2 serum was provided by Dr. H. von Boehmer, Institute of Immunology, Basle, Switzerland. The serum killed 99 percent of C_3H thymocytes at a dilution of 1:256 and a cell concentration of 10^7 cells/ml. It was found to abrogate T-cell-mediated cytotoxicity *in vitro* as well as T-helper function *in vivo*. B-lymphocytes were not affected.

Aliquots of 2.5×10^7 bone marrow cells were incubated for 30 min at 0°C in 1 ml anti-Thy 1.2 serum (final dilution 1:4). The cells were then washed and incubated for 40 min at 37°C, with 1 ml guinea pig complement (final dilution 1:4).

IMMUNIZATION PROCEDURES

The BDF_1 mice were inoculated (i.p.) with 10^7 x-irradiated L-1210 cells in 0.5 ml TC 199. Their marrow cells or macrophages were used 6 and 15 days after immunization.

X-IRRADIATION

Tumor cells were irradiated with 6000 R, at a rate of 100 R/min, using a Muller X-ray apparatus (250 kVp, 30 ma; Filter 0.2 mm copper). Cell tubes were placed at a target distance of 60 mm.

STATISTICAL ANALYSIS

The significance of the results was determined by the Wilcoxon Rank Test.

EXPERIMENTAL DESIGN

Experimental BDF_1 and CBA mice (L-1210 and Moloney YBA lymphoma) were inoculated (s.c.) on day 0 with two fragments of solid tumors containing about 3×10^6 tumor cells. After randomization, the mice were injected (i.p.) on day 1 with a noncurative dose of AAFC, DMM, or CY followed by transplantation of bone marrow or fetal liver cells 10 hr (L-1210) or 24 hr (Moloney YBA) later. The BALB/c mice received no chemotherapy, but fetal liver cells were injected on day 7 after s.c. tumor inoculation with 3×10^5 METH-A ascites cells. The tumor volume was measured every second day. In L-1210, all mice were sacrificed on day 6 or 8, i.e., the time of dying of the tumor-bearing control animals. The tumors were excised, and the weights of the tumors of mice treated chemotherapeutically was expressed as a percent of the tumor weight of the untreated control mice. In some experiments, survival time was registered after i.p. inoculation of either 1×10^5 or 5×10^5 leukemic ascites cells. The mice were divided into the following groups after inoculation of tumor cells:

CONTROL, tumor-bearing mice—no treatment
CT, chemotherapy (noncurative dose)—no cell infusion
BMC/FLC, bone marrow cells/fetal liver cells—no chemotherapy
CT + BM/FLC, chemotherapy plus bone marrow cells/fetal liver cells

The syngeneic transplantation of hemopoietic cells were modified by different kind of cells:

BMC, bone marrow cells of normal donors (spleen-m lymph node-cells)
BMC:Immun., bone marrow cells of immunized donors
BMC (a θ + C^1), bone marrow cells treated with anti-Thy 1.2 serum + complement
PEC[Immun.], peritoneal exudate cells of immunized donors

B-LY, B-lymphocytes of normal nude mice ($C_{57}BL/6$—congeneic)
FLC, fetal liver cells (day 12 to 17 of pregnancy)

RESULTS

Chemotherapy and Syngeneic Bone Marrow Transplantation

Adult female BDF_1 mice were inoculated (s.c.) with approximately 3×10^6 leukemic cells on days 0. On day 1, the mice received a noncurative dose of AAFC (500 mg/kg, i.p.) and were injected 10 hr later with 3×10^7 syngeneic bone marrow cells (i.v.), nonimmune or immune to L-1210. Tumor weights were recorded on day 8, and the tumor regression or tumor growth calculated.

The results are summarized in Table 15-1. Control mice without treatment showed progressive tumor growth. Chemotherapy alone had an antitumor effect on nearly 60 percent of the tumors. If bone marrow cells of immunized donors were given as an adjunct to chemotherapy, a significant increase of tumor regression was observed. Surprisingly, bone marrow cells of nonimmunized donors also had an antitumor effect. In fact, there was no significant difference between the marrow cells from the two different groups of donors. With syngeneic marrow cells (day 1 after tumor inoculation) a 60 percent increase in the median survival time was observed. Clear-cut additive effects between chemotherapy and subsequent transfer of 3×10^7 non-

TABLE 15-1. Subcutaneous Growth of L-1210 Cells in BDF Mice Engrafted Either with Syngeneic Bone Marrow Cells or Fetal Liver Cells Subsequent to Therapy with AAFC[a]

TREATMENT	PERCENT TUMOR REGRESSION[b,c]	
None	—	(68)
AAFC	59.1	(63)
AAFC + bone marrow	89.5	(35)
AAFC + immunized bone marrow	95.2	(17)
— bone marrow	37.8	(24)
— Immunized bone marrow	64.2	(15)
None	—	(23)
AAFC	53.2	(28)
AAFC + fetal liver cells	89.4	(29)
— Fetal liver cells	79.2	(6)

[a]Adult BDF_1 mice were inoculated (s.c.) on day 0 with 3×10^6 leukemic cells. On day 1 after tumor cell implantation, some of the mice were injected with 500 mg/kg AAFC; 10 hr after either experimental or control treatment, the mice were engrafted (i.v.) with either 3×10^7 syngeneic bone marrow cell or fetal liver cells.

[b]The tumor regression in experimentally treated mice was expressed as a percentage of the mean tumor weights of control mice.

[c]Numbers in parentheses indicate the number of mice used in each group.

immune syngeneic bone marrow cells were also observed in several other systems: The L-1210 ascites tumor in AAFC-treated BDF$_1$ and DBA/2 mice receiving bone marrow cells by the i.p. route (data not shown) and the subcutaneous growing Moloney YBA lymphoma in CBA mice receiving a single dose of either 200 mg/kg cyclophosphamide or 12.5 mg/kg DMM (i.p.) followed by intravenous transfusion of bone marrow cells [Fig. 15-1(A,B)]. In the case of the cyclophosphamide-treated animals, the tumor partially regressed, and this was which was most marked by day 12, whereas in mice receiving combination therapy with bone marrow cells, the tumors regressed rapidly and completely, though only a temporary remission occurred. The antitumor effect of DMM—a very mild cytostatic drug characterized by a slight immunosuppressive activity (10)—was significantly increased in combination with cells that were not immune to lymphoma YBA.

Bone Marrow Transplantation after Treatment with Anti-Thy 1.2 Serum plus Complement

To determine if the antitumor effect was mediated by T-lymphocytes, bone marrow cells of unimmunized and immunized donors were treated with anti-Thy 1.2 serum and complement prior to transfer. The results are shown in Table 15-2. Bone marrow cells, from which T-cells had been eliminated, were still effective against the L-1210 leukemia. There was no difference between the additive effect of a normal T-cell-depleted bone marrow cell suspension and the effect of similarly treated cells from immunized donors given in combination with chemotherapy. Thus, the antitumor effect of bone marrow cells can not be attributed solely to thymus-derived lymphocytes.

TABLE 15-2 Subcutaneous Growth of L-1210 Cells in BDF Mice Engrafted with Anti-theta, Serum-treated, Syngeneic Bone Marrow Cells Subsequent to Therapy with AAFC[a]

TREATMENT	PERCENT TUMOR REGRESSION[b]	
None	—	(6)
AAFC	46.0	(6)
AAFC + normal bone marrow (anti-theta + complement)	81.9	(6)
AAFC + immunized bone marrow (anti-theta + complement)	81.9	(6)

[a]Adult BDF$_1$ mice were inoculated (s.c.) on day 0 with 3 × 10^6 leukemic cells. On day 1 after tumor cell implantation, some of the mice were injected with 500 mg/kg AAFC; 10 hr after either experimental or control treatment, the mice were engrafted (i.v.) with either 3 × 10^7 syngeneic bone marrow cell or fetal liver cells.

[b]The tumor regression in experimentally treated mice was expressed as a percentage of the mean tumor weights of control mice.

Chemotherapy and Transplantation of Syngeneic Fetal Liver Cells

Since it appeared possible that the efficacy of the bone marrow grafts was mediated by stem cells, syngeneic fetal liver cells were also used. Indeed, when fetal liver cells were injected following a single dose of AAFC, tumor regression was comparable to that observed with chemotherapy and bone marrow cells in the L-1210 leukemia (Table 15-1). This effect was also observed in mice bearing the Moloney lymphoma after treatment with DMM in combination with fetal liver cells [Figure 15-1(B)].

Syngeneic Transplantation of Bone Marrow Cells and Fetal Liver Cells without Chemotherapy

Without chemotherapy, a remarkable tumor regression could be induced by transfusion of bone marrow or fetal liver cells into mice (Table 15-1). The antitumor effect was also observed in L-1210 ascites tumor-bearing DBA/2 mice injected with a single dose of either 5 × 10^7 fetal liver or bone marrow cells. Repeated treatment of DBA/2 mice with bone marrow cells increased the median survival time (MST) to 76 percent. There was no difference in the survival of mice injected with marrow cells obtained from immune or nonimmune donor mice. Repeated treatment with fetal liver cells led to peritonitis, with a MST of only 10 days (Table 15-3). Very few leukemic ascites cells were found, and the liver and spleen were of normal

TABLE 15-3 Intraperitoneal Growth of L-1210 Cells in DBA/2 Mice Subsequent to Single or Multiple Transfusions of Either Bone Marrow or Fetal Liver Cells[a,b]

TREATMENT	MEDIAN SURVIVAL TIME	
	Days	Percent Prolongation
None	5.8	—
Bone marrow (day 1)	7.6	31
None	7.5	—
Bone marrow (days 1, 2, 4, 6)	13.2	76
Immunized bone marrow (days 1, 2, 4, 6)	13.2	76
Fetal liver cells (day 1)	9.9	32
None	6.9	—
Fetal liver cells (day 1)	10.2	47.9
None	7.5	—
Fetal liver cells (days 1, 2, 4, 6)	10.0	33.3

[a]DBA/2 mice inoculated (i.p.) on day 0 with 1.5 × 10^5 L-1210 ascites cells. On day 1 after tumor cell implantation, the mice were treated with 5 × 10^7 bone marrow with 3 × 10^7 bone marrow or fetal liver cells on days 1, 2, 4, and 6. The results were expressed as the median survival time of untreated, tumor cell-injected mice.

[b]Five mice were used in each group.

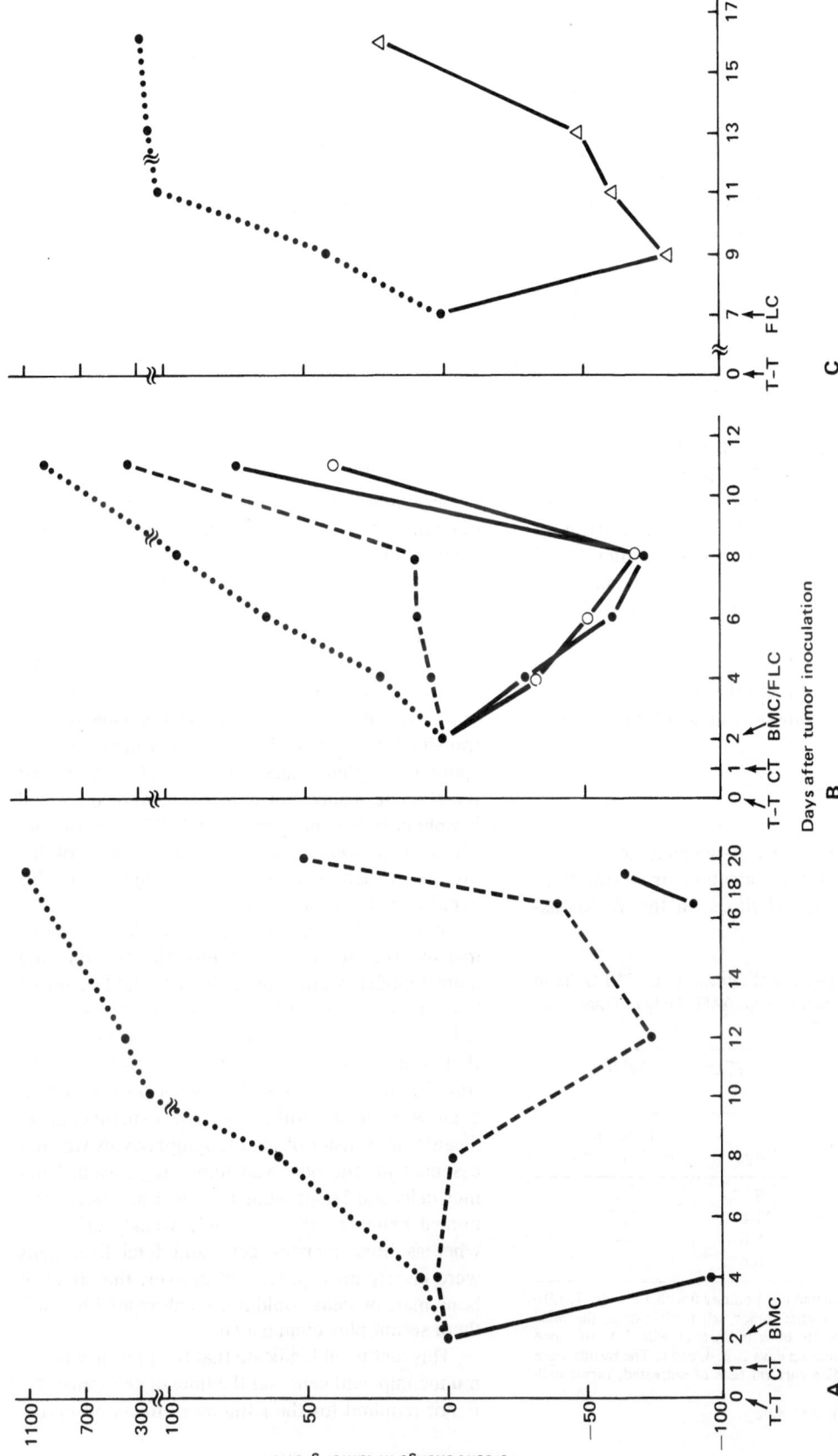

FIGURE 15-1. (A) Adult CBA mice were inoculated (s.c.) on day 0 with approximately 3×10^4 Moloney lymphoma cells. Two groups of these mice were treated on day 1 with 200 mg of cyclophosphamide (CY)/kg (BMC) and one group of mice received 3×10^7 CBA bone marrow cells (BMC) (i.v.) chemotherapy 10 hr later. (B) The same experiment (A) was performed with 12.5 mgCy/kg. In this experiment, an additional group of mice was treated with DMM + fetal liver cells (FLC); bone marrow cells (BMC) (C) Adult BALB/c mice were inoculated (s.c.) with 3×10^5 methylcholonthiene-induced fibrosarcoma cells on day 0. On day 7, these mice were injected with 3×10^7 fetal liver cells (FLC). These mice were not given chemotherapy. The tumor volume was measured every second day. Results are expressed as a percentage change of the control tumor growth. Tumor transplanted (T-T), chemotherapy (CT), control (●.....●), chemotherapy (●—●), chemotherapy + bone marrow cells (●—●), chemotherapy + fetal liver cells (○—○); fetal liver cells (○—○).

weight. Moreover, in two experiments, the MST after fetal liver cells administration without chemotherapy was 30 percent longer than the MST of groups with chemotherapy alone. Reduced tumor dissemination in livers and spleens was observed histologically after treatment with fetal liver cells alone, in comparison to widespread metastases in the control groups.

The antitumor effect was also observed in fibrosarcoma–METH-A-implanted BALB/c mice given no chemotherapy and a single injection of 3×10^7 fetal liver cells i.v. on day 7 after tumor inoculation. A transitory regression of nearly 80 percent was noted [Fig. 15-1(C)]. Thus, the antitumor effect of hemopoietic tissue grafts appears to be mediated by cells present in both bone marrow and fetal liver (Table 15-4).

Effects of Hemopoietic Cells from Various Organs

Cell suspensions from other tissues were examined for their anti L-1210 tumor effect by transfer of peritoneal exudate cells from immunized donors. This was less effective than bone marrow or fetal liver cells. Thymocytes (day 4 after birth) lymph node cells, or spleen cells from nonimmune donors, lymph node cells of congenitally athymic nu/nu-$C_{57}BL/6$ mice, cells of other fetal organs (lung, kidney), as well as the supernatant of killed fetal liver cells (frozen and thawed), were no more effective than chemotherapy alone (Table 15-5).

DISCUSSION

Some studies indicate that T-lymphocytes play an essential role in tumor rejection or destruction. Studies of immunoprophylaxis, in the WINN-as-

TABLE 15-4 Intraperitoneal Growth of L-1210 Cells in DBA/2 after Either Therapy with AAFC Only or Fetal Liver Cell Transplantation Only[a,b]

TREATMENT	MEDIAN SURVIVAL TIME	
	Days	Percent Prolongation
AAFC	9	—
Fetal liver cells	11.8	31.3
AAFC	8.6	—
Fetal liver cells	8.6	0

[a]DBA/2 mice inoculated (i.p.) on day 0 with 1.5×10^5 L-1210 ascites cells. On day 1 after tumor cell implantation, the mice were treated with 5×10^7 bone marrow or with 3×10^7 bone marrow or fetal liver cells on days 1, 2, 4, and 6. The results were expressed as the median survival time of untreated, tumor cell-injected mice.
[b]AAFC:325 mg/kg, effective dose$_{50}$.

TABLE 15-5 Subcutaneous Growth of L-1210 Cells in BDF$_1$ Mice Engrafted with Various Hemopoietic Cells Subsequent to Therapy with AAFC[a]

TREATMENT	PERCENT TUMOR REGRESSION	
None	—	(6)
AAFC	56.8	(6)
AAFC + 5×10^6 sensitized peritoneal cells	69.7	(5)
AAFC + nu/nu lymphocytes	47.2	(5)
AAFC + thymocytes	60.4	(6)
AAFC + fetal organs	49.2	(5)
AAFC + supernatant-killed fetal liver cells	48.6	(5)
None	—	(6)
AAFC	62.8	(6)
AAFC + spleen cells	64.2	(6)
AAFC + lymphocytes	66.1	(6)

[a]See footnotes to Table 15-1.

say, and of the adoptive immunotherapy and chemoimmunotherapy models, to determine the effector cells of syngeneic tumor rejection in vivo, suggest that immune, theta-bearing lymphocytes are essential for this reaction (1,6,7,12,13,17, 22–24,28–31).

Initial clinical studies on tumor patients treated with chemotherapeutic agents and autologous bone marrow transplantation indicated that lymphopoiesis could be stimulated. Repeated bone marrow transfer increased the numbers of lymphocytes responding to phytohemagglutinin, and in the mixed lymphocyte reaction. The mechanism of increased lymphopoiesis is not understood. Whether the observed improvement in the clinical course of the disease in such patients was a consequence of this event or not is also not clear.

In order to analyze better the effect of bone marrow transfer on tumor growth, experimental animal models were chosen. It was established that bone marrow transfer led to tumor regression in the animal systems tested. In addition, it was shown that transfer of syngeneic bone marrow and the transfer of fetal liver cells have a positive effect even without chemotherapy. The experiments indicate that transfer of mature lymphocytes was not essential for the observed tumor regression. Thymus cells and lymph node cells had no effect; peritoneal exudate cells were only weakly effective, whereas bone marrow cells and fetal liver cells were clearly most potent. Moreover, the effect of bone marrow cells could not be abrogated by antitheta serum plus complement.

This fact would indicate that the presence of immunocompetent cells—at the time of cell transfer—is not required for the antitumor effects observed.

This is in agreement with the findings that immunization of the cell donors was not required for tumor regression in the recipients. The present data do not allow identification of the cell type(s) involved in this reaction. However, there appears to exist a positive correlation between the stem cell content of the cell populations transferred and their efficacy in the tumor regression models.

Nevertheless, the fate and function of the cells after transfer, their influence on immune competence and on nonspecific defense mechanism of the host, as well as the role of the chemotherapeutic regimen beyond its possible mere cytotoxic effect, all remain to be determined. Indeed, the synergism between chemotherapy and hemopoietic cell transplantation suggests the possibility that tumor cells exposed to only small doses of cytotoxic drugs become more susceptible to destruction from either donor or host cells. The transfer of hemopoietic cells increase the colony-stimulating activity in host compartments, which might in turn facilitate the proliferation and differentiation of myeloid killer cells capable of tumor destruction.

SUMMARY

In our previous work, autologous bone marrow transplantation was performed as an adjunct to chemotherapy in tumor patients. The clinical course showed a prolongation of relapse-free intervals and an increase of tumor regression after the chemotherapy and bone marrow cells than after the same chemotherapy without bone marrow cells. We had previously reported that a disseminated Moloney lymphoma in CBA mice could be eradicated by a lethal dose of a chemotherapeutic agent in combination with syngeneic hemopoietic cells. The experiments reported here were designed to establish an experimental basis for synergistic effects between noncurative chemotherapy and hemopoietic cells and to determine which cells mediate the antitumor effect in syngeneic bone marrow and fetal liver cell transplantation. Three murine tumor models were used: leukemia L-1210, Moloney YBA lymphoma, and methylcholanthrene-induced fibrosarcoma (METH-A). Adult mice were grafted with a lethal dose of tumor cells on day 0, treated on day one with a noncurative dose of a chemotherapeutic agent followed by transfusion of bone marrow or fetal liver cells 10 hr later. The percentage change of tumor growth was measured, and in the L-1210 tumor, weights were recorded on day 8 and tumor regression was determined. The effects of bone marrow transfer on tumor regression could be reproduced in the other animal tumor models. Fetal liver cells had the same antitumor effect. The synergism between chemotherapy and hemopoietic cell transfer was evident, but surprisingly, the transfer of bone marrow or fetal liver cells had a positive effect even without chemotherapy. The antitumor effect was independent of thymus-derived lymphocytes. Bone marrow cells—immune or not immune to tumor antigens and treated with anti-Thy 1.2 serum + complement—were as effective as untreated cells. The addition of syngeneic cells to the chemotherapy protocol prolonged survival time and reduced tumor dissemination. The present data do not allow identification of the cell type(s) involved in this reaction. However, the stem cell content of the cell populations transferred and their effects on tumor regression seem to be positively correlated.

ACKNOWLEDGMENTS

The authors wish to thank Dr. H. von Boehmer for his interest and advice in the experiment with anti-theta serum. Mrs. Ch. Steiner is acknowledged for her excellent technical assistance.

REFERENCES

1. Berenson, J.R., Einstein, A.B., and Fefer, A. Syngeneic adoptive immunotherapy and chemoimmunotherapy of a friend leukemia: Requirement for T cells. *J. Immunol., 115*:234, 1975.
2. Cerottini, J.C., Nordin, A.A., and Brunner, K.T. Specific *in vitro* cytotoxicity of thymus-derived lymphocytes sensitized to alloantigens. *Nature (Lond), 228*:1308, 1970.
3. Cerottini, J.C., Nordin, A.A., and Brunner, K.T. *In vitro* cytotoxic activity of thymus cells sensitized to alloantigens. *Nature (Lond)., 227*:72, 1970.
4. Einstein, A.B. Jr., Fass, L., and Fefer, A. Suppression of secondary cellular immunity to a tumor allograft by cyclophosphamide and 1,3-*Bis*(2-chloroethyl)-1-nitrosurea. *Cancer Res., 35*:492, 1975.
5. Evans, R., and Alexander, P. Mechanism of immunologically specific killing of tumour cells by macrophages. *Nature (Lond), 236*:168, 1972.
6. Evans, R., and Alexander, P. Cooperation of immune lymphoid cells with macrophages in tumour immunity. *Nature (Lond), 228*:620, 1970.
7. Fefer, A., McCoy, J.L., Perk, K., and Glynn, J.P. Immunologic, virologic, and pathologic studies of regression of autochthonous Moloney sarcoma virus-induced tumors in mice. *Cancer Res., 28*:1577, 1968.
8. Fefer, A., Einstein, A.B., Cheever, M.A., and Berenson, J.R. Models for syngeneic adoptive chemoimmunotherapy of murine leukemias. *Ann. N.Y. Acad. Sci., 276*:573, 1976.
9. Floersheim, G.L. Treatment of Moloney lymphoma with lethal doses of dimethyl-myleran combined with injections of haemopoietic cells. *Lancet, i*:228, 1969.

10. Floersheim, G.L., and Ruszkiewicz, M. Bone-marrow transplantation after antilymphocyte serum and lethal chemotherapy. *Nature (Lond), 222*:854, 1969.

11. Fox, J.J., Falco, E. A., Wempen, J., Pomeroy, D., Dowling, M.D., and Burchenal, J.H. Oral and parenteral activity of 2,2'-anhydro-1-*b*-D-arabinofuranosyl-5-fluorocytosin against both intraperitoneally and intracerebrally inoculated mouse leukemia. *Cancer Res., 32*:2269, 1972.

12. Friedman, L.R., Cerottini, J.C., and Brunner, K.T. *In vivo* studies of the role of cytotoxic T cells in tumor allograft immunity. *J. Immunol., 109*:1371, 1972.

13. Gorczynski, R.M. Evidence for *in vivo* protection against murine-sarcoma virus-induced tumors by T lymphocytes from immune animals. *J. Immunol., 112*:533, 1974.

14. Grant, C.K., Evans, R., and Alexander, P. Multiple effector roles of lymphocytes in allograft immunity. *Cell. Immunol., 8*:136, 1973.

15. Grant, C.K., Currie, G.A., and Alexander, P. Thymocytes from mice immunized against an allograft render bone marrow cells specifically cytotoxic. *J. Exp. Med., 135*:150, 1972.

16. Heberman, R.B., Nunn, M.E., Lavrin, D.H., and Asofsky, R.J. Effect of antibody to O antigen on cell-mediated immunity induced in syngeneic mice by murine sarcoma virus. *Nat. Cancer Inst., 51*:1509, 1973.

17. Howell, S.B., Dean, J.H., Esber, E.C., and Law, L.W. Cell interactions in adoptive immune rejection of a syngeneic tumor. *Int. J. Cancer, 14*:662, 1974.

18. Klein, G., Clifford, P., Klein, E., and Stjernswärd, J. Search for tumor-specific immune reactions in Burkett lymphoma patients by the membrane immunofluorescence reaction. *Proc. Nat. Acad. Sci. (USA), 55*:1628, 1966.

19. Lamon, W.E., and Wigzell, H., Klein, E., Andersson, B., and Skurzak, H.M. The lymphocyte response to primary Moloney sarcoma virus tumors in BALB/c mice. Definition of the active subpopulations at different times after infection. *J. Exp. Med., 137*:1472, 1973.

20. Plata, F., Gomard, E., Leclerc, J.C., and Levy, J.P. Comparative *in vitro* studies on effector cell diversity in the cellular immune response to murine sarcoma virus (MSV)-induced tumors in mice. *J. Immunol., 112*:1477, 1974.

21. Rouse, B.T., Wagner, H., and Harris, A.W. *In vivo* activity of *in vitro* immunized lymphocytes. I. Tumor allograft rejection mediated by *in vitro* activated mouse thymocytes. *J. Immunol., 108*:1353, 1972.

22. Rouse, B.T., Röllinghoff, M., and Warner, N.L. Anti-θ-serum induced suppression of the cellular transfer of tumour specific immunity to a syngeneic plasma cell tumour. *Nature (New Biol.), 238*:116, 1972.

23. Rouse, B.T., Röllinghoff, M., and Warner, N.L. Tumor immunity to murine plasma cell tumors. II. Essential role of T lymphocytes in immune response. *Eur. J. Immunol., 3*:218, 1973.

24. Röllinghoff, M., and Warner, N.L. Specificity of *in vivo* tumor rejection assessed by mixing immune spleen cells with target and untreated tumor cells (37688). *Proc. Soc. Exp. Biol. Med., 144*:813, 1973.

25. Stinner, G., Rose, A., and Steenbeck, L. Untersuchungen über die Funktionen der Granulopoese nach zytostatischer Therapie und Knochenmarktransfusionen. *Folia Haemat., 91*:318, 1969.

26. Stinner, G., Boden, B., and Fünfhausen, G. Serologische Untersuchungen nach homologer Knochenmarktransfusion. *Folia Haemat., 92*:326, 1970.

27. Stinner, G., and zur Horst-Meyer, H. Immunologische Untersuchungen in der Lymphozytenkultur nach Strahlenintensivtherapie oder hochdosierter zytostatischer Stossbehandlung und Knochenmarktransplantationen. *Z. Ges. Inn. Med., 24*:(Suppl):116, 1969.

28. Tevethia, S.S., Blasecki, J.W., Waneck, G., and Goldstein, A.L. Requirement of thymus-derived O-positive lymphocytes for rejection of DNA virus (SV 40) tumors in mice. *J. Immunol., 113*:1417, 1974.

29. Winn, H.J. Immune mechanism in homotransplantation. II. Quantitative assay of the immunologic activity of lymphoid cells stimulated by tumor homografts. *J. Immunol., 86*:228, 1961.

30. Zarling, J.M., and Tevethia, S.S. Transplantation immunity to simian virus -40-transformed cells in tumor-bearing mice. II. Evidence for macrophage participation at the effector level of tumor cell rejection. *J. Nat. Cancer Inst., 50*:149, 1973.

31. Zighelboim, J., Bonavida, B., and Fahey, J.L. Heterogeneous populations of cytotoxic cells in the peritoneal cavity of BALB/c mice with allogeneic EL 4 leukemia cells. *Cell. Immunol., 12*:280, 1974.

PART IV

Therapeutic Approaches to Aplastic Anemia

B. Speck

16

On the Pathogenesis and Treatment of Aplastic Anemia

B. Speck, P. Cornu, C. Nissen, P. Groff, W. Weber, and M. Jeannet

INTRODUCTION

Progress in the treatment of aplastic anemia has been hampered by our lack of understanding of its pathogenesis. Even with maximal supportive care and the use of anabolic steroids, aplastic anemia continues to take an almost uniformly fatal course in its severe form (SAA) (3,5). Results of bone marrow grafts from siblings matched at the major histocompatibility complex (MHC) aiming at complete hemopoietic chimerism have been considered to be the treatment of choice for SAA (3,5,24). It was assumed that the primary defect is located at the level of the pluripotent hemopoietic stem cell itself (24). More recently, there has been rapidly increasing direct and indirect evidence that autoimmune mechanisms directed against hemopoietic precursor cells may be operative in SAA (1,2,6,8,17–20,23,25).

We have studied some basic aspects of benzene-induced marrow aplasia in rabbits and mice. Studies were then extended to experimental treatment with histoincompatible marrow grafts and their clinical application.

MATERIAL AND METHODS

Marrow aplasia was induced in 20 Dutch rabbits weighing 1.5 to 2 kg by daily injections of 0.2 ml of pure benzene/kg (s.c.). The concentration of their peripheral cells was ascertained by complete blood counts twice weekly. Two weeks after the first injection, radioautographic studies were performed *in vivo* in 10 animals using [^{3}H]-thymidine and in 10 animals using [^{3}H]-cytidine, as previously described (10). At the same time, cytogenetic studies of marrow cells were performed using the slightly modified short-term culture method of Moorhead et al. (13).

Ten CBA mice were given the same dose of benzene. Complete blood counts were performed before injection, and at 1 and 2 weeks after injection. The number of colony-forming units (CFU-s) was determined at 2 weeks by transferring either 1/250 or 1/500 of the marrow cells of one femur into syngeneic recipients that were exposed to 750 rad total-body irradiation (TBI).

Sixty rabbits of a cross between Vienna white and Alaska weighing 1 to 2 kg were given daily (s.c.) injections of benzene for a period of 16 to 20 weeks (0.5 to 1.0 ml/kg) in an attempt to produce permanent marrow aplasia. Forty animals of the same strain received a single (i.v.) injection of 1.4 mC of ^{32}P/kg. At the nadir of pancytopenia and marrow hypoplasia, the animals were assigned to one of four study groups: (i) 3 ml/kg of s.c. antilymphocyte serum (ALS) injected on 4 successive days followed by a histoincompatible rabbit marrow transfusion. (ii) The same dose of ALS, no marrow. (iii) No ALS, administration of histoincompatible marrow only. (iv) Controls, not any treatment. For all experiments, the same batch of equine antirabbit ALS was used (21,22).

Twenty-six Dutch rabbits were sensitized against Burgundy rabbits by a skin allograft and by twice-weekly blood transfusions from the prospective marrow donor for 3 weeks. They were then given a 1,200 rad TBI midline tissue dose from a ^{60}Co source at a dose rate of 30 rad/min in a single session. Four groups of animals were examined: (i) Immunization + radiation + Burgundy rabbit marrow. (ii) Immunization + radiation. (iii) Radiation + Burgundy rabbit marrow. (iv) Radiation alone.

Twenty-four patients with SAA (3,5) were treated according to one of the three following protocols: (i) Commercially available equine antihuman thoracic duct ALG (4 × 40 mg/kg, i.v.) followed by one HLA haplotype-mismatched, MLC-positive family marrow (N =9). (ii) The same ALG, same dose, no marrow (N = 5). (iii) Donor buffy coat followed by cyclophosphamide (CY) (4 × 50 mg/kg, i.v.) and MHC-matched marrow (N = 10). The three groups were virtually identical in all factors known to influence the prognosis of SAA and the outcome of hemopoietic grafts.

RESULTS

Pancytopenia was seen in all animals. Within two weeks, median granulocyte counts dropped from 3,600 to 120/mm³, reticulocytes from 50,000 to 600/mm³, and platelets from 450,000 to 95,000/mm³. Radioautographs of the marrow cells at 2 weeks are illustrated in Figure 16-1. Labeling indices were only slightly subnormal at the level of the pronormoblast but markedly decreased at the basophilic normoblast and in the myeloid precursors, indicating a profoundly disturbed proliferation and maturation. In the marrow, the most striking cytogenetic aberrations at 2 weeks were breaks and gaps that were observed in all animals. Polyploid mitoses were seen in one animal (10).

The pancytopenia in CBA mice was comparable to that of rabbits. The number of colony-forming units of the animals with benzene-induced pancytopenia and normal controls were virtually identical (Table 16-1).

Figure 16-2 shows the survival data of 52 rabbits living for more than 10 days after induction of aplasia. In group 1, 18 out of 22 rabbits were long-term survivors (81.9 percent). By means of genetic markers, we were able to document that, for a period of 6 weeks to 6 months, these rabbits were split chimeras. Marrow cells and B-lymphocytes were predominantly of the donor type, while T-lymphocytes remained of host type explaining the

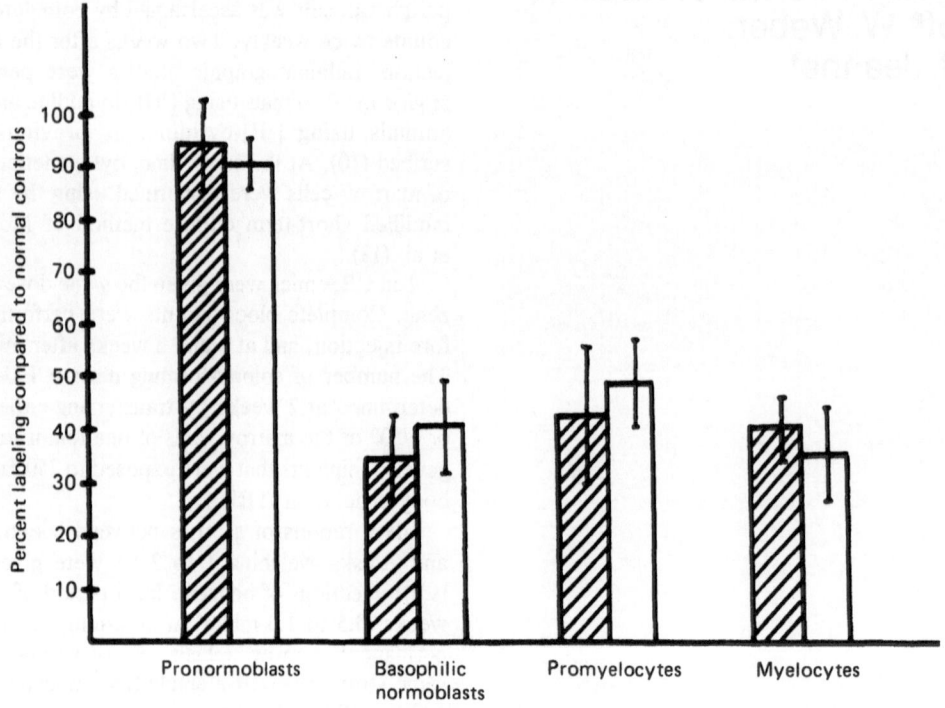

FIGURE 16-1. Radioautographic studies of rabbit marrow cells after 2 weeks' exposure to benzene. Mean values and standard deviations of 10 animals are expressed in percent of normal controls. Labeling by [³H] thymidine (■); by [³H] cytidine (□).

TABLE 16-1 Colony-forming units (spleen) in CBA mice. Mean values and standard deviations of 10 normal animals after 2 weeks' exposure to benzene. Transfusion of 1/250 or 1/500 of a femur bone marrow into syngeneic recipients exposed to 750 rad TBI.

FEMUR BONE MARROW	NORMAL ANIMALS	ANIMALS WITH BENZENE APLASIA
1/250	69±5	72±4
1/500	35±4	37±4

absence of graft-versus-host disease (GvHD). Split hemopoietic chimerism was regularly followed by complete autologous reconstitution. We were unable to induce marrow aplasia that was uniformly fatal in the control animals, yet it is obvious that animals on ALS alone or on bone marrow alone do poorly.

Immunization against the marrow donor followed by 1,200 rad TBI and transfusion of histoincompatible cells from Burgundy rabbits results in two distinctly different patterns of all repopulation. Eight animals repopulated their marrow and normalized their peripheral blood counts. Three of them died within 4 weeks from infectious complications, and five became long-term survivors. In every instance, these reconstitutions could be shown to be autologous by genetic markers, and GvHD was never seen. Another eight animals in this group died without any evidence of hemopoietic reconstitution. Immunization alone did not protect against radiation, and there was no difference in survival compared to the radiation controls. All animals given TBI and marrow without previous immunization died from GvHD (Figure 16-3).

At the present time, five out of nine patients in group 1 are in complete remissions (CR), from over 10 months to over 4½ years. Two patients showed a transient hemopoietic improvement, whereas in two others, no amelioration was seen (Figure 16-4). These four died of hemopoietic failure. In the five successful cases, we were unable to document hemopoietic chimerism, and GvHD was never

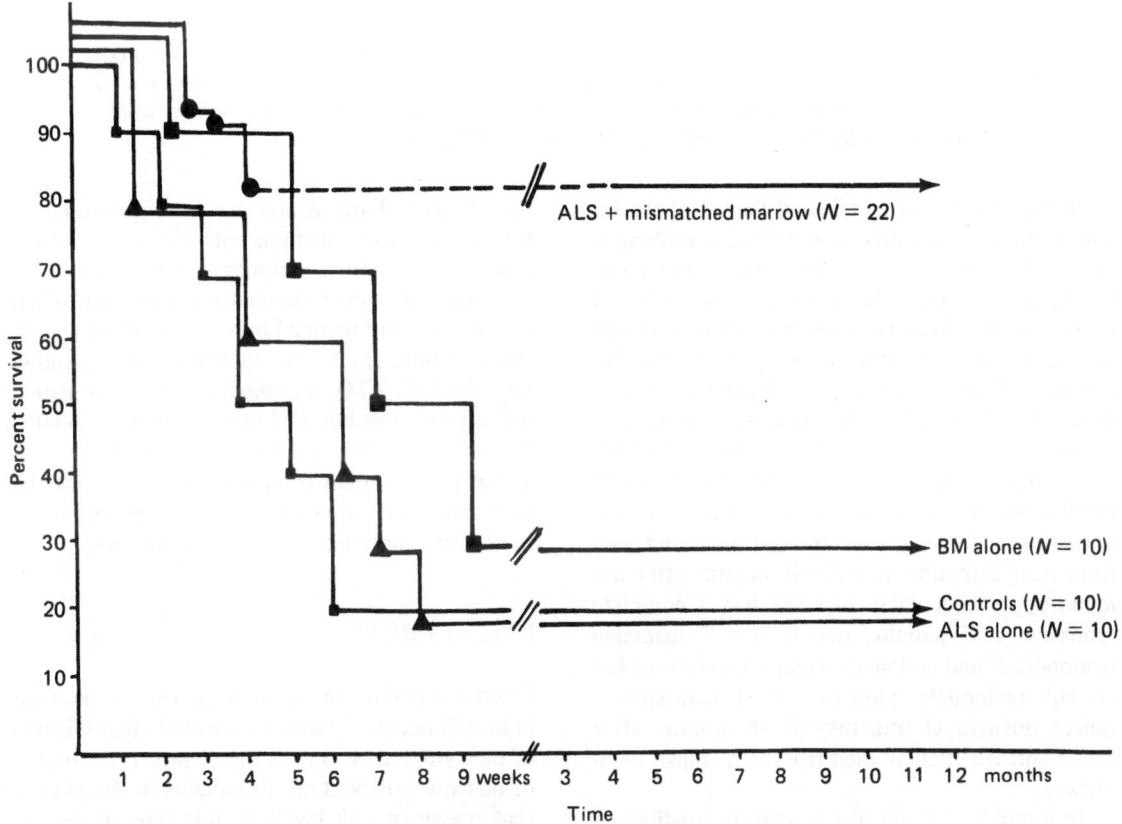

FIGURE 16-2. Survival of rabbits with benzene- and ^{32}P-induced marrow aplasia treated with ALS plus transfusion of histoincompatible marrow, and three control groups. Animals receiving a bone marrow cell transfusion have a significantly higher rate of survival ($p < 0.0005$).

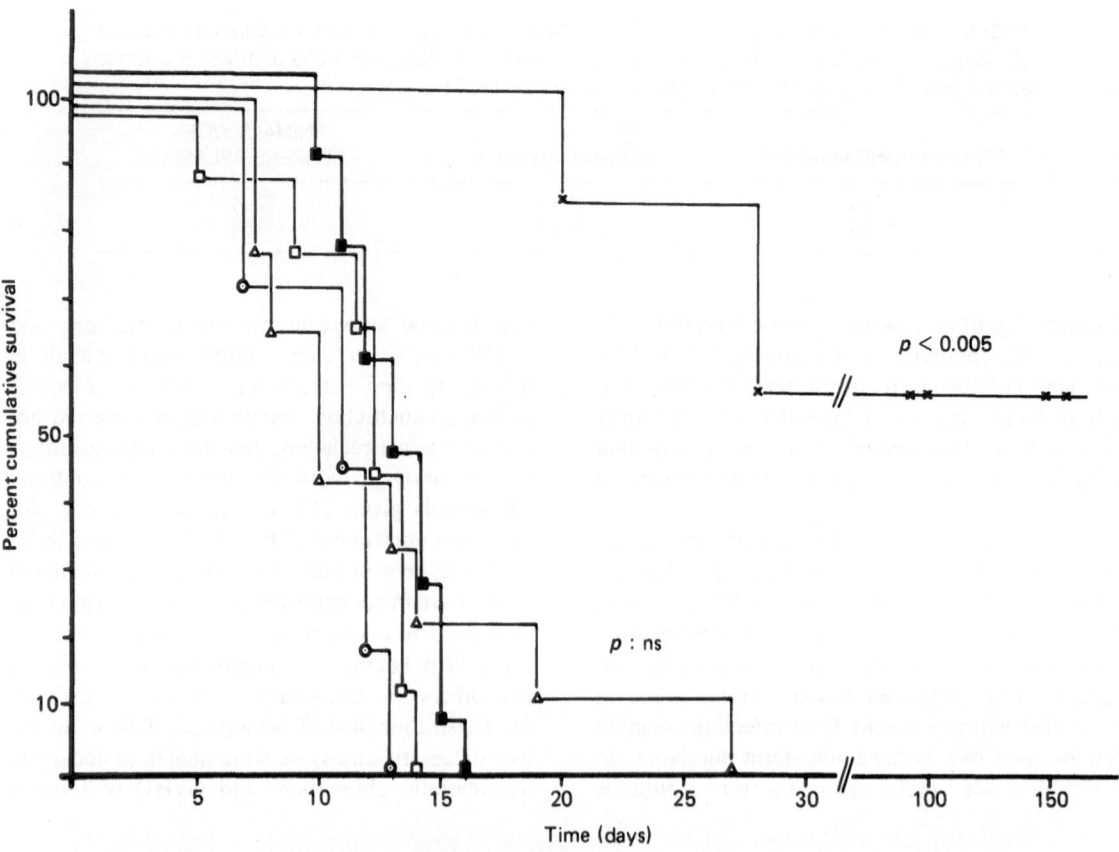

FIGURE 16-3. Survival of rabbits after 1,200 rad TBI midline tissue dose (MTD). Only the animals previously sensitized to donor antigens and given a marrow graft became long-term survivors (x). Their marrow recoveries were autologous. Radiation controls (~), immunized radiation controls (■) and animals with marrow grafts but no previous immunization (△) do not show a significant difference in survival rate. Animals immunized without reconstitution (□).

seen. Genetic markers indicated that all these reconstitutions were autologous. It is of considerable interest to note that all five successful patients had been given androgen therapy before the graft and proved to be refractory. After immunosuppression and transfusion of marrow, they became responsive and two of them even became dependent on low doses of androgen. Complete remissions were seen in three of the idiopathic cases, one probably chloramphenicol induced and one posthepatitic SAA. Of the five patients in group 2 treated with ALG followed by androgens, four are still alive and well from over 2 months to over 18 months after immunosuppression. Only one has had a complete remission at 7 months, two have self-sustaining hemopoiesis and no longer require transfusions but are still moderately cytopenic. One patient still requires occasional transfusions 18 months after treatment. One patient died from intracranial hemorrhage.

In group 3, five out of ten patients conditioned with CY became long-term survivors. They are at home leading a normal life, from over 10 months to over 3 years after treatment. Three patients are

hemopoietic chimeras and two have recovered autologous marrow function, both after clear-cut evidence of early transient donor marrow engraftment. Two patients never showed any evidence of engraftment nor of resumption of autochthonous marrow function. Both were regrafted after conditioning with CY, ATG, and procarbazine. One patient had a good graft but died of interstitial pneumonia, the other one rejected the second graft as well. Three patients died from bacterial infections that were refractory to combined broad-spectrum antibiotics and daily granulocyte transfusions.

DISCUSSION

From our studies on the pathogenesis of benzene-induced marrow aplasia, we conclude that this type of pancytopenia is due to the disturbed maturation of marrow cells and not to a defect at the pluripotent precursor cell level. In this type of marrow aplasia, and in ^{32}P-induced aplasia, it is possible to keep over 80 percent of the experimental rabbits alive after ALS conditioning and transfusion of

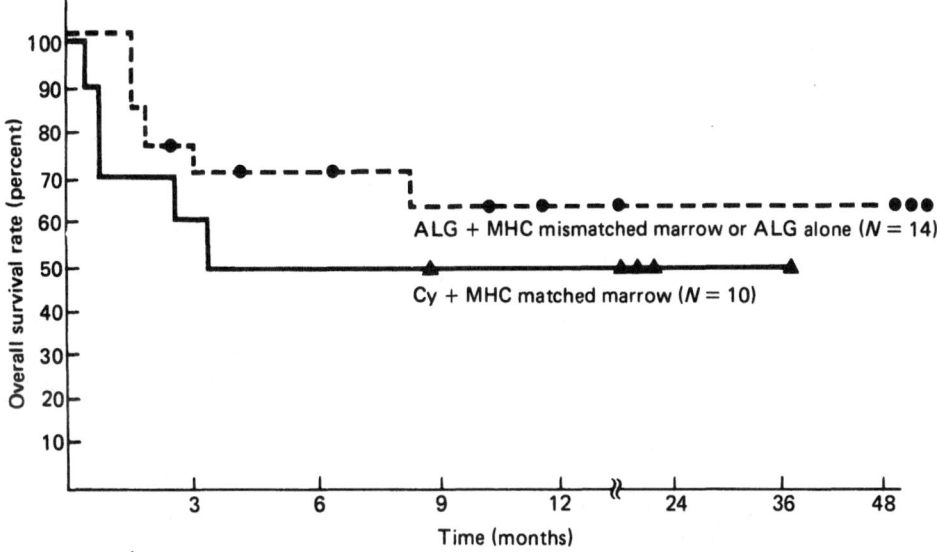

FIGURE 16-4. Comparison of survival rates of 14 patients with severe aplastic anemia on ALG plus mismatched marrow or ALG alone, with 10 patients on cyclophosphamide and matched marrow (1973 through September 1977).

mismatched allogeneic marrow. Transient split hemopoietic chimerism is ultimately followed by complete autologous reconstitution. It is possible that the split hemopoietic chimerism is only a means of overcoming a period of severe cytopenia.

But it is conceivable that, in addition to the documented split take, there is a therapeutic synergism between the ALS and the bone marrow (4,9,11,15). In rabbits exposed to 1,200 rad TBI, autologous marrow reconstitutions are only possible if the animals are sensitized to donor antigens and transfused with allogeneic marrow. Autologous hemopoietic reconstitution after 1,200 rad TBI is a unique phenomenon in the immunobiology of bone marrow transplantation. Here again we must postulate a synergistic effect between the state of immunization and the transfusion of allogeneic marrow. Possibly, there is a transient split take or a microchimerism (11), which stimulates endoreduplication and differentiation of some residual radioresistant pluripotent hemopoietic stem cells; this seems to lead, ultimately, to complete reconstitution of the autochthonous marrow function.

These experimental studies may have important implications for the pathogenesis and treatment of aplastic anemia in man, but we are aware that benzene-induced marrow aplasia is probably not identical with human SAA. As a matter of fact, we are convinced that, in SAA, certain pathogenetic mechanisms may modify the course of the disease. The fact that 10 out of 24 patients in our series recovered autologous marrow function after immunosuppression with or without marrow infusion favors the concept that an autoimmune process directed against the hemopoietic precursor cell com-

partment occurs in a large number of patients. Considerable *in vitro* culture experiments support our clinical data (1,6,8). We have largely confirmed these experiments in our laboratory (14), finding that ALG has a major advantage over CY in suppressing this immune mechanism against hemopoietic precursor cells. Autologous reconstitutions seem to occur much more frequently than after CY and there is no risk of GvHD. Thus, ALG offers the great advantage that no MHC-identical sibling is required. Instead of treatment being for only a small number of patients, effective treatment is available for virtually everyone with SAA. We are not yet sure if transfusion of marrow after immunosuppression is essential, but the experimental evidence strongly favors its use (18,21,22). In the two patients who recovered autologous marrow function after CY conditioning and transfusion of allogeneic MHC-matched marrow, there was direct genetic evidence of transient donor marrow engraftment. After ALG treatment, followed by mismatched marrow transplantation, we could only provide indirect evidence that partial split engraftment had actually taken place (7). Yet, we think that some kind of transient split take or microchimerism may provide an essential factor for the marrow microenvironment or that the lymphocytes transfused with the marrow graft may influence favorably the disturbed interaction between helper and suppressor lymphocytes that is thought to be responsible for SAA. Moreover, all the remissions we have seen after marrow administration are complete and permanent. Three patients have been followed over 4½ years and continue to do well. Our experience is confirmed by the original work of

Mathé et al. (12,16). All their patients who reacted well to ALS and marrow treatment have been in continuous remission for over 7 years. But our experience with ALG alone is insufficient to draw any conclusions, and we have not randomized patients for comparative studies with ALG and bone marrow. However, only one of five patients treated with ALG alone is in remission versus five of nine on ALG plus marrow transfusion. It will be crucial to develop further *in vitro* tests that can predict optimal treatment for individual patients. While we await these developments, randomized cooperative multicenter clinical trials seem to be the most promising approach to the problem.

In our mind, one question is settled. Severe aplastic anemia is not uniformly a defect at the level of the earliest hemopoietic progenitor cell, and, therefore, hemopoietic chimerism is not the only possible cure for this disorder. Our clinical findings have been confirmed at other institutions (20), and we think that there is hope for the cure of all SAA patients not only for the few patients with syngeneic or MHC-matched siblings.

SUMMARY

It has been generally assumed that aplastic anemia is a defect at the level of the pluripotent stem cell. Marrow grafts to achieve hemopoietic chimerism were considered the treatment of choice for this disorder. So far this was only possible for patients with sibling donors matched at the major histocompatibility complex (MHC). Based on our marrow transplant studies in experimental (benzene-induced) aplasia in rabbits and mice, we tried to produce split hemopoietic chimerism without GvHD by grafting one HL-A haplotype-mismatched, mixed leukocyte culture (MLC)-positive marrow after ALG conditioning in patients. Five of nine patients with severe aplastic anemia (SAA) had a sustained recovery of hemopoiesis. Hemopoietic chimerism could never be documented, however, and all these restorations could clearly be shown to be autologous. Five of 10 patients grafted with marrow from MHC-matched siblings after cyclophosphamide (CY) conditioning had a sustained recovery of hemopoiesis. In three patients, the hemopoietic cells were shown to be donor cells, in two, they were host cells. The fact that partial to complete reconstitution of autochthonous marrow function were documented after immunosuppression with ALG and CY implies that, at least in some instances, aplastic anemia is an immunological disorder. Recently, repressor lymphocytes were demonstrated in a patient with this disorder. We confirmed these findings in our laboratory. From our

experimental studies, it seems clear that the allogeneic marrow transfusion is needed for the autochthonous marrow reconstitution but in clinical SAA we do not know if a comparably vigorous immunosuppression alone could give similar results.

ACKNOWLEDGMENTS

We wish to thank Miss B. Rubin and Miss M. Kissling for expert technical assistance, Mrs. J. Lehnherr for the editorial work, and Schweizerisches Serum- und Impfinstitut, Bern for supplying the ALG.

Supported by the Swiss National Science Foundation 3.3320.74 and by the Swiss Cancer League FOR.101.AK.77 (2).

REFERENCES

1. Ascensao, J.G., Pahwa, R., Kagan, W.A., Moore, M.A.S., Hansen, J., and Good, R.A. Aplastic anaemia: Evidence for an immunological mechanism. *Lancet, i*:669, 1976.
2. Baran, D.T., Griner, P.F., and Klemperer, M.R. Recovery from aplastic anemia after treatment with cyclophosphamide. *New Eng. J. Med., 295*:1522, 1976.
3. Camitta, B.M., Thomas, E.D., Nathan, D.G., Santos, G.W., Gordon-Smith, E.C., Gale, R.P., Rapaport, J.M., and Storb, R. Severe aplastic anemia: A prospective study of the effect of early marrow transplantation on acute mortality. *Blood, 48*:63, 1976.
4. Chertkov, J.L., Lemenova, L.N., Mendelevitch, O.A., and Udalov, G.A. Stimulation of haemopoietic colony formation by antilymphocyte serum. *Cell Tissue Kinet., 5*:387, 1972.
5. Gale, R.P., Cline, M.J., Fahey, J.L., Feig, S., Opelz, G., Young, L., Territo, M., Golde, D., Sparkes, R., Naeim, N., Juillard, G., Haskell, C., Fawzi, F., Sarna, G., and Falk, P. Bone marrow transplantation in severe aplastic anemia. *Lancet, ii*:921, 1976.
6. Hoffman, R., Zanjani, E.D., Lutton, J.D., Zalusky, R., and Wasserman, L.R. Suppression of erythroid colony formation by lymphocytes from patients with aplastic anemia. *New Eng. J. Med., 296*:10, 1977.
7. Jeannet, M., Speck, B., Rubinstein, A., Pelet, B., Wyss, M., and Kummer, H. Autologous marrow reconstitutions in severe aplastic anemia after ALG pretreatment and HLA semi-incompatible bone marrow cell transfusion. *Acta Haemat., 55*:129, 1976.
8. Kagan, W.A., Ascensao, J.G., Pahwa, R.N., Hansen, J.A., Goldstein, G., Valera, E.B., and Good, R.A. Aplastic anemia: Presence in human bone marrow of cells that suppress myelopoiesis. *Proc. Nat. Acad. Sci. (USA), 73*:2890, 1976.
9. Kanamaru, A., Kitamura, Y., Kanamaru, A., and Naggi, K. Synergism between lymph node and bone marrow cells for production of granulocytes. I: Requirement for immunocompetent cells. *Exp. Hematol., 2*:35, 1972.

148

10. Kissling, M., and Speck, B. Further studies on experimental benzene induced aplastic anemia. *Blut,* *15*:97, 1972.

11. Liégois, A., Escourrou, J., Ouvré, E., and Charreire, J. Microchimerism: A stable state of low-ratio proliferation of allogeneic bone marrow. *Transpl. Proc.,* *9*:273, 1977.

12. Mathé, G., Amiel, J.L., Schwarzenberg, L., Choay, J., Trolard, P., Schneider, M., Hayat, M., Schlumberger, J.R., and Jasmin, C. Bone marrow grafts in man after conditioning by antilymphocyte serum. *Brit. Med. J.,* *2*:131, 1970.

13. Moorhead, P.S., Nowell, P.C., Mellman, W.J., Battips, D.W., and Hungerford, D.A. Chromosome preparations of leukocytes cultured from human peripheral blood. *Exp. Cell Res.,* *20*:613, 1960.

14. Nissen, C., and Speck, B. *In vitro* studies on the hemopoietic precursor compartment in severe aplastic anemia (in preparation), 1978.

15. Rajewsky, K., Roelants, G.E., and Askonas, B.A. Carrier specificity and allogeneic effect in mice. *Eur. J. Immunol.,* *21*:107, 1972.

16. Schwarzenberg, L., and Mathé, G. Bone marrow transplantation after antilymphocyteglobulin conditioning: Split lymphocyte chimerism. *Behring Res. Commun.: ALG-therapy and Standardization Workshop*, pp. 163-1975, 1972.

17. Sensenbrenner, L.L., Steele, A.A., and Santos, G.W. Recovery of hematologic competence without engraftment following attempted bone marrow transplantation in a patient with severe aplastic anemia. *Exp. Hematol.,* *5*:51, 1977.

18. Speck, B., Buckner, C.D., Cornu, P., and Jeannet, M. Rationale for the use of ALG as sole immunosuppressant in allogeneic bone marrow transplantation for aplastic anemia. *Exp. Hematol.,* *4*:131, 1976.

19. Speck, B. Cornu, P. Jeannet, M., Nissen, C., Burri, H.P., Groff, P., Nagel, G.A., and Buckner, C.D. Autologous marrow recovery following allogeneic marrow transplantation in a patient with severe aplastic anemia. *Exp. Hematol., 4*:131, 1976.

20. Speck, B., Gluckman, E., Haak, H.L., and van Rood, J.J. Treatment of aplastic anemia by ALG with and without allogeneic bone marrow. *Lancet* (submitted), 1978.

21. Speck, B., and Kissling, M. Studies on bone marrow transplantation in experimental ^{32}P-induced aplastic anemia after conditioning with antilymphocyte serum. *Acta Haemat., 50*:193, 1973.

22. Speck, B., and Kissling, M. Successful bone marrow grafts in experimental aplastic anemia using antilymphocyte serum for conditioning. *Eur. J. Clin. Biol. Res., 10*:1047, 1971.

23. Territo, M.C., for the UCLA Bone Marrow Team. Autologous bone marrow repopulation following high dose cyclophosphamide and allogeneic marrow transplantation in aplastic anaemia. *Brit. J. Haemat., 36*:305, 1977.

24. Thomas, E.D., Storb, R., Clift, R.A., Fefer, A., Johnson, F.L., Neiman, P.E., Lerner, K.G., Glucksberg, G., and Buckner, C.D. Marrow transplantation. *New Eng. J. Med., 292*:832, 895, 1975.

25. Thomas, E.D., Storb, R., Giblett, E.R., Longpre, B., Weiden, P.L., Fefer, A., Witherspoon, R., Clift, R.A., Buckner, C.D. Recovery from aplastic anemia following attempted marrow transplantation. *Exp. Hematol., 4*:97, 1976.

17

Engraftment of HLA Identical Marrow in Treatment of Severe Aplastic Anemia: Results of 110 Consecutive Transplants

R. Storb, E. D. Thomas,
P. L. Weiden, C. D. Buckner,
R. A. Clift, A. Fefer,
B. W. Goodell,
F. L. Johnson, P. E. Neiman,
J. E. Sanders, and J. Singer

INTRODUCTION

We report here our experience with 110 consecutive cases of severe aplastic anemia (SAA) treated with marrow grafts from HLA-identical siblings (109 patients) or from an HLA-identical mother (one patient) after the recipient had been conditioned with high doses of cyclophosphamide (Cy) and/or total-body irradiation (TBI). Detailed descriptions on the courses of the first 73 of these patients have been given previously (20,21,23).

MATERIALS AND METHODS

The patients were 2 to 67 (median 18) years old. Seventy-nine had SAA of unknown causes. In eighteen, SAA was associated with drugs, in eight with hepatitis, in two it occurred following a history of acquired paroxysmal nocturnal hemaglobinuria, and in three it was part of a Fanconi syndrome. The SAA had lasted for 0.5 to 96 (median 3) months before transplantation. Seventy-nine patients had been treated with androgenic steroids without effect, and forty-eight patients had infections at admission. All but 10 patients had received many random red blood cell (RBC) and platelet transfusions before transplantation. In addition, 12 had received transfusion from parents or siblings; 40 were refractory to platelets from random donors. Granulocyte counts ranged from 0 to 3,800 (median 200)/ mm^3. Platelet counts before transfusion ranged from 0 to 29,000 (median 5,000)/mm^3.

In 109 cases, the marrow donor was a sibling, and in one case, the mother served as the marrow donor. Serological histocompatibility typing of the patient and family showed inheritance of the same HLA haplotypes in both donor and recipient. In all cases, patient and donor were mutually nonreactive in mixed leukocyte culture (MLC). Seventy-one patients were prepared for transplantation by CY, 50 mg/kg intravenously (i.v.) on each of 4 successive days. Nineteen patients were prepared by a combination of procarbazine, antithymocyte globulin (ATG), and Cy (21). Twenty patients were prepared by 1,000 rad midline tissue exposure of TBI. Seven of these patients were given TBI only, eight were given additional CY and five, additional procarbazine and ATG. Thirty-six hours after CY or within 24 hr of TBI, marrow from the donor was infused (i.v.), 0.9 to 14.2 (median 3.0) × 10^8 marrow cells/kg body weight of recipient. The day of marrow infusion was designated day 0. Following transplantation, methotrexate was administered to prevent or ameliorate graft-versus-host disease (GvHD), 10 to 15 mg/m^2 (i.v.) on day 1 and 10 mg/m^2 on days 3, 6, 11, and weekly thereafter for no

151

TABLE 17-1 Surviving Patients[a]

GENERAL ACTIVITY	NUMBER OF PATIENTS	PROBLEMS
Fully rehabilitated	41	None
Back to school or work	6	Chronic GvHD in skin
Reduced	3	and/or liver
Total	50	

[a]As of August 26, 1977, the survival in months of the 50 patients was as follows: 4, 4.5, 5, 5, 5.5, 6.5, 7.5, 8.5, 10.5, 11.5, 11.5, 11.5, 11.5, 13, 13, 13.5, 15, 17.5, 18, 18, 20, 22, 22.5, 22.5, 24, 25, 27.5, 28, 29, 31, 31.5, 33.5, 35, 39.5, 40, 41.5, 44, 47, 48.5, 53, 53, 54, 54.5, 54.5, 56.5, 59.5, 64, 68, 72, 75.

more than 100 days. Some patients were given rabbit, goat, or horse ATG or prednisone in an attempt to combat life-threatening GvHD (15,31). After day 100, all immunosuppressive therapy was discontinued. Details on isolation procedures, supportive care, and consent forms have been described (20,21).

RESULTS

Fifty patients were alive with normal marrow function 4 to 75 months after marrow grafting (Table 17-1). Forty-one (including five less than 6 months after grafting) are fully rehabilitated, whereas nine show evidence of chronic GvHD. Sixty patients are dead (Table 16-2). Five died early: one on day 0, of heart failure; four on days -1 to 8, of bacterial infections. One patient who had received earlier transfusions from his mother and the marrow donor died on day 24 without engraftment, presumably as a result of acute graft rejection. Twenty-six patients died after rejection of the initial marrow transplant. Twenty-two patients died with active GvHD, or soon after the resolution of

GvHD, of infection. Six patients died without GvHD, two of fungal infection on days 18 and 64 and four of interstitial pneumonia on days 58 to 156. One patient who rejected his graft is alive after more than 3½ years, with complete recovery of his own marrow (26).

In 21 patients, second transplants were carried out after rejection of the first transplant. In sixteen cases, the same HLA-identical sibling served as the marrow donor, and in five cases, a different HLA-identical sibling was used. Seven of the 21 attempts at second transplantation were successful. Only one of these patients, however, is now alive with a functioning graft more than 2½ years after transplantation, whereas six died of GvHD and/or infection.

A multifactorial regression analysis recently carried out on the first 73 patients (17) indicated that two factors strongly correlated with marrow graft rejection: (a) The presence of recipient sensitization against donor cells as determined by *in vitro* tests of cell-mediated immunity before transplantation ($p < 0.01$); tests included the relative response index (RRI) in MLC and the ^{51}Cr release test (10,28–30). (b) A low (less than 3×10^8 cells/kg) marrow cell dose ($p < 0.05$).

Recent attempts to prevent marrow graft rejection in patients with positive *in vitro* tests of sensitization included the use of 1,000 rad TBI in combination either with CY (50 mg/kg on days -6, -5, and -4) or with procarbazine (15 mg/kg on days -8, -6, and -5) and ATG (15 mg IgG/kg on days -7, -5, and -3). Eleven patients underwent one of the TBI regimens and only one patient rejected his transplant. Survival was poor, however, mainly because of death secondary to interstitial pneumonitis, either associated with or following an episode of GvHD.

A second approach to reduce marrow graft rejection in sensitized patients has involved the infusion of unirradiated buffy coat cells from the mar-

TABLE 17-2 Patients who Died

EVENTS ASSOCIATED WITH DEATH	NUMBER OF PATIENTS	DAYS AFTER TRANSPLANTATION (RANGE)
Graft rejection (and infection)[a]	27	24–206
GvHD and		
(a) Bacterial or fungal infection	12	19–719
(b) Interstitial pneumonia	10	54–350
No GvHD		
(a) Bacterial or fungal infection	6	-1–64
(b) Interstitial pneumonia	4	58–156
Heart failure	1	0
Total	60	

[a]Two additional patients with graft rejection are alive, one by a successful second graft from another HLA identical sibling, one by recovery of his own marrow. Five additional patients of the 27 patients with rejection had successful second grafts but died of GvHD and/or infection.

TABLE 17-3 ABO Incompatible Transplants

| | NUMBER OF PATIENTS | | | | |
| | | | DIED | | |
INCOMPATIBILITY	Studied	Alive	Early	Rejec- tion	GvHD
Minor	16	6	2	7	1
Major	5	4	0	0	1
Total	21	10	2	7	2

row donor on days 1, 2, and 3 following the infusion of bone marrow. An average of 4.8×10^8 buffy coat cells/kg was administered. This was done in the hope of infusing additional hemopoietic stem cells that were present among the peripheral leukocytes, as suggested from studies in rodents, dogs, and primates (2,4,8,9,11,13,15). Eight sensitized patients were given combined marrow and buffy coat infusions following CY (50 mg/kg) on each of four successive days. Three of the eight patients rejected, and five sustained, engraftment. All five are alive without evidence of GvHD.

A second multifactor analysis was carried out for those 47 patients among the first 73 who did not reject their initial graft (18). Two factors were significantly associated with GvHD and death: (a) sex mismatch of donor and recipient ($p < 0.01$) and (b) refractoriness to random donor platelets at the time of transplantation ($p < 0.05$). The latter factor (b) adversely influenced the survival of those patients with a sex-mismatched donor.

Table 17-3 illustrates the results of ABO-incompatible transplants. No specific precautions were taken for "minor" incompatibilities ($O \rightarrow A$; $O \rightarrow B$, $O \rightarrow AB$). Six of the sixteen patients with minor incompatibility are currently alive, and GvHD was not a major problem in these patients. Five patients had a major incompatibility with their donor ($A \rightarrow O$; $B \rightarrow O$); and four of these patients are alive with good graft function and no late evidence of hemolysis. To avoid a transfusion reaction at the time of transplantation, i.e., hemolysis of the RBC in the infused marrow suspension, plasma exchanges were carried out to lower the patients isohemagglutinin titers (21). Results indicate that an ABO barrier is not a contraindication to marrow transplantation and that the A and B antigens are not involved in GvHD or rejection.

DISCUSSION

Of our 110 consecutive patients with SAA treated by marrow transplantation from HLA-identical family members, 50 have now survived up to more than 6 years, with complete hematological resto-

ration and sustained engraftment (15,20,21,23). The period of greatest mortality was the first 3 months after grafting. Despite the still high early mortality, marrow transplantation was recently found to be more effective than conventional therapy of SAA in a prospective trial (3). Results similar to ours, with smaller numbers of patients, have been reported by other groups of investigators (1,27).

Further improvement of the results of marrow transplantation can only be achieved by overcoming the three major obstacles responsible for most of the mortality: marrow graft rejection, GvHD, and infection. Studies in our laboratory in dogs have indicated that future marrow graft recipients can be sensitized to "minor" histocompatibility antigens by blood transfusions, resulting in rejection of subsequent marrow transplants from DLA-identical littermates (12,19). We believe that some of the human patients with SAA have also been sensitized by random or family donor transfusions to minor antigens present on marrow cells of the HLA-identical donor, resulting in marrow graft rejection. Only 10 of the 110 patients in this summary had not been transfused until just before transplantation. All 10 had successful engraftment and none rejected the graft. All 10 are alive with sustained grafts for up to 5 years. Although the number of non-transfused patients is still small, these results support our hypothesis that transfusion-induced sensitization is a major cause of marrow graft rejection in man. Additional support comes from the results of in vitro tests, the RRI in MLC and the ^{51}Cr release assay (10,17,28–30). Positive tests presumably are due to the presence of specifically altered lymphocytes in some patients with SAA. Results of the in vitro tests predict the fate of the marrow graft, provide evidence that rejection is immunologically mediated, and allow us to distinguish nonsensitized patients with a low chance of marrow graft rejection from sensitized patients with a high chance of rejection. The antigens detected by the in vitro tests, and presumably involved in rejection, appeared to be independent of ABO or HLA antigens, sex, or refractoriness to random donor platelets.

The second factor predicting the fate of the marrow graft was the marrow cell dose, i.e., with a decreasing marrow cell dose, the incidence of graft rejection increased and vice versa (17). A dose of marrow cells below 3×10^8 cells/kg of body weight increased the incidence of rejection even in the absence of positive in vitro tests. The mechanism of rejection of a transplant with a low marrow cell dose in the absence of sensitization is obscure. It could be that genetic determinants similar to the genes of hemopoietic histocompatibility described in mice are involved (5,6).

Subsequent to the analysis, we implemented two changes for the conditioning of sensitized patients with SAA for marrow transplantation. The first change consisted of more intensive immunosuppression, either by a combination of CY and TBI, a regimen used to condition leukemia patients for transplantation (22), or a combination of procarbazine-ATG and TBI, a regimen found to be effective in overcoming sensitization by blood products in the canine marrow graft model (14,32). Both regimens were very effective in preventing marrow graft rejection, but survival was poor mainly because of more severe and more often fatal GvHD and infection than usually seen after CY. The use of the TBI regimens was, therefore, suspended. The second change involved the use of larger numbers of marrow cells by harvesting marrow not only from the anterior and posterior iliac crests, but also from the sternum. However, there are limits to the amount of marrow that can be obtained from a donor (25). Therefore, eight sensitized patients conditioned by CY were given additional peripheral blood buffy coat cells from the marrow donor. Three of the eight rejected the graft, but five had sustained engraftment and are still alive. In contrast, 13 of 16 previous sensitized patients given CY and marrow, but no additional buffy coat cells rejected their transplants. The number of sensitized patients given buffy coat cells in addition to the marrow inoculum is still too small to draw firm conclusions regarding their effectiveness. However, the decrease in the incidence of marrow graft rejection is encouraging. Furthermore, the incidence of GvHD was not increased by infusions of viable buffy coat cells from the marrow donor. In fact, none of the five patients with sustained engraftment had GvHD. Therefore, the infusion of additional buffy coat cells might be a reasonable approach to reduce the high incidence of graft rejection in patients with positive *in vitro* tests of sensitization.

Graft-versus-host disease has remained a major problem; it is likely that GvHD is due to differences between donor and recipient for minor histocompatibility systems outside of HLA. Our results have suggested a role for X and Y chromosome-associated transplantation antigens in the outcome of marrow grafts between HLA-identical siblings (18). Such antigen systems have been described in mice (see review by Gasser and Silvers, reference 7). The significance of a second factor, refractoriness to random donor platelets at the time of transplantation, which lowers the survival rate of sex-mismatched patients, is currently unclear. The problem of GvHD will remain until more effective means are found to prevent it. However, matching of donor and recipient for sex may decrease the incidence of GvHD.

Infections are more frequent in patients with GvHD than in those without clinical evidence of GvHD. Patients pass through a period of severe combined immunodeficiency following allogeneic marrow transplantation (for review, see reference 33). This immunodeficiency seems to be more pronounced and prolonged in patients with GvHD (33). The immunodeficiency is likely to be a major factor in the high incidence of fatal infections among marrow transplant recipients. In time, patients regain immunological reactivity and, concomitantly, are less susceptible to infection. Efforts should therefore be directed at accelerating immunological reconstitution, thus shortening the time of risk for fatal infections.

Results in these 110 patients given allogeneic marrow grafts and our previous successful results in patients given grafts from their monozygous twins (24) suggest that, in most instances, SAA is due to a stem cell defect, which can be corrected by the infusion of normal hemopoietic stem cells. Treatment of SAA by marrow transplantation has proven to be more effective than conventional therapy (3). Early transplantation to avoid transfusion-induced sensitization and, thus, graft rejection should, therefore, be a prime consideration in any patient with SAA who has an HLA-identical sibling.

SUMMARY

Seventy-three patients, 2 to 67 years old, with severe aplastic anemia were given marrow transplants from HLA-identical siblings following 200 mg of cyclophosphamide/kg or 1,000 rad total-body irradiation. Forty-seven patients had aplastic anemia of unknown etiology; in 15 it was associated with drugs, in six with hepatitis, in three it was a Fanconi syndrome, and in two it occurred after a history of paroxysmal nocturnal hemoglobinuria. The duration of aplastic anemia ranged from 0.5 to 95 months, with a median of 3 months. Fifty-one patients had received androgens. After transplantation, all patients were given intermittent methotrexate for 100 days to obviate graft-versus-host disease (GvHD). Five patients died too early to evaluate success or failure of engraftment: one at day 0, of heart failure; five at days 1 to 8, of bacterial infection. One patient died on day 28 without engraftment. Sixty-seven patients had successful marrow engraftment, but thirty-six of these died: 16 with marrow graft rejection, days 33 to 183; 16 with active GvHD or soon after the resolution of GvHD, days 19 to 350 (eight with bacterial or fungal infection, eight with interstitial pneumonia); one with interstitial pneumonia but no GvHD, day 58;

two with bacterial infection, days 18 and 206; one of unknown causes, day 427. One patient rejected his graft and is alive after more than 2½ years, with complete autologous marrow recovery. Thirty additional patients are surviving with normally functioning marrow grafts between 200 and 802 days (median 740 days) after grafting, and 28 have returned to normal activities. Although the mortality is still high, the 42 percent survival in this group of patients indicates that marrow grafting is a most effective treatment for severe aplastic anemia.

The most frequent cause of failure was marrow graft rejection, which occurred in 21 of 68 evaluable patients (three of these were successfully retransplanted, and one survived, with autologous marrow recovery). A multifactor regression analysis indicated that two factors were strongly correlated with graft rejection: (a) detection of recipient sensitization against donor cells by *in vitro* tests of cell-mediated immunity performed before transplantation ($p < 0.01$) and (b) a low (3×10^8 cells/kg) marrow cell dose ($p < 0.05$). In those patients not rejecting their grafts, sex match of donor and recipient predicted better survival ($p < 0.01$). Death in patients with grafts was most often related to GvHD. This suggests that "minor" transplantation antigen systems, one not sex linked and involved in rejection and the other associated with the X and Y chromosomes and involved in GvHD, are important in determining survival of the patient with aplastic anemia given a marrow graft from an HLA-identical sibling.

ACKNOWLEDGMENT

This investigation was supported by Grant Numbers CA 18029, CA 17117, CA 18579, CA 15704, CA 05231, and CA 18221, awarded by the National Cancer Institute, DHEW, and Contract AI 52515 from the National Institute of Allergy and Infectious Diseases.

REFERENCES

1. A report from the ACS/NIH bone marrow transplant registry: Bone marrow transplantation from histocompatible, allogenic donors for aplastic anemia. *JAMA, 236*:1131, 1976.
2. Calvo, W., Fliedner, T.M., Herbst, E., Hugl, E., and Bruch, C. Regeneration of blood-forming organs after autologous leukocyte transfusion in lethally irradiated dogs. II. Distribution and cellularity of the marrow in irradiated and transfused animals. *Blood, 47*:593, 1976.
3. Camitta, B.M., Thomas, E.D., Nathan, D.G., Santos, G., Gordon-Smith, E.C., Gale, R.P., Rappeport, J.M., and Storb, R. Severe aplastic anemia: A pro- spective study of the effect of early marrow transplantation on acute mortality. *Blood, 48*:63, 1976.
4. Cavins, J.A., Scheer, S.C., Thomas, E.D., and Ferrebee, J.W. The recovery of lethally irradiated dogs given infusions of autologous leukocytes preserved at −80°C. *Blood, 23*:38, 1964.
5. Cudkowicz, G., and Bennett, M. Peculiar immunobiology of bone marrow allografts. I. Graft rejection by irradiated responder mice. *J. Exp. Med., 134*:33, 1971.
6. Cudkowicz, G., and Rossi, G.B. Hybrid resistance to parental DBA/2 grafts: Independence from the H-2 locus. I. Studies with normal hematopoietic cells. *J. Nat. Cancer Inst., 48*:131, 1972.
7. Gasser, D.L., and Silvers, W.K. Genetics and immunology of sex-linked antigens. *Adv. Immunol., 15*:215, 1972.
8. Goodman, J.W., and Hodgson, G.S. Evidence for stem cells in the peripheral blood of mice. *Blood, 19*:702, 1962.
9. Malinin, T.I., Perry, V.P., Kerby, C.C., and Dolan, M.F. Peripheral leukocyte infusion into lethally irradiated guinea pigs. *Blood, 25*:693, 1965.
10. Mickelson, E.M., Fefer, A., Storb, R., and Thomas, E.D. Correlation of the relative response index with marrow graft rejection in patients with aplastic anemia. *Transplantation, 22*:294, 1976.
11. Storb, R., Epstein, R.B., Ragde, H., Bryant, J., and Thomas, E.D. Marrow engraftment by allogeneic leukocytes in lethally irradiated dogs. *Blood, 30*:805, 1967.
12. Storb, R., Epstein, R.B., Rudolph, R.H., and Thomas, E.D. The effect of prior transfusion on marrow grafts between histocompatible canine siblings. *J. Immunol., 105*:627, 1970.
13. Storb, R., Epstein, R.B., and Thomas, E.D. Marrow repopulating ability of peripheral blood cells compared to thoracic duct cells. *Blood, 32*:662, 1968.
14. Storb, R., Floersheim, G.L., Weiden, P.L., Graham, T.C., Kolb, H.J., Lerner, K.G., Schroeder, M.L., and Thomas, E.D. Effect of prior blood transfusions on marrow grafts: Abrogation of sensitization by procarbazine and antithymocite serum. *J. Immunol., 112*:1508, 1974.
15. Storb, R., Gluckman, E., Thomas, E.D., Buckner, C.D., Clift, R.A., Fefer, A., Glucksberg, H., Graham, T.C., Johnson, F.L., Lerner, K.G., Neiman, P.E., and Ochs, H. Treatment of established human graft-versus-host disease by antithymocyte globulin. *Blood, 44*:57, 1974.
16. Storb, R., Graham, T.C., Epstein, R.B., Sales, G.E., and Thomas, E.D. Demonstration of hemopoietic stem cells in the peripheral blood of baboons by cross circulation. *Blood, 50*:537, 1977.
17. Storb, R., Prentice, R.L., and Thomas, E.D. Marrow transplantation for treatment of aplastic anemia. An analysis of factors associated with graft rejection. *N. Eng. J. Med., 296*:61, 1977.
18. Storb, R., Prentice, R.L., and Thomas, E.D. Treatment of aplastic anemia by marrow transplantation from HLA identical siblings. Prognostic factors associated with graft versus host disease and survival. *J. Clin. Invest., 59*:625, 1977.

19. Storb, R., Rudolph, R.H., Graham, T.C., and Thomas, E.D. The influence of transfusions from unrelated donors upon marrow grafts between histocompatible canine siblings. *J. Immunol., 107*:409, 1971.

20. Storb, R., Thomas, E.D., Buckner, C.D., Clift, R.A., Johnson, F.L., Fefer, A., Glucksberg, H., Giblett, E.R., Lerner, K.G., and Neiman, P. Allogeneic marrow grafting for treatment of aplastic anemia. *Blood, 43*:157, 1974.

21. Storb, R., Thomas, E.D., Weiden, P.L., Buckner, C.D., Clift, R.A., Fefer, A., Fernando, L.P., Giblett, E.R., Goodell, B.W., Johnson, F.L., Lerner, K.G., Neiman, P.E., and Sanders, J.E. Aplastic anemia treated by allogeneic bone marrow transplantation: A report on 49 new cases from Seattle. *Blood, 48*:817, 1976.

22. Thomas, E.D., Buckner, C.D., Banaji, M., Clift, R.A., Fefer, A., Flournoy, N., Goodell, B.W., Hickman, R.O., Lerner, K.G., Neiman, P.E., Sale, G.E., Sanders, J.E., Singer, J., Stevens, M., Storb, R., and Weiden, P.L. One hundred patients with acute leukemia treated by chemotherapy, total body irradiation, and allogeneic marrow transplantation. *Blood, 49*:511, 1977.

23. Thomas, E.D., Buckner, C.D., Storb, R., Neiman, P.E., Fefer, A., Clift, R.A., Slichter, S.J., Funk, D.D., Bryant, J.I., and Lerner, K.G. Aplastic anemia treated by marrow transplantation. *Lancet, i*:284, 1972.

24. Thomas, E.D., Rudolph, R.H., Fefer, A., Storb, R., Slichter, S., and Buckner, C.D. Isogeneic marrow grafting in man. *Exp. Hematol., 21*:16, 1971.

25. Thomas, E.D., and Storb, R. Technique for human marrow grafting. *Blood, 36*:507, 1970.

26. Thomas, E.D., Storb, R., Giblett, E.R., Longpre, B., Weiden, P.L., Fefer, A., Witherspoon, R., Clift, R.A., and Buckner, C.D. Recovery from aplastic anemia following attempted marrow transplantation. *Exp. Hematol., 4*:97, 1976.

27. UCLA Bone Marrow Transplant Team. Bone-marrow transplantation in severe aplastic anemia. *Lancet, ii*:921, 1976.

28. Warren, R.P., Storb, R., Weiden, P.L., Mickelson, E.M., and Thomas, E.D. Direct and antibody-dependent cell-mediated cytotoxicity against HLA-identical sibling lymphocytes. Correlation with marrow graft rejection. Brief communication. *Transplantation, 22*:631, 1976.

29. Warren, R.P., Storb, R., Weiden, P.L., Su, P.J., and Thomas, E.D. Detection of lymphocyte-mediated, antibody-dependent cell-mediated and complement-dependent cytotoxicity against HLA-identical sibling lymphocytes: prediction of marrow graft rejection. *Transplant. Proc.* (in press).

30. Warren, R.P., Storb, R., Weiden, P.L., Su, P.J. and Thomas, E.D. Cytotoxicity against HLA-identical sibling lymphocytes: Genetic determination and independent segregation of the antigenic system. *Transplant. Proc.* (in press).

31. Weiden, P.L., Doney, K., Storb, R., and Thomas, E.D. Antihuman thymocyte globulin (ATG) for prophylaxis and treatment of graft-vs.-host disease in recipients of allogeneic marrow grafts. *Transplant. Proc.* (in press).

32. Weiden, P.L., Storb, R., Slichter, S., Warren, R.P., and Sale, G.E. Effect of six weekly transfusions on canine marrow grafts: Tests for sensitization and abrogation of sensitization by procarbazine and antithymocyte serum. *J. Immunol., 117*:143, 1976.

33. Witherspoon, R., Noel, D., Storb, R., Ochs, H.D., and Thomas, E.D. The effect of graft-vs.-host disease on reconstitution of the immune system following marrow transplantation for aplastic anemia or leukemia. *Transplant. Proc.* (in press).

18

The Role of Cyclophosphamide and Total-body Irradiation in Marrow Transplantation for Severe Aplastic Anemia

Robert Peter Gale for the UCLA Bone Marrow Transplant Team

INTRODUCTION

Graft failure is a major obstacle to successful bone marrow transplantation in aplastic anemia with an incidence of 30 to 40 percent in several large series (1,3,14,17,21). Graft failure, i.e., either initial failure of engraftment or engraftment with the subsequent breakdown of the graft, is thought to be immunologically mediated and related to graft rejection in most instances.

Important data regarding mechanisms of marrow graft failure come from studies in dogs by Storb and coworkers. These investigators have demonstrated the importance of the major histocompatibility complex (DLA) in determining graft survival (16); they have also shown that pre-transplant sensitization either to DLA and non-DLA antigens of the donor can increase the incidence of graft rejection (13,24). In the dog, the adverse effects of sensitization can be overcome with more intense pre-transplant immunosuppression (14,25). Thomas et al. (19) have reported a relationship between engraftment and marrow dose in the dog. These observations suggest that immunological factors may be important in graft failure.

In man, evidence of an immune mechanism of graft failure is indirect. First, the onset of graft failure is usually delayed. Second, the occasionally successful second transplants tend to exclude microenvironmental defects as a common cause of graft failure. Third, as Storb and coworkers have reported, there is a correlation between marrow cell dose, *in vitro* test of anti-donor immunity, and graft failure (11,15,23). Fourth, correlation between graft failure and pre-transplant transfusions has been reported in one large series (1). Taken together, these observations suggest that an immune mechanism may be responsible for graft failure in man.

In view of this evidence, we and others have investigated the hypothesis that more intensive pre-transplant immunosuppression might be effective in preventing graft failure. Weiden and coworkers (25) reported a randomized trial comparing cyclophosphamide (Cy) to the combination of Cy, procarbazine, and anti-thymocyte globulin (ATG). No significant difference in the incidence of graft rejection was observed. A similar trial using total-body irradiation, procarbazine, and ATG is in progress. Because of the low incidence of graft rejection in leukemic patients conditioned for transplantation with Cy and total body irradiation (TBI) (18,20), we evaluated this regimen in patients with aplastic anemia. Results of this study suggest that, although Cy and TBI is effective in preventing graft rejection, overall survival is not improved, due to an increased incidence and severity of graft-versus-host disease (GvHD) and interstitial pneumonitis.

MATERIALS AND METHODS

Patients

Twenty patients with severe aplastic anemia (SAA) were evaluated. Criteria for diagnosis included a granulocyte count of less than $0.5 \times 10^9/$ liter, a platelet count of less than $20 \times 10^9/$liter, a corrected reticulocyte count of less than 1.0 percent, and two or more bone marrow biopsy specimens indicating moderate to severe aplasia, with less than 70 percent non-myeloid cells. Eleven patients received Cy and nine received Cy and TBI (Cy-TBI). Clinical data and the pretransplant hematological parameters of the two groups are indicated in Table 18-1.

Histocompatibility

Marrow donors were selected by HLA and mixed lymphocyte culture (MLC) testing of available siblings and was performed by standard techniques. All 20 pairs were HLA-A, B, and MLC (HLA-D) identical. Relative response indexes (RRI) between donor and recipient were calculated by dividing the MLC of the recipient against the donor by the recipient's mean response to two or more unrelated controls. The ABO typing was performed using standard techniques. Lymphocytotoxins were determined using standard techniques against a panel of 80 to 100 unrelated control cells. One patient in the Cy-TBI group was a major mismatch (A→O) and required plasma exchange prior to transplantation (6).

Transplantation

Informed consent, approved by the UCLA Human Subject Protection Committee, was obtained. Two conditioning regimens were evaluated: (a) cy-clophosphamide: donor buffy coat rich plasma was given on day −5, 50 mg of Cy/kg was injected intravenously (i.v.) on days −4, −3, −2, and −1 (19) and (b) Cy-TBI: 60 mg of Cy/kg was injected (i.v.) on days −4 and −3 and TBI (1,000 rad) was given on day −1 (Fig. 18-1). Donor bone marrow was infused intravenously on day 0. Methotrexate was given post-transplant to modify graft-versus-host disease (GvHD) (20). Engraftment was documented by an increase in the peripheral blood cell count and marrow cellularity, peripheral blood and bone marrow chromosome analysis (12 patients), and by red blood cell and leukocyte antigens and isoenzyme gene markers (19 patients) (9).

Graft Rejection

Failure of engraftment was defined as a granulocyte count of less than $0.05 \times 10^9/$liter, a platelet count of less than $10 \times 10^9/$liter a reticulocyte count of less than 0.1 percent, and the absence of identifiable hemopoietic precursors and of spontaneously dividing donor cells in the bone marrow of recipients surviving 21 days or more following transplantation. Graft rejection was defined as the development of marrow aplasia following evidence of engraftment as defined above. Engrafted patients surviving 28 days or more were considered at risk for graft rejection.

Graft-versus-Host Disease

Criteria for diagnosis and staging of GvHD have been reported (20). Patients with at least stage II GvHD were treated with corticosteroids. Patients with grafts surviving 30 days or more post-transplant were considered at risk for GvHD.

Interstitial Pneumonitis

The diagnosis of interstitial pneumonitis was based on typical radiological findings, significant

TABLE 18-1 Pre-transplant Characteristics of Patients Undergoing Marrow Transplantation

	CYCLOPHOSPHAMIDE	CYCLOPHOSPHAMIDE–TOTAL-BODY IRRADIATION
Number	11	9
Age (years)	24 (12–48)	16 (5–56)
Sex (male/female)	2.7	3.5
Symptoms to First Clinic Visit (days)	10 (0–167)	13 (3–22)
Presenting Symptoms		
Hemorrhage	9	9
Anemia	11	6
Infection	1	3
Etiology		
Unknown	7	6
Hepatitis	4	2
Drug (?)	0	1
Therapy		
Androgen	5	5
Steroids	8	4

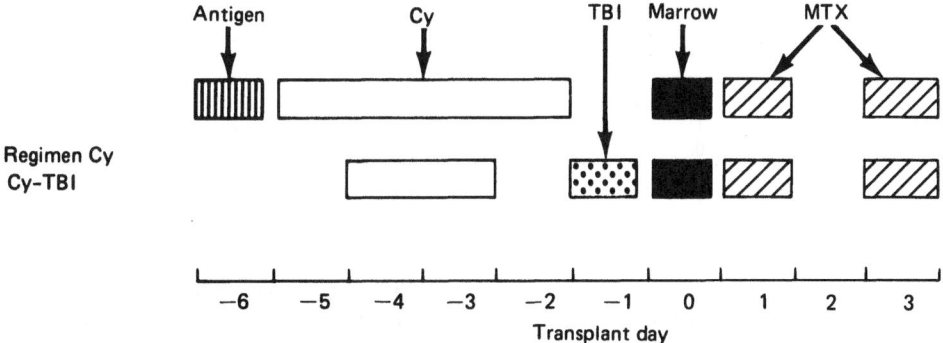

FIGURE 18-1. Treatment scheme. Cyclophosphamide, CY; total-body irradiation, TBI; methotrexate, MTX.

hypoxemia, recovery of the suspected agent from the pulmonary tissue, identification of intranuclear inclusions typical of cytomegalovirus (CMV), or a twofold rise in antibody titer temporarily related to the interstitial process. Routine evaluation included fiberoptic bronchoscopy with brushings and washings, transbronchial biopsy, and open-lung biopsy, when indicated. Patients dying before day 40 of unrelated causes were not considered at risk for interstitial pneumonitis.

Clinical Management

Details regarding general clinical management have been reported (21,22). HLA-unmatched platelets and granulocytes were used pre-transplant, and recipients did not receive blood products from related donors. Following transplantation, the marrow donor and family members frequently provided platelet and granulocyte support. Blood products were irradiated with 1,500 rad to prevent engraftment. Transplant recipients received cotrimoxazole twice weekly from day +50 to +150 as prophylaxis for *Pneumocystis carinii* (10).

Data Analysis

Data were analyzed using a two-tailed Fisher's Exact T-Test. Survival data were analyzed by means of a life table analysis and survival rate (BMD 11S of the UCLA Health Science Computing Facility) (4).

RESULTS

Twenty patients entered the study; 11 received cyclophosphamide and nine received Cy and TBI. Pre-transplant clinical (Table 18-1) and hematological parameters for the two groups were comparable. Most patients in the study were young, had aplasia of unknown etiology, had failed to respond to androgens and/or corticosteroid therapy, and had received between 10 to 100 transfusions pre-

transplant. The groups were also comparable for factors thought to possibly influence graft outcome, including sex and ABO matching, interval from diagnosis to transplant, and marrow dose. Relative response indices (RRI) for the two groups were -0.49 ± 0.32 and -0.05 ± 0.26, and no recipient had an RRI greater than 1.5. Eight grafts were between donors and recipients of the same sex, and twelve were mismatched—eight in the Cy group and four in the Cy-TBI group. Median marrow dose was $2.51 \pm 0.31 \times 10^8$ cells/kg body weight in patients receiving Cy versus $2.81 \pm 0.82 \times 10^8$ cells/kg in patients receiving Cy-TBI. Six patients in the Cy group and four in the Cy-TBI group received more than 2.9×10^8 marrow cells/kg.

Graft rejection occurred in four of 10 patients at risk who received Cy. None of the patients receiving Cy-TBI rejected their grafts ($p = 0.087$). There was no correlation between either marrow cell dose or RRI and graft rejection.

Clinically significant (stage II or over) GvHD was observed in eight of 14 patients at risk, including two of six patients in the Cy group and six of eight patients in the Cy-TBI group. No correlation was observed between sex or ABO matching and GvHD. The RRI of the donor against the recipient did not correlate with GvHD.

Interstitial pneumonitis was observed in 12 of 15 patients at risk, including five of seven patients in the Cy group and seven of eight patients in the Cy-TBI group. Four cases were probably related to CMV and one each to herpes simplex zoster, adenovirus, and *Pneumocystis carinii*. No etiology was determined in five cases. One patient receiving Cy died of interstitial pneumonitis. In contrast, interstitial pneumonitis was a primary cause of death in one patient and a significant contributing factor (with GvHD) in three additional patients in the Cy-TBI group. The overall incidence of interstitial pneumonitis between the two groups did not differ significantly. The incidence of fatal pneumonitis is significant at the $p = 0.03$ level.

TABLE 18-2 Cause of Treatment Failure

	CY	CY-TBI
Rejection	3	0
Infection	1	0
GvHD	0	1
Interstitial Pneumonia	0	2
GvHD and pneumonia	1	2
Other	1	2
	7	6

Survival data is indicated in Figure 18-2. Seven patients are currently alive, 130 days to over 3 years following transplantation. Actuarial 3-year survival is 30 percent. One-year survival in the Cy group is 36 percent compared to 20 percent in the Cy-TBI group. This difference is not significant. Causes of treatment failure are indicated in Table 18-2. Graft rejection was a major cause of death in patients receiving Cy alone, whereas GvHD and interstitial pneumonitis were major problems in those receiving Cy-TBI ($p = 0.03$).

DISCUSSION

The role of allogeneic bone marrow transplantation in patients with SAA has been well established (1,3,14,17,21). Results with marrow grafting, however, should be improved, since the overall survival rates are 30 to 45 percent in most large series.

Graft rejection is a major cause of treatment failure despite matching at the HLA gene complex. The high incidence of graft rejection suggests that non-HLA antigens are probably involved. Other factors that may contribute to graft rejection include inherent differences in immune reactivity, extent of pre-sensitization, and variable response to immunosuppressive therapy.

Ideally, one would like to determine both the pathogenetic mechanism of graft rejection and to identify patients at greatest risk so that immunosuppressive therapy could be individualized. Recently, Storb and coworkers (11,15,23) have reported a correlation between *in vitro* immune tests and graft rejection. Although these data require confirmation, this correlation offers the potential of restricting intensive immunosuppression to the 30 to 40 percent of patients at highest risk of rejection without escalating immunosuppression therapy for the remaining 60 to 70 percent.

In our relatively small series, we were unable to identify clinical or *in vitro* prognostic factors of graft rejection and therefore evaluated the role of more intensive immunosuppression with Cy and TBI. This combination was effective in eliminating graft rejection but was unfortunately associated with a higher incidence and/or severity of GvHD and interstitial pneumonitis. The overall effect was that survival was not significantly improved.

Although immunologically mediated graft rejection appears to be the most likely etiology of graft failure, other potential mechanisms, such as microenvironmental abnormalities, must be considered. The successful hematological reconstitution of most syngeneic and allogeneic marrow graft recipients militates against this possibility in most patients. The temporal course of marrow graft failure also makes a microenvironmental defect unlikely. We and others have, however, observed the rare failure of restoration of normal hemopoiesis in recipients of marrow grafts from identical twins (5,6). Recently, several investigators have suggested that an

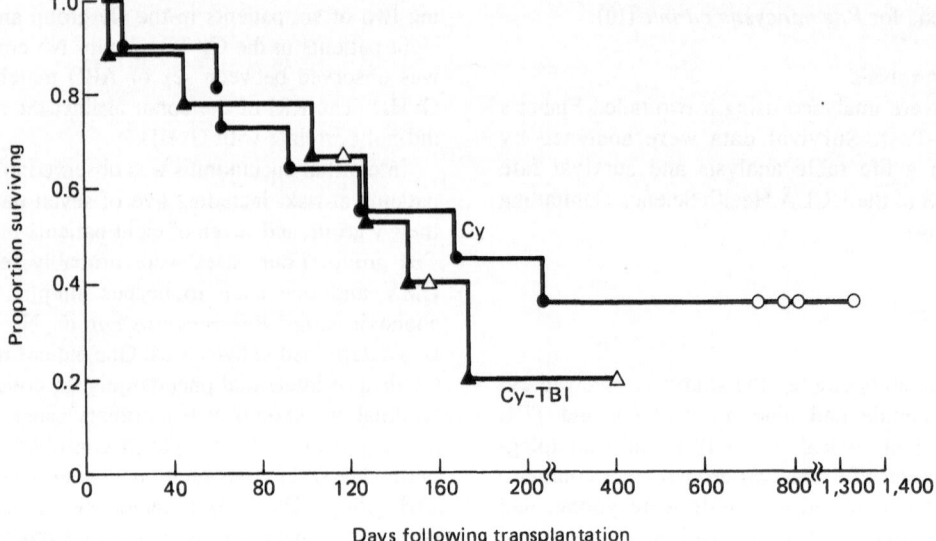

FIGURE 18-2. Survival of bone marrow transplant patients (transplants January through April 1977). Cyclophosphamide, CY; total-body irradiation, TBI.

immune mechanism may be operative in aplasia and related disorders (2,8). This could explain graft failure in identical twins and allograft recipients. We, however, have repeatedly failed to find *in vitro* evidence of anti-donor immunity in such patients (6,12).

In summary, we approached the problem of graft rejection with an intensive immunosuppressive conditioning regimen of Cy and TBI. Although this approach was successful in preventing graft rejection, survival was not improved. Treatment failure was related primarily to an increased incidence and/or severity of GvHD and interstitial pneumonitis. Optimally, one would like to reserve intensive immunosuppression for those patients at high risk of rejection and to control more effectively GvHD and interstitial pneumonitis. Studies directed toward these aims are currently underway.

SUMMARY

Graft failure is observed in 30 to 40 percent of patients with SAA receiving HLA-identical bone marrow transplants. Rejection is thought to be the etiology, in most instances. Graft rejection probably represents an immunological response of the recipient against non-HLA antigens of the marrow donor. One approach to this problem is the use of more intensive immuno-suppressive conditioning regimens. In the present study, we compared Cy alone to Cy-TBI in patients undergoing transplantation for SAA. Although the incidence of rejection was markedly reduced, overall survival was not improved. This was due to an increased incidence and severity of graft-versus-host disease and interstitial pneumonitis. Although the incidence of graft failure can be decreased with intensive immunosuppression, alternative approaches to this problem are clearly needed.

ACKNOWLEDGMENTS

This work could not have been possible without the dedicated teamwork of the physicians, nurses, and staff of the UCLA Bone Marrow Transplant Unit.

Supported by Grants CA12800, CA15688, and RR00865 from the National Institutes of Health, Bethesda, Maryland.

REFERENCES

1. ACS/NIH Bone Marrow Transplant Registry. Bone marrow transplantation from histoincompatible allogeneic donors with aplastic anemia. *JAMA, 236*:1131, 1976.

2. Ascensao, J., Kagan, W., Moore, M., Pahwa, R., Hansen, J., and Good, R. Aplastic anemia. Evidence for an immunologic mechanism. *Lancet, i*:669, 1976.

3. Camitta, B.M., Thomas, E.D., Nathan, D.G., Santos, G., Gordon-Smith, E.C., Gale, R.P., Rappeport, J.M., and Storb, R.: Severe aplastic anemia: A prospective study of the effect of early marrow transplantation on acute mortality. *Blood, 48*:63, 1976.

4. Dixon, W.J. (ed.).: *BMD Biomedical Computer Programs, X-series Supplement.* University of California at Los Angeles, Berkeley, 1969.

5. Fernbach, D.J., and Trentin, J.J. Isologous bone marrow transplantation in an identical twin with aplastic anemia. *Proceed VIII Internat. Congr. Hematol., Tokyo,* 1960, pp. 150–155.

6. Gale, R.P., Falk, P., Feig, S.A., Cline, M.J. for the UCLA Bone Marrow Transplant Team. Failure of recovery following syngeneic marrow grafting in aplastic anemia. *Exp. Hematol.,* (in press) (Abstr.) 5(Suppl. 2):103, 1977.

7. Gale, R.P., Feig, S.A., Ho, W., Falk, P., Rippee, C., and Sparkes, R.S. ABO blood group system and bone marrow transplantation. *Blood, 50*:185, 1977.

8. Hoffman, R., Zanjani, E.D., Villa, J., Zalasky, R., Lutton, J.D., and Wasserman, L.R. Diamond-Blackfan syndrome: Lymphocyte-mediated suppression of erythropoiesis. *Science 193*:899, 1976.

9. Sparkes, M.C., Crist, M.L., Sparkes, R.S., Gale, R.P., Feig, S.A., and the UCLA Bone Marrow Transplant Team. Gene markers in human bone marrow transplantation. *Vox. Sang.* (In Press) 1978.

10. Lau, W.K., and Young, L.S. Trimethoprim-sulfamethoxazole treatment of *Pneumocystis carinii* pneumonia in adults. *New Eng. J. Med., 295*:716, 1976.

11. Mickelson, E.M., Fefer, A., Storb, R., and Thomas, E.D. Correlation of the relative response index with marrow graft rejection in patients with aplastic anemia. *Transplantation, 22*:294, 1976.

12. Opelz, G., Gale, R.P., and the UCLA Bone Marrow Transplant Team: Absence of specific mixed leukocyte culture reactivity during graft-versus-host disease and following bone marrow rejection. *Transplantation, 22*:474, 1976.

13. Storb, R., Epstein, R.B., Rudolph, R.H., and Thomas, E.D. The effects of prior transfusions on marrow grafts between histocompatible canine siblings. *J. Immunol., 105*:627, 1970.

14. Storb, R., Florsheim, G.L., Weiden, P.L., Graham, T.C., Kolb, H., Lerner, K.G., Schroeder, M., and Thomas, E.D. Effects of prior blood transfusion in marrow graft: Abrogation of sensitization by procarbazine and anti-thymocyte serum. *J. Immunol., 112*:1508, 1974.

15. Storb, R., Prentice, R.L., and Thomas, E.D. Marrow transplantation for treatment of aplastic anemia. An analysis of factors associated with graft rejection. *New Eng. J. Med., 296*:61, 1977.

16. Storb, R., Thomas, E.D., Buckner, C.D., Clift, R.A., Johnson, F.L., Fefer, A., Glucksberg, H., Giblett, E.R., Lerner, K.G., and Neiman, P.E.: Allogeneic marrow grafting for treatment of aplastic anemia. *Blood, 43*:157, 1974.

17. Storb, R., Thomas, E.D., Weiden, P.L., Buckner,

C.D., Clift, R.A., Fefer, A., Fernando, L.P., Giblett, E.R., Goodell, B.W., Johnson, F.L., Lerner, K.G., Neiman, P.E., and Sanders, J.E. Aplastic anemia treated by allogeneic bone marrow transplantation: a report of 49 new cases from Seattle. *Blood, 48*:817, 1976.

18. Thomas, E.D., Buckner, C.D., Banaji, M., Clift, R.A., Fefer, A., Fluornoy, N., Goodell, B.W., Hickman, R.O., Lerner, K.G., Neiman, P.E., Sale, G.E., Sanders, J.E., Singer, J., Stevens, M., Storb, R., and Weiden, P.L. One hundred patients with acute leukemia treated by chemotherapy, total body irradiation, and allogeneic marrow transplantation. *Blood, 49*:511, 1977.

19. Thomas, E.D., Collins, J.A., Herman, E.D., Jr., and Ferrebee, J.W. Marrow transplants in lethally irradiated dogs given methotrexate. *Blood, 19*:217, 1962.

20. Thomas, E.D., Storb, R., Clift, R.A., Fefer, A., Johnson, F.L., Neiman, P.E., Lerner, K.G., Glucksberg, H., and Buckner, C.D. Bone marrow transplantation. *New Eng. J. Med., 292*:832, 895, 1975.

21. UCLA Bone Marrow Transplant Team. Bone marrow transplantation in severe aplastic anemia. *Lancet, ii*:921, 1976.

22. UCLA Bone Marrow Transplant Team. Bone marrow transplantation with intensive combination chemotherapy-radiation therapy (SCARI) in acute leukemia. *Ann. Int. Med. 86*:155, 1977.

23. Warren, R.P., Storb, R., Weiden, P.L. Mickelson, E.M., and Thomas, E.D. Direct and antibody-dependent cell-mediated cytotoxicity against HLA identical sibling lymphocytes: Correlation with marrow graft rejection. *Transplantation, 22*:631, 1976.

24. Weiden, P.L., Storb, R., Schlicter, S., Warren, R.P., and Sale, G. Effect of six weekly transfusions in canine marrow grafts: Tests for sensitization and abrogation of sensitization with procarbazine and anti-thymocyte serum. *J. Immunol., 117*:143, 1976.

25. Weiden, P.L., Storb, R., Thomas, E.D., Graham, T.C., Lerner, K.G., Buckner, C.D., Fefer, A., Neiman, P.E., and Clift, R.A. Proceeding transfusions and marrow graft rejection in dogs and man. *Transplant Proc., 8*:551, 1976.

19

Histocompatibility Testing for Bone-marrow Grafting

M. Jeannet and B. Speck

INTRODUCTION

At the present time, the three main indications for bone marrow transplantation are severe combined immunodeficiency (SCID), bone marrow aplasia, and acute leukemia. In these diseases, the success of bone marrow transplantation is hampered by many difficulties. These difficulties include the lack of a suitable compatible donor, the frequent development of graft-versus-host disease (GvHD), and the occurrence of graft rejection. The latter is mainly seen in patients with severe aplastic anemia (SAA). Despite considerable recent advances in our knowledge of the major histocompatibility complex (MHC), and in the use of HLA-identical sibling donors for marrow transplantation, the problems of GvHD and graft rejection remain unsolved.

In this review, we summarize the present status of knowledge in the field of histocompatibility testing for bone marrow transplantation and discuss the various tests that are thought to be useful for the selection of a compatible donor.

THE SELECTION OF AN HLA-COMPATIBLE DONOR

Aside from the rarely occurring monozygotic twin, who represents of course the ideal donor, an HLA-identical sibling has always been considered to be virtually the only acceptable donor for a bone marrow graft. Sometimes an HLA-identical sibling is incompatible with the recipient for the ABO system, but this ABO barrier has been successfully ignored in a number of cases (26). When recipients have "minor" ABO incompatibilities with their donors (recipient A, B, or AB, donor O), no specific precautions need be taken; however, when there is a "major" ABO incompatibility (recipient O, donor A, B, or AB) plasma-exchange transfusion must be carried out to remove anti-A or anti-B antibodies before transplantation, in order to prevent hemolysis of the red blood cells (RBC) in the infused marrow suspension. In some cases, *in vivo* absorption of antibodies has been attempted, using blood group substances and incompatible RBC. Often there are few children in the patient's family, and an HLA-identical sibling is not available. Although the search for a HLA-compatible donor can be further carried on within the family, chances of finding such a donor are small. Moreover, the high frequency of GvHD and of graft rejection, even with HLA-identical donors, have made many clinicians reluctant to undertake bone marrow transplantation with donors who were not fully HLA identical. In SCID, however, it has been shown that immunological reconstitution can be obtained with bone marrow grafts from family donors other than an HLA-identical sibling (3,5). In these cases, selection of donors was based on nonreactivity in mixed lymphocyte culture (MLC). One patient has been successfully grafted with marrow from an MLC-identical, but A-locus nonidentical sibling. The HLA-A recombination between the paternal HLA-A and B determinants led to HLA-B and D identity between patient and donor (5). In two instances, the donor was the father, who was HLA identical and MLC negative with the recipient, because the father and the mother happened to share the same HLA haplotype. In another case, the donor was a maternal uncle who was incompatible with the recipient at the A and B loci, but compatible at the D locus, as shown by MLC and HLA-D typing (3).

Recently, Hansen et al. (9) have reported a number of instances of weak reactions in MLC between patients with SCID and other hemopoietic diseases and at least one parent. They have suggested a possible relationship between the sharing of histocompatibility determinants between mother and child and the pathogenesis of these diseases. In our experience, also, sharing of HLA-D determinants and MLC compatibility between the parents of patients with hemopoietic diseases is much more frequent than in ''Normal'' families (10). We have observed several examples of nonstimulation in MLC between a male patient and his mother. In a number of these cases, the sharing of HLA-D determinants could be confirmed by the serological typing of B-lymphocytes (Table 19-1).

In the absence of an HLA-identical sibling, a related or even an unrelated HLA-D compatible donor could theoretically be selected even if the donor has other different HLA loci (A, B, or C). Studies in experimental animals have provided evidence that the MLC is an *in vitro* correlate of the proliferative phase of GvHD (15). However, more experience is needed before one can recommend the use of donors who are only compatible at the HLA-D locus for clinical bone marrow transplantation. It seems that MLC reactivity is not always an infallible predictor of graft rejection or severe GvHD. A successful graft has recently been achieved in a patient with acute myeloblastic leukemia using an HLA-identical, MLC-reactive sibling donor (4). This unusual situation may result from a crossover between the HLA-B and HLA-D loci in the maternal or paternal chromosomes. The importance of assessing the role of each locus of the HLA complex in matching for bone marrow transplantation is best illustrated by the controversial issue of the use of unrelated donors when no HLA-identical sibling is available. So far, only one successful case has been reported (3). No family donor was available for this male patient with the autosomal recessive form of SCID. A prospective unrelated female donor who was compatible with the patient for HLA-B and -D, but not compatible for HLA-A was identified. After several bone marrow transplants from the same donor, complete hematological and lymphoid reconstitution was achieved. However, other patients with aplastic anemia have been grafted using unrelated MLC-compatible donors without success (7,24). The future of bone marrow transplants from unrelated donors remains uncertain. If further experience confirms that only HLA-D compatibility and MLC non-reactivity are necessary for successful transplantation, the task of finding suitable donors for patients with SCID, aplastic anemia, or even leukemia would be much easier. On the other hand, if full genotypic compatibility is necessary, the polymorphism of the HLA complex might make the finding of donors matched with the recipient for more than two or three loci impossible.

Recently, another approach has been initiated for patients having no HLA-identical sibling. Mathé and his coworkers (18) have reported that grafts from HLA-non-identical, MLC-positive family do-

TABLE 19-1 Weak Stimulation in MLC between Members of the Family of a Patient with Acute Leukemia Sharing B-Lymphocyte, HLA-D Antigens

| | STIMULATING CELLS | | | | |
RESPONDING CELLS	Father	Mother	Patient	Sister	Control
Father	—	0.9[a]	1.2	0.5	7.8
Mother	2.5	—	1.4	0.9	5.5
Patient	2.4	2.9	—	0.5	6.4
Sister	1.5	4.1	1.5	—	4.6
Control	3.1	5.7	0.8	TF[b]	10.7

Father, A2, B12, *DW2* / A11, B7, *DW7*
Mother, AW24, BW18, *DW2* / AW30, B13, *DW7*
Patient, A2, B12, *DW2* / AW30, B13, *DW7*
Sister, A2, B12, *DW2* / AW30, B13, *DW7*
[a]Stimulation index.
[b]TF, technical failure.

nors could induce remissions in aplastic anemias provided anti-lymphocyte globulin (ALG) was given before engraftment. This protocol has now been used in 12 aplastic patients (13,23). Eight are still alive, and five of these patients have had complete hemopoietic reconstitution. An additional aplastic patient who was treated with ALG alone, without a bone marrow graft, has also had complete hemopoietic reconstitution. The exact mechanism of these remissions is still unknown, but convincing evidence has been accumulating in recent years that aplastic anemia may have an immune pathogenesis, explaining, why immunosuppressive therapy by itself could have induced the remission. Although this protocol cannot be used for patients with SCID or leukemia, it provides an alternative form of therapy for patients with bone marrow aplasia who lack matched sibling donors.

DETECTION OF PRIOR SENSITIZATION

In spite of vigorous treatment with immunosuppressive drugs, the frequency of rejection in patients with aplastic anemia grafted with an HLA-identical sibling marrow is as high as 30 percent (26). Observations in both animal and human bone marrow transplantation have indicated that graft rejection is most likely related to prior exposure of the recipient to donor histocompatibility antigens by earlier transfusions (26,31). Detection of such sensitization may be possible by using a variety of in vitro tests of cell-mediated and humoral immunity.

Cellular Sensitization

Mickelson et al. (19) recently reported that the relative response index (RRI), that is the response of the patient's cells to donor's cells compared with the response to unrelated cells in MLC, can detect individuals who are most likely to reject their marrow grafts. Although these observations have not yet been confirmed, we have also found, in cases where patients have been previously transfused, a weakly positive response in MLC of patient cells to stimulation by lymphocytes of the HLA-identical sibling (10). These positive MLC tests (among HLA-identical siblings) cannot usually be explained by a single crossover between the HLA-B and D loci. The most likely explanation for these anomalous MLC tests would be the immunization of the patient against non-HLA antigens by previous blood transfusions. Thorsby and Helgesen (28) reported a significant MLC response of lymphocytes from a thrombopenic patient when confronted with stimulating cells from his HLA-iden-

tical brother, after numerous platelet transfusions from the brother. In addition, it has been shown in the rat (32) and in man (1) that lymphocytes immunized in vitro with MHC antigens react more promptly to the presence of cells from the immunizing donor. It appears, therefore, that the MLC test measures not only the overall amount of tissue antigenic disparity between graft donor and recipient, but probably also the extent of cell-mediated immunity. In fact, in a patient who had received a bone marrow graft from her HLA-identical MLC-negative sibling, we observed that the recipient's lymphocytes responded to stimulation by irradiated lymphocytes from the donor after graft rejection, whereas the reverse one-way test remained negative (10,22). The maximum level of response was observed 48 hr after culturing, whereas, with non-sensitized lymphocytes stimulated by allogeneic cells, the maximum response usually occurs between 120 to 144 hr (Table 19-2). In this patient, the indirect cell-mediated lympholysis (CML) test, performed at the same time, indicated that, after in vitro sensitization of the recipient's lymphocytes by the irradiated cells from the HLA-identical donor, the former could cause significant lysis of the donor's lymphocytes when used as target cells (10). Similar findings have been recently reported by Goldmann in a patient with aplastic anemia who rejected a bone marrow graft from an unrelated HLA-matched donor (7). These findings suggest that both the MLC and CML tests can sometimes detect cellular sensitization against non-HLA antigens that might be responsible for the rejection of the graft.

Recently, a similar test of cell-mediated immunity, the direct CML test, was shown to detect in vivo sensitization of patients with aplastic anemia to the transplantation antigens of HLA-identical and MLC-nonreactive sibling donors. A variation of the CML test, the so-called antibody-dependent cell-mediated lympholysis (ADCML) or lymphocyte-dependent antibody (LDA) assay, has also been investigated and will be discussed in the section on humoral sensitization. Parkman et al. (21) reported a positive CML test in four patients with aplastic anemia and a positive ADCML in one additional patient. Two of these patients subsequently rejected their marrow grafts from HLA-identical siblings. Warren et al. (30) investigated the ability of the CML assay to predict graft rejection in 26 patients with aplastic anemia. They found that this assay correlated significantly with graft outcome and could be used to identify patients who were likely to reject marrow grafts from HLA-identical siblings.

Recently, Goulmy et al. (8) reported an interesting phenomenon observed in a female patient

TABLE 19-2 Mixed Lymphocyte Culture Tests in Patient M.A. 1 Year after Bone Marrow Transplantation for Aplastic Anemia

RESPONDING CELL	STIMULATING CELL	24 HOUR		48 HOUR		120 to 144 HOUR	
		cpm[a]	Index	cpm[a]	Index	cpm[a]	Index
Recipient	Donor	635	3.9	298	2.8	1169	2.4
Donor	Recipient	249	1.9	126	1.7	375	1.4
Recipient	Unrelated control 1	165	1.0	120	1.1	2372	4.8
Recipient	Unrelated control 2	187	1.1	212	1.9	3660	7.5
Recipient	Recipient	163	—	107	—	490	—

[a]cpm, counts per minute.

with aplastic anemia who rejected a bone marrow graft from her HLA-identical brother. Using both the direct and indirect CML tests, they demonstrated a cytotoxicity of the patient's lymphocytes against the donor's lymphocytes 31 weeks after grafting. They were able to show, in addition, that the specificity of the cytotoxicity was not directed only against HLA antigens. Lysis occurred when target cells were derived from male donors carrying the HLA-A2 antigen of the original sensitizing cell, but not when the target cells were from female donors, even when they carried A2. This strong association with maleness suggests that the non-HLA specificity of the cytotoxic reaction was coded for by the Y chromosome. The clinical relevance of the Y-associated killing phenomenon is that it is probably an *in vitro* reflection of the known influence of sex on the incidence of GvHD and perhaps on rejection in bone marrow transplantation between HLA-identical siblings (25).

As shown in Table 19-3, we have also found the CML test useful in detecting sensitization of the patient against donor antigens. We used the CML assay in 14 patients with aplastic anemia, of which 10 received a bone marrow transplant. Seven of these transplants were from an HLA-identical brother (group 1). Another three, conditioned with ALG (group 2), were grafted with marrow from an HLA haplotype identical related donor. The CML test was positive, using donor lymphocytes as target cells in two out of seven patients of group 1. Two additional patients had a positive CML test, using the lymphocytes of another family member or those of unrelated individuals as target cells. Three of these four patients either had no engraftment or rejected their graft soon after transplantation. The fourth patient had engraftment but died from GvHD. Three of four patients had a sex-mismatched donor. In contrast, the three patients with a negative direct CML test had engraftment from a sex-matched donor. One of these patients later rejected her graft, but the two other patients have been full chimeras for more than a year. Two patients from group 2 had a negative CML test and reconstitution of their own marrow. The third, as

shown by the CML test, was strongly sensitized against his donor and other members of his family and had no remission. In summary, the CML test appears to be the best test presently available to predict the outcome of bone marrow grafts.

Humoral Sensitization

In human kidney transplantation, the presence in the recipient of humoral antibodies specific for donor lymphocytes has been shown to be associated with a hyperacute rejection of the graft (11). In bone marrow graft recipients, HLA-specific, complement-dependent lymphocytotoxic antibodies are frequently found at the time of transplantation. They are most probably produced in response to the numerous blood or platelet transfusions these patients have received because of their profound pancytopenia. In our experience, the presence of these lymphocytotoxic antibodies is not correlated with an increased risk of graft rejection, probably because these antibodies are rarely directed against non-HLA antigens. In fact, we have occasionally observed a positive crossmatch test, using the patient's serum and the lymphocytes of the HLA-identical donor. In the patient mentioned above, who rejected a graft from her HLA-identical sister, we found after rejection a strong lymphocytotoxic antibody that was reactive at 15 and 22 but not at 37°C with the donor cells and the patient's autologous lymphocytes (22). However, this cold antibody (Table 19-4) the activity of which was shown to reside in the IgM, was already demonstrable in the patient's serum before grafting, although the titer was lower. Its exact specificity remains unknown, but, since it was reactive with almost all the allogeneic lymphocyte suspensions tested, it does not appear to be directed against antigens belonging to the HLA system. Similar cold lymphocytotoxins, also reactive with the patient's autologous lymphocytes, have been observed in patients with viral and autoimmune diseases (27) and in patients waiting for a kidney transplant (16). In the latter, the antibodies do not appear to be associated with an increased risk of graft rejection. In our experience, they are

TABLE 19-3 Direct CML Activity before Bone Marrow Transplantation for Aplastic Anemia

PATIENT	HLA HAPLO-TYPE SHARED	SEX Donor	SEX Recipient	PANEL/FAMILY (MAXIMUM SPECIFIC LYSIS, PERCENT)	DONOR (SPECIFIC LYSIS, PERCENT)	COURSE
MAU	2	F	F	1.2	−0.8	Take, rejection, remission survival >4 years
JUL	2	F	F	5.2	7.7	No take, death 41 days
SCH	2	M	M	0.1	1.3	Take survival > 1 year
RIN	2	F	M	13.5	4.2	No take, death 28 days
PEL	2	M	M	2.3	3.2	Take survival > 1 year
ABD	2	M	F	40.3	18.4	Take, GvH, death 86 days
STO	2	M	F	5.3	−2.6	Take, rejection, remission survival > 250 days
BEN	1	M	M	23.0	7.0	No remission, death 76 days
CHA	1	M	M	3.6	−3.1	Remission survival > 240 days
MOR	1	F	F	7.6	0.9	Remission ? survival > 180 days

also occasionally detected in patients with aplastic anemia, but their significance and their possible role in bone marrow graft rejection is still ambiguous. Gluckman et al. (26) detected similar cold antibodies both before and after bone marrow grafting, the greater proportion of which were directed against B- rather than T-lymphocytes. The anti-B antibodies appear to increase in number after bone marrow rejection and during GvHD. We also observed a non-HLA, cytotoxic, complement-dependent antibody during GvHD in a patient grafted for severe combined immunodeficiency (12). This IgG antibody reacted with lymphocytes from the patient's father, mother, sisters, and with 70 percent of unrelated allogeneic lymphocytes. It was not reactive with the donor lymphocytes, suggesting that it was formed during the GvH reaction by the donor lymphocytes against a non-HLA-antigen present on the recipient tissues but absent on the donor cells. Recently, Opelz et al. (20) observed a patient who, following the rejection of marrow transplants from two HLA-identical sibling donors, developed antibodies that reacted strongly against donor cells but were thought not to be cold autocytotoxins. Because the non-HLA antigens that might be detected by these antibodies may be important in bone marrow transplantation, careful searching for donor-specific antisera in patients who have rejected grafts from HLA-identical sibling or who have had GvHD seems warranted. This may yield valuable typing reagents for non-HLA histocompatibility antigens.

Among the various techniques that have been used for the detection of humoral antibodies in human transplantation immunology, the lymphocyte-dependent antibody (LDA) assay (14) is certainly the most sensitive. Although the cells effecting the killing of target cells in both the previously discussed CML and the LDA assay appear to be lymphocytes, they possess different characteristics. The effectors in the CML reaction are specific immune cytotoxic T-lymphocytes (CTL), which act directly on target cells in the absence of antibody. In contrast, the cytotoxic mechanism of the LDA depends on an effector-antibody-target association and the effectors (Killer or K-cells) possess receptors for the F_c portion of the IgG antibody coating the target, Thus, the characteristics of the effector cell, as well as the mechanism of cytolysis, is quite different in these two assays.

Recently, Warren et al. (30) reported their results using two variations of the LDA assay. Twenty-one recipients were studied using patient lymphocytes as effector cells and their HLA-identical donor cells as target cells. They found that eight recipients were positive, and seven rejected their grafts, whereas thirteen were negative and five had a graft rejection. Twenty-four recipients were tested with the LDA assay using recipient plasma and unrelated lymphocytes as effector cells. Seven recipients were positive and six rejected their grafts, whereas seventeen were negative and five had a graft rejection. The correlation of both combinations of the test with graft outcome was significant. In addition, most of the patients also had positive direct CML tests (see the section on cellular sensitization). Using the LDA technique with unrelated lymphocytes as effectors, we have investigated 17 patients with aplastic anemia prior to bone marrow grafting.

Antibodies were detected using family or unrelated lymphocytes as target cells in five cases, whereas in only one case was the activity directed against the HLA-identical donor lymphocytes. This patient, nevertheless, had successful engraftment. The LDA assay seems to represent a useful technique for the detection of humoral sensitization against donor non-HLA antigens in patients with

TABLE 19-4 Mixed Lymphocyte Culture 4-Hour Test with Donor and Panel Lymphocytes in Patient M.A. before and after Bone Marrow Transplantation for Aplastic Anemia

	15°C		22°C		37°C		PERCENT PANEL		
	Donor	Recipient	Donor	Recipient	Donor	Recipient	15°C	22°C	37°C
Before transplantation serum	1:4	1:2	1:2	—	—	—	86	78	—
One month after transplantation									
serum	1:16	1:8	1:16	1:4	—	—	100	100	—
IgG fraction	1:1	1:1	—	—	—	—	—	—	—
IgM fraction	1:16	1:4	1:16	1:8	—	—	100	84	—
Two months after transplantation									
serum	1:8	1:8	1:8	1:1	—	—	100	100	—
IgG fraction	—	—	—	—	—	—	—	—	ND[a]
IgM fraction	1:4	1:2	1:4	—	—	—	70	70	—
Five months after transplantation serum	1:1	—	—	—	—	—	30	—	—
8 months after transplantation serum	1:2	—	—	—	—	—	30	—	—

[a]ND, no data.

aplastic anemia, especially when patient serum and effector cells are used in combination. It remains to be seen whether the demonstration of humoral sensitization against the donor by this assay, in the absence of cellular sensitization, will predict graft rejection. Bone marrow graft recipients may be sensitized by blood transfusions against non-HLA antigens present on blood cells other than lymphocytes. It is well established, for instance, that granulocytes (17) and platelets (2) carry specific antigens that are not carried by other cells. The possible role of such antigens in graft rejection and GvHD has not been investigated. Recently, Warren et al. (29) failed to demonstrate an involvement of human leukocyte group 5 antigens in GvHD and graft rejection. It is clear that further efforts are needed before the immunogenetic basis of bone marrow transplantation is fully understood.

SUMMARY

Bone marrow transplantation from an HLA-identical sibling is an effective form of therapy for patients with severe aplastic anemia and immunodeficiency. However, in spite of HLA compatibility with the donor, approximately one-third of the patients with aplastic anemia reject their graft and graft-versus-host disease occurs frequently. Histocompatibility antigens outside the HLA complex may therefore be important in bone marrow grafting. Among the tests commonly used for the selection of a compatible donor, HLA typing and the mixed lymphocyte culture test appear to be the most important at the present time. *In vitro* tests

of humoral and cellular immunity, such as the lymphocyte-dependent antibody and the cell-mediated lympholysis assays, may also yield valuable information on the state of sensitization of the recipient to donor non-HLA antigens. In the future, it may become necessary to match the donor and recipient for non-HLA antigens present on granulocytes, lymphocytes, or platelets, in order to avoid marrow rejection or severe graft-versus-host reactions.

ACKNOWLEDGMENTS

We are grateful to Karen Binder for reviewing the manuscript and to Nadia Carnal for secretarial help.

This work was supported by the Swiss National Fund for Scientific Research Grant 3.407.074.

REFERENCES

1. Bondevik, H., and Thorsby, E. Still a role for HLA in the MLC interaction. In *Lymphocyte Recognition and Effector Mechanisms*. New York, London: Academic Press, 1974, p. 225.

2. Colombani, J., and Colombani, M. Serologic recognition of histocompatibility antigens using complement fixation. *Semin. Hematol.*, 2:273, 1974.

3. Dupont, B., Hansen, J.A., Good, R.A., and O'Reilly, R. Histocompatibility testing for clinical bone marrow transplantation. In Ferrara, G.B., ed., *HLA System - New Aspects*. Amsterdam: Elsevier/North Holland Biomedical Press, 1977, p. 153.

4. Feig, S.A., Opelz, G., Winter, H.S., Falk, P.M.,

Neerhout, R.C., Sparkes, R., and Gale, R.P. Successful bone-marrow transplantation against mixed lymphocyte culture barrier. *Blood, 48*:385, 1975.

5. Gatti, R.F., Meuwissen, H.J., Terasaki, P.I., and Good, R.A. Recombination within the HLA-A locus. *Tissue Antigens, 1*:239, 1971.

6. Gluckman, E., Andersen, E., Lepage, U., and Dausset, J. Non-HLA lymphocytotoxic antibodies during GVHD after bone marrow transplantation (BMT). *Transpl. Proc., 11*:761, 1977.

7. Goldmann, S.F. Selection of related and unrelated bone marrow donors. *Transplant. Proc., 9*:507, 1977.

8. Goulmy, E., Termijtelen, A., Bradley, B.A., and Van Rood, J.J. Y-antigen killing by T cells of women is restricted by HLA. *Nature (Lond.), 266*:544, 1977.

9. Hansen, J.A., Good, R.A., and Dupont, B. HLA-D compatibility between parent and child. *Transplantation, 23*:366, 1977.

10. Jeannet, M., Klouda, P.T., Vassalli, P., Ramirez, E., Legendre, C., and Speck, B. Anomalous MLC and CML Tests in Human Bone Marrow Transplantation in Histocompatibility Testing. Copenhagen: Munksgaard, 1975, p. 885.

11. Jeannet, M., Pinn, V.W., Flax, M.H., Winn, H.J., and Russell, P.S. Humoral antibodies in renal allotransplantation in man. *N. Eng. J. Med., 3*:282, 1970.

12. Jeannet, M., Rubinstein, A., and Pelet, B. Studies on non-HLA cytotoxic and blocking factor in a patient with immunological deficiency successfully reconstituted by bone-marrow transplantation. *Tissue Antigens, 3*:411, 1973.

13. Jeannet, M., Rubinstein, A., Pelet, B., and Kummer, H. Prolonged remission of severe aplastic anemia after ALG pretreatment and HLA semi-incompatible bone-marrow cell transfusion. *Transplant. Proc., 6*:359, 1974.

14. Jeannet, M., and Vassalli, P. The role of lymphocyte-dependent antibody in kidney transplantation. *Transplantation, 22*:493, 1976.

15. Klein, J. Relative importance of H2-regions in the development of graft-versus-host reactions. *Transplant. Proc., 8*:335, 1975.

16. Klouda, P.T., and Jeannet, M. Cold and warm antibodies and graft survival in kidney allograft recipients. *Lancet, i*:876, 1976.

17. Lalezari, P., and Radel, E. Neutrophil-specific antigens: Immunology and clinical significance. *Semin. Hematol., 2*:281, 1974.

18. Mathe, G., Amiel, J.H., Schwarzenberg, H., Choay, J., Trolard, P., Hayar, M., Schlumberger, J.R., and Jasmin, C. Bone-marrow graft in man after conditioning by antilymphocytic serum. *Transplant. Proc., 3*:325, 1971.

19. Mickelson, E.M., Fefer, A., Storb, R., and Thomas, E.D. Correlation of the relative response index with marrow graft rejection in patients with aplastic anemia. *Transplantation, 22*:294, 1976.

20. Opelz, G., Territo, M.C., Gale, R.P., and Sparkes, R. Sensitization against non-HLA antigens following bone marrow graft rejection. *Tissue Antigens, 9*:209, 1977.

21. Parkman, R., Rosen, F.S., Rappeport, J., Camitta, B., Levey, R.L., and Nathan, D.G. Detection of genetically determined histocompatibility antigen differences between HLA identical and MLC non-reactive siblings. *Transplantation, 21*:110, 1976.

22. Speck, B., Cornu, P., Jeannet, M., Nissen, C., Burri, H.P., Groff, P., Nagel, G.A., and Buckner, C.D. Autologous marrow recovery following allogeneic marrow transplantation in a patient with severe aplastic anemia. *Exp. Hematol., 4*:131, 1976.

23. Speck, B., Cornu, P., Sartorius, J., Nissen, C., Groff, G., Burri, H.P., and Jeannet, M. Immunologic aspects of aplasia. *Transplant. Proc.* (in press).

24. Speck, B., Zwaan, F.E., Van Rood, J.J., and Eernisse, J.G. Allogeneic bone-marrow transplantation in a patient with aplastic anemia using a pheno-typically HLA-identical unrelated donor. *Transplantation, 16*:24, 1973.

25. Storb, R., Prentice, R.L., and Thomas, E.D. Treatment of aplastic anemia by marrow transplantation from HLA identical siblings. Prognostic factors associated with graft-versus-host disease and survival. *J. Clin. Invest., 9*:625, 1977.

26. Storb, R., Thomas, E.D., Weiden, P.L., Buckner, C.D., Clift, R.A., Fefer, A., Fernando, L.P., Giblett, E.R., Goodell, B.W., Johnson, F.L., Lerner, K.G., Neiman, P.E., and Sanders, J.E. Aplastic anemia treated by allogeneic bone-marrow transplantation: a report of 49 new cases from Seattle. *Blood, 48*:817, 1976.

27. Terasaki, P.I., Mottironi, V.D., and Barnett, E.V. Cytotoxins in disease. Autocytotoxins in lupus. *N. Eng. J. Med., 283*:724, 1970.

28. Thorsby, E., and Helgesen, A. Possible detection of sensitization against non-HLA histocompatibility antigens *in vitro*. *Symp. Ser. Immunobiol. Stand., 18*:141, 1973.

29. Warren, R.P., Storb, R., Nguyen, P.D., and Thomas, E.D. The failure to demonstrate an involvement of human leukocyte group 5 antigens in graft-versus-host disease and marrow graft rejection. *Transplantation, 24*:89, 1977.

30. Warren, R.P., Storb, R., Weiden, P.L., Mickelson, E.M., and Thomas, E.D. Direct and antibody-dependent cell-mediated cytotoxicity against HLA identical sibling lymphocytes. *Transplantation, 22*:631, 1976.

31. Weiden, P.L., Storb, R., Schlichter, S., Warren, R.P., and Sale, G.E. Effects of six weekly transfusions on canine marrow grafts: Tests for sensitization and abrogation of sensitization by procarbazine and antithymocyte serum. *J. Immunol., 117*:143, 1976.

32. Wilson, D.B., and Nowell, P.C. Quantitative studies on the mixed lymphocyte interaction in rats. V. Tempo and specificity of the proliferative response and the number of cells from immunized donors. *J. Exp. Med., 133*:442, 1971.

20

Recent Developments in Histocompatibility Testing in Bone Marrow Transplantation

J. J. van Rood,
A. van Leeuwen, E. Goulmy,
A. Munro, A. Termijtelen,
and B. A. Bradley

INTRODUCTION

One of the main goals in histocompatibility research ought to be the identification of non-HLA loci important in graft-versus-host disease (GvHD) and graft rejection. Even if this were achieved, it would only help the clinician identify those patients with HLA-identical siblings who have a minimal risk of GvHD or graft rejection. Although an important contribution, it would not solve the problem of what to do for those patients who do not have an HLA-identical sibling donor who is also identical at these non-HLA loci. In fact, with increasing selectivity based on better typing, the number of these problem patients increases. For such patients who have no related donor, help could come from two directions:

1. The use of unrelated donors compatible for HLA and non-HLA loci. However, we need to know whether matching for HLA between unrelated donor-recipient pairs can reach the same degree of compatibility as that between siblings.
2. More effective immunosuppression.

WHAT PROGRESS HAS BEEN MADE IN RECOGNIZING NON-HLA ANTIGENS?

Successful platelet transfusion holds a key position in bone marrow transplantation, especially in sustaining the patient during and after bone marrow transplantation. In hyperimmunized patients, only about 60 percent of the platelet transfusions will be successful if HLA-matched platelets are used (3).

The Seattle bone marrow transplant group has shown that GvHD is more serious in patients who are resistant to platelet transfusions (26). The possibility of matching for platelet antigens and preventing immunization to them are thus important goals in bone marrow transplantation and will be discussed first. Following the lead of the Seattle group, several studies have shown that the prognosis of bone marrow transplantation, especially as determined by GvHD, is poorer when the donor is a female and the recipient a male (Table 20-1) (1,11,26). The question one is confronted with is whether this is due to an *in vitro* detectable immunity against the H-Y antigen, as has been described in rodents. This question is especially relevant because not all sex-mismatched grafts fail because of GvHD, and such an *in vitro* test would make it possible to identify which donor-recipient pairs were at risk for GvHD.

TABLE 20-1 Successful Bone Marrow Transplantation in Males

DONOR/RECIPIENT	FRACTION OF SURVIVORS		
	Seattle series[a]	E.B.M.T. series[b]	S.C.I.D.[c] International Bone Marrow Transplant Registry Report[d]
M → M	8/9	17/24	15/32
F → M	7/18	5/21	5/25
p Value	0.02	0.002	0.03

[a]Storb et al. (26).
[b]European Bone Marrow Transplant (EBMT) Cooperative Group (11).
[c]Severe combined immunodeficiency disease.
[d]JAMA (1).

PLATELET ANTIGENS

Brand et al. have recently shown that a near 100 percent success rate can be achieved if HLA matching is supplemented by a sensitive platelet cross-match by an immunofluorescence technique (3). With this technique, a set of non-HLA antigens can be recognized.

It can be anticipated, however, that the incidence of hyperimmunized patients will diminish. Eernisse has shown that sensitization by platelet transfusion can be prevented in over 70 percent of the patients by the use of platelets that have been centrifuged three times and that contain virtually no lymphocytes (8). Eernisse's contribution is thus of direct importance in clinical bone marrow transplantation because of the already-mentioned fact that GvHD is more severe in platelet-resistant patients (26).

H-Y IMMUNITY

Goulmy et al. studied six hyperimmunized women for evidence of anti-H-Y immunity using the cell-mediated lympholyses (CML) test (16). They found significant evidence for such immunity in two patients, both of whom were suffering from aplastic anemia and one of whom had received a bone marrow transplant. It is of interest that this immunity is only evident when both the female effector cells and the male target cells carry HLA-A2 (Table 20-2). It remains to be determined whether this test really predicts H-Y dependent GvHD. Nevertheless, its relevance in allograft reactions is emphasized by the finding that male donor-female recipient kidney grafts that share HLA-A2 have a 2-year graft survival of 38 percent, whereas if they both lack HLA-A2, it is 58 percent (15).

The HLA-A2 restricted H-Y immunity is the first example in man of the dual recognition phenomen that has for some time been recognized in the mouse (33). In mouse systems, the restriction

TABLE 20-2 Sex-related Killing in CML of HLA-A2 Positive Target Cells

	KILL (PERCENT)	SEX OF HLA-A2 POSITIVE TARGET	
		M	F
CML	> 30	15	0
by Mrs. R.'s	10–20	0	2
effector cells[a]	< 10	0	17
CML	> 30	6	0
by Mrs. K.'s	10–20	0	5
effector cells[a]	< 10	0	5

[a]The effector cells were HLA-A2 positive.

phenomenon has been described not only for H-Y but also for microbial and artificial antigens (2,14). The precise mechanism of the dual recognition phenomenon is not known.

The HLA-A2 restricted H-Y immunity can be demonstrated for more than a year after immunization, but it eventually wanes. It can, however, be reactivated in vitro by stimulation with unrelated HLA-A, HLA-B, and HLA-C identical, but HLA-D different lymphocytes.

The serum from one of these patients (Mrs. R.) was also examined for serological activity against H-Y. The standard, two-step complement-dependent cytotoxicity (CdC) technique was negative, but van Leeuwen was able to show that a two-color fluorescence technique, which enabled the recognition of CdC directed to identifiable subsets of mononuclear cells showed antibodies that reacted with HLA-A2 male donors (Table 20-3) (23). Only

TABLE 20-3 HLA-A2 Restricted Anti-H-Y Immunity as Detected by Serology

TWO-COLOR FLUORESCENCE CYTOTOXICITY TEST	CELL-MEDIATED LYMPHOLYSIS	
	Positive	Negative
+	8	1
−	0	13

a fraction of the mononuclear cells that were not typical B-cells were killed.

The interpretation of these two sets of data is difficult. In the case of the CML reaction, one can conceive of two hypotheses. In the first, the killer cell carries two receptors, one for HLA-A2 and one for H-Y. In the second, the killer cell carries one receptor directed to a neoantigen generated in an interaction between HLA-A2 and H-Y.

The serological reactions must be considered separately. Here, there are three possibilities. First, we would be dealing with two antibodies produced by two clones of B-cells directed to HLA-A2 and H-Y; second, we could be dealing with an antibody directed to an neoantigen produced by HLA-A2 and H-Y; and third, it could be that all three antibodies, HLA-A2, H-Y, and the antibody raised by the neoantigen, reacting together are needed for killing to occur. It should be possible to determine which of these situations exists.

It can be said that progress is being made in two areas that are relevant to bone marrow transplant survival. These are platelet alloantigens and immunity to sex-linked transplantation antigens. There are, of course, many more non-HLA systems. The compatibility of some of these systems, such as the ABO blood group system and the tissue system group Five, have been shown not to be relevant to bone marrow transplantation (4,21). For others, such as the granulocyte systems (NA, NB, 14) and the endothelial and monocyte system E, relevancy has not been tested for (18,19,24). For none of these systems, with the exception of the ABO system, has it been ascertained by cross-matching procedures whether circulating antibodies against these systems in the recipient could cause the rejection of the bone marrow graft. It appears to us to be a point of some urgency to find out whether this is so. The Seattle group and Jeannet have provided further evidence in that ^{51}Cr release studies, antibody-dependent CdC, as a sensitive indicator of cellular and humoral immunity of the recipient against the donor, correlated with graft rejection (32; Jeannet, personal communication). It is not known for which non-HLA system incompatibility was detected. Several other systems in animals have been shown to be of possible relevance for bone marrow transplantation: the Hh and Mli systems in the mouse and the colonic secretion system W/Z in the dog (7,12,34). Whether analogous systems exist in man has not been determined. In this context, it is of significance that, for the detection of all these determinants, techniques other than the CdC technique are necessary. Furthermore, the target cells in these techniques will also be different from the mononuclear cell suspension routinely used in the CdC technique. The

European Bone Marrow Transplant Cooperative Group plans to make the technical facilities necessary to carry out these tests available to its members.

CAN HLA-IDENTICAL, UNRELATED DONORS BE USED IN BONE MARROW TRANSPLANTATION?

The question can be answered, in all probability, in the affirmative. But only a few of these transplants have been attempted. With the possible exception of one case, none was successful (10). But encouragement is given by a small number of mixed leukocyte culture (MLC) negative haplo-identical transplants (parent-child, siblings) from non-inbred families, several of which have been reported to be successful (6). It must be assumed that if it is possible to match successfully for one "unrelated" haplotype, it should also be possible to match for two. The lesson here is that if no HLA-identical sibling is available, one should not despair but look for other MLC-negative family members. An added advantage is that parents have a good chance of being identical for non-HLA loci.

The question then can be asked whether one should match for the SD or HLA-A, -B, and -C antigens or the MLC-stimulating determinants or both. In man, sufficient data are of course not available. Furthermore, no data are available for outbred mice and only limited data for the dog (27,31). Most critical would be studies in monkeys, which will probably soon be forthcoming.

Because bone marrow transplant data in experimental animals are lacking, it seems reasonable to ask whether other transplants in man may indicate what to expect. Table 20-4 summarizes experimental skin transplants in man, and these are a much more severe test of compatibility than are bone marrow transplants. It is clear that a negative MLC is a more effective way to improve skin graft survival than SD matching, but the combination of the two is still better. The 17-day mean graft survival that can be obtained in this manner is not much shorter than that obtained in HLA-identical siblings in our laboratory (19 days). Similar data are avail-

TABLE 20-4 Skin Graft Survival and Matching for HLA

HLA-A and HLA-B	MLC	MEAN SURVIVAL Time ± S.E. (days)	NUMBER
Nonidentical	+	10.0 ± 1.08	13
Identical	+	11.8 ± 0.59	20
Nonidentical	−	14.4 ± 2.07	5
Identical	−	17.3 ± 2.5	4

TABLE 20-5 The Recognition of HLA-D Determinants by B-Cell Serology

HLA-D	ANTI-B-CELL SERUM	+ +	− +	+ −	− −	HTC	Anti-B cell sera
		HLA-D ANTI-B-CELL SERUM				GENE FREQUENCIES DEFINED BY	
DW1	He	8	0	4	19	0.0889	0.1339
	Du	11	0	0	21		
DW2	CB	13	0	1	47	0.1287	0.1229
	RD	9	0	4	44		
DW3	Moa	27	0	0	34	0.1462	0.2033
	WH	16	2	4	39		
DW4 (+ 7)	Smith (Bodmer)	20	10	1	30	0.0669	0.1229
DW5	Pichon (Dausset)	8	5	2	45	0.0412	0.0906
DW6 (+ 2)	Po	14	4	8	35	0.0458	0.1229
LD 107	Si	15	1	0	45	0.0672	0.1451
LD 108	TL (Thorsby)	8	0	7	26	0.1039	0.0594
						0.6888	1.0010

The panel was first typed for HLA-D by HTC and PLT, and was then studied by sera recognizing B-cell determinants. The fit of the B-cell typing, with PLT-corrected HTC typing, is remarkably good for some determinants. It is clear that the sum of the gene frequencies of the HLA-D determinants as established by HTC alone is much smaller than if the HLA-D determinants had been studied by Bcell serology. The names of the investigators who made these sera available are given in parentheses. The remainder of the sera were obtained from Leiden.

able for kidney grafts (5). The question is, How does one obtain unrelated donors who are identical for the HLA-A, -B, and -C antigens and MLC negative with the recipient. Large numbers of donors have been HLA typed, with the intention of using them for platelet and granulocyte support therapy, and some of these donors are willing to donate bone marrow. In Europe alone, the files already exceed 50,000 HLA-typed donors. Thus, it would be possible to select, in many instances, for a given patient a donor who is at least HLA-A and -B matched. Those donors who are MLC negative with the recipient could be further selected by direct MLC testing. This is, however, a laborious procedure and when the patient has only a few lymphocytes, it will be difficult if not impossible to realize.

Recently, it has become possible to type for the alleles of the strong MLC-stimulating HLA-D locus using serological techniques (22). (Table 20-5). There are only eight alleles for HLA-D (30). It is quite possible that some of the HLA-D specificities will be "split" in two or more determinants, as was the case for the HLA-A and -B locus antigens. Nevertheless, as Table 20-6 shows, matching for even this small number of alleles increases MLC negativity from practically 0 percent in the HLA-D different combinations through 25 percent in the SD-nonidentical, HLA-D-identical combination, to nearly 80 percent in the SD- and HLA-D-identical combinations. These data confirm and extend earlier observations and prove that MLC reactivity is governed not only by loci in the HLA-D region, but also by loci near HLA-A and/or -B. These findings have furthermore a number of theoretical and prac-

TABLE 20-6 Matching for HLA-A, -B, and -D and the Outcome of the MLC Test

HLA		MLC STIMULATION INDEX		
-A and -B	-D	< 2	2–8	> 8
≠	≠	1	65	148
=	≠	0	11	11
≠	=	15	26	16
=	=	10	2	0

tical implications. From the practical point of view, the findings are clear cut. Identity at HLA-A, -B, and -D implies a 80 percent chance of a negative MLC. The loci responsible for the remaining 20 percent could be outside HLA and should be identified. From the theoretical point of view, these findings are less simple. The one MLC negative HLA-A, -B, and -D disparate combination found conflicts with most accepted theories on MLC testing. Short of a technical error, which is unlikely, it can only be explained by non-reactive clones, as was first described by Svejgaard's group (29). Individuals who are mismatched for HLA-A and HLA-B, but matched for HLA-D, produce a significant stimulation presumably because of differences for loci identical with or near to HLA-A and/or -B.

It is clear that, until now, our view on the mechanism of strong MLC stimulation has been simplistic. It is even possible that an equivalent of the H-2 I-J locus coding for determinants on suppressor cells exists in man and that disparity for such a locus can suppress MLC activation (28). It is, of course, too early to know whether these findings can be used in the selection of unrelated donors for bone marrow transplantation, but it is equally clear that we ought to find out. For the time being, our attitude should be one of extreme caution. If at all possible, we should wait until data from Rhesus monkeys are available. Until that time, only SD-identical, MLC-negative donors should be used and the CML test should also be negative.

THE INFLUENCE OF SENSITIZATION ON THE EFFECTIVENESS OF IMMUNOSUPPRESSION

Gluckman et al. have shown that the responsiveness of the patient, i.e., his or her capacity to form immunity against major histocompatibility complex (MHC) determinants and other antigens, and the effect immunosuppression has on this immunity, correlates with bone marrow graft prognosis and especially rejection (13). The immunological capacity of both patient and donor are thus variables to reckon with, and the question whether we can influence it by other means than chemotherapy alone is thus a relevant one. Although we know that anti-HLA-A, and -B antibodies in the recipient will destroy most kidney grafts from incompatible donors, it is equally certain that kidney grafts in recipients who had received no blood transfusions before transplantation have a graft prognosis that is extremely poor: less than 20 to 30 percent of the grafts function at 1 year in the non-transfused group, versus 70 percent in the transfused group (17,20). Balner and his colleagues have confirmed these findings in a randomized prospective study in Rhesus monkeys (9). The unexpected observation was that whereas immunosuppression alone (standard doses of Imuran and prednisone) *after* transplantation had no effect on graft survival, the combination of immunosuppression after transplantation with blood transfusions *before* transplantation had a very significant effect (Figure 20-1).

The mechanism for this very effective graft prolongation is completely unclear (competition, suppression cell induction, enhancing antibodies?). But even if the mechanism is unknown, it might still be of interest to test whether it could be used to suppress the GvH reactivity of the donor, especially against non-HLA determinants. It might be a relatively small price for the donor to pay if this could prevent GvHD in the recipient.

SUMMARY

With the help of the immunofluorescence test, platelet antibodies can be detected and compatible platelets selected. Such platelets, if HLA compat-

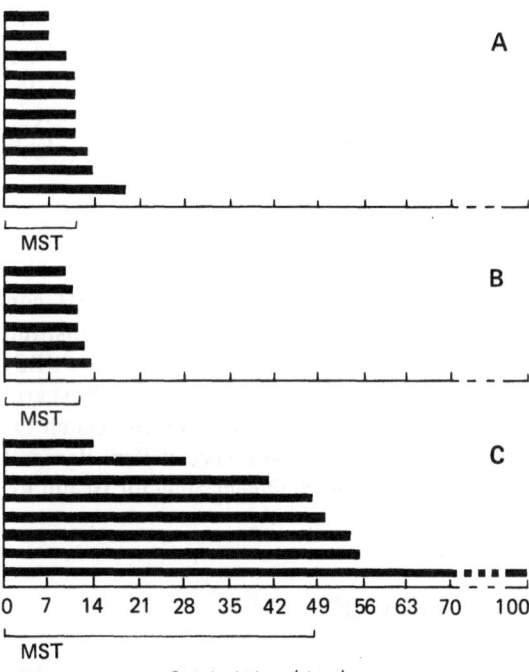

FIGURE 20-1. The influence of blood transfusion on kidney allograft survival in unrelated Rhesus monkeys. (A) Non-transfused/no immunosuppression, mean survival time (MST), 11.4 days; (B) non-transfused/immunosuppression, MST, 11.0 days; transfused/immunosuppression, MST, 48.8 days. Reprinted with permission of *Lancet* (see reference 9).

ible, have a nearly 100 percent survival rate, even in hyperimmunized patients. The use of lymphocyte-free platelets for transfusion, which can be easily obtained by three-step differential centrifugation, prevents immunization in over 70 percent of the recipients.

The cell mediated lympholysis (CML) test and a complement-dependent cytotoxicity (CdC) test can be used to detect HLA-restricted, anti-H-Y immunity in some immunized women. This might explain, in part, the poorer prognosis of the sex-mismatched bone marrow grafts.

In the future, if HLA-identical siblings of patients are not available or suitable for bone marrow transplantation, unrelated HLA-identical donors might take their place. Clinical data obtained with other organ transplants indicate that prognosis when the donor is unrelated might be comparable or better than that of HLA-identical sibling donors. Thus, HLA-D-identical donors can be found relatively easily because it is now possible to type for HLA-D serologically. The HLA-A, -B, and -D-identical, donor-recipient pairs are in mixed leukocyte culture (MLC) negative about 80 percent of the cases.

If blood transfusions are combined with chemotherapy (Imuran and prednisone), very effective "immunosuppression" can be obtained in kidney transplantation. It would be of interest to test whether such a procedure could be used to increase the number of successful bone marrow transplants.

ACKNOWLEDGMENT

This work was in part supported by NIH contract NO-AI-4-2508, the J. A. Cohen Institute for Radipathology and Radiation Protection (IRS), the Dutch Foundation for Medical Research (FUNGO), which is subsidized by the Dutch Organization for the Advancement of Pure Research (ZWO), and the Dutch Organization for Health Research (TNO).

REFERENCES

1. Advisory Committee of the International Bone Marrow Transplant Registry. Severe combined immunodeficiency disease—characterization of the disease and results of transplantation. *JAMA* (in press) 1977.
2. von Boehmer, H., Fatham, C.G., and Haas, W. H-2 gene complementation in cytotoxic I cell responses of female against male cells. *Eur. J. Immunol., 7*:443, 1977.
3. Brand, A., van Leeuwen, A., Eernisse, J.G., and van Rood, J.J. Platelet transfusion therapy. Optimal donor selection with a combination of lymphocytotoxicity and platelet fluorescence tests. *Blood* (submitted)
4. Bussel, A., Gluckman, E., Benbunan, M., Marty, M., Dausset, J., and Bernard, J. Bone marrow grafting and major ABO incompatibility. *Proc. IV Ann. Conf. Int. Soc. Exp. Hematol. Medena Bay*, Sept 21-24, 1975, p. 71.
5. Cochrum, K., Perkins, H., Payne, R., Kountz, S., and Belzer, F.O. The correlation of MLC with graft survival. *Transplant. Proc., 5*:391, 1973.
6. Copenhagen Study Group on Immunodeficiencies. Bone marrow transplantation from an HLA non-identical but mixed lymphocyte culture identical donor. *Lancet, i*:1146, 1973.
7. Cudkowicz, G., and Lotzova, E. Hemopoietic cell-defined components of the major histocompatibility complex of mice: Identification of responsive and unresponsive recipients for bone marrow transplants. *Transplant. Proc., 5*:1399, 1973.
8. Eernisse, J.G., and Brand, A. Postponement of sensitization by blood and random platelet transfusions (Abstr). *Exp. Hematol., 5* (Suppl 2):83, 1977.
9. van Es, A.A., Marquet, R.L., van Rood, J.J., Kalff, M.W., and Balner, H. Blood transfusions induce prolonged kidney allograft survival in rhesus monkeys. *Lancet, i*:506, 1977.
10. De L'Espérance, P., Hansen, J.A., Jersild, C., O'Reilly, R., Good, R.A., Thomsen, M., Nielsen, L.S., Svejgaard, A., and Dupont, B. Bone marrow donor selection among unrelated four-locus identical individuals. *Transplant. Proc., 9*:823, 1977.
11. European Bone Marrow Transplant Cooperative Group. Unpublished data.
12. Festenstein, H., and Halim, K. HLA-D-locus determinants detected by sperm-lymphocyte culture. *Transplant. Proc., 9*:1239, 1977.
13. Gluckman, E., Devergie, A., Marty, M., Bussel, A., Rottembourg, J., Dausset, J., and Bernard, J. Allogeneic bone marrow transplantation in aplastic anemia. Report of 25 cases. *Transplant. Proc.* (in press), 1977.
14. Gordon, R.D., Simpson, E., and Samuelson, L.E.: *In vitro* cell-mediated immune responses to the male specific (H-Y) in mice. *J. Exp. Med., 142*:1108, 1975.
15. Goulmy, E., Bradley, B.A., Lansbergen, Q., and van Rood, J.J. The importance of H-Y incompatibility in human organ transplantation. In preparation.
16. Goulmy, E., Termijtelen, A., Bradley, B.A., and van Rood, J.J. Y-antigen killing by women is restricted by HLA. *Nature (Lond.), 266*:544, 1977.
17. van Hooff, J.P., Kalff, M.W., van Poelgeest, A.E., Persijn, G.G., and van Rood, J.J.: Blood transfusions and kidney transplantation. *Transplantation, 22*:306, 1976.
18. Lalezari, P., and Bernard, G.E. An isologous antigen-antibody reaction with human neutrophiles, related to neonatal neutropenia. *J. Clin. Invest., 45*:1741, 1966.
19. Moraes, J.R., and Stastny, P. A new antigen system expressed in human endothelial cells. *J. Clin. Invest., 60*:449, 1977.
20. Opelz, G., Mickey, M.R., and Terasaki, P.I. Identification of unresponsive kidney transplant recipients. *Lancet, i*:868, 1972.
21. van Rood, J.J. Major and minor histocompatibility systems in man and importance in bone marrow transplantation. *Transplant. Proc., 8*:429, 1976.
22. van Rood, J.J., van Leeuwen, A., Keuning, J.J., and Termijtelen, A. Evidence for two series of B cell antigens in man and their comparison with HLA-D. *Scand. J. Immunol., 6*:373, 1977.
23. van Rood, J.J., van Leeuwen, A., and Ploem, J.S. Simultaneous detection of two cell populations by two colour fluorescence and application to the recognition of B cell determinants. *Nature (Lond.), 262*:795, 1976.
24. van Rood, J.J., van Leeuwen, A., Schippers, A.M.J., Pearce, R., van Blankenstein, M., and Volkers, W. Immunogenetics of the group four, five and nine systems. In *Histocompatibility Testing 1967*. Baltimore: Williams & Wilkins, 1967, p. 203.
25. Shearer, G.M. Cell-mediated cytotoxicity to trinitro-phenyl-modified syngeneic lymphocytes. *Eur. J. Immunol., 4*:527, 1974.
26. Storb, R., Prentice, R.L., and Thomas, E.D. Treatment of aplastic anemia by marrow transplantation from HLA identical siblings. Prognostic factors associated with graft-versus-host disease and survival. *J. Clin. Invest., 59*:625, 1977.
27. Storb, R., Weiden, P.L., Graham, T.C., Lerner,

K.G., and Thomas, E.D. Marrow grafts between unrelated dogs homozygous and identical for DLA antigens. *Transplant. Proc., 9*:281, 1977.

28. Suciu-Foca, N., and Rubinstein, P.: Genetic fine structure of the HLA region. *Transplant. Proc., 9*:385, 1977.

29. Thomsen, M., Morling, N., Platz, P., Ryder, L.P., Staub-Nielsen, L., and Svejgaard, A. Specific lack of responsiveness to certain HLA-D (MLC) determinants with notes on primed lymphocyte typing (PLT). *Transplant. Proc., 8*:455, 1976.

30. Thorsby, E., and Piazza, A. Joint report from the VIth Int Histocompatibility Workshop Conference. II. Typing for HLA-D (LD-1 or MLC) determinants. In *Histocompatibility Testing 1975*. Copenhagen: Munksgaard, 1975, p. 414.

31. Vriesendorp, H.M., Bijnen, A.B., Westbroek, D.L., and van Bekkum, D.W. Genetics and transplantation immunology of the DLA complex. *Transplant. Proc., 9*:293, 1977.

32. Warren, R.P., Storb, R., Weiden, P.L., Mickelson, E.M., and Thomas, E.D. Direct and antibody-dependent cell-mediated cytotoxicity against HLA identical sibling lymphocytes. Correlation with marrow graft rejection. *Transplantation, 22*:631, 1976.

33. Zinkernagel, R.M., and Doherty, P.C. Restriction of *in vitro* T cell-mediated cytotoxicity in lymphocytic choriomeningitis within a syngeneic or semi-allogeneic system. *Nature (Lond.), 248*:701, 1974.

34. Zweibaum, A., Oriol, R., Dausset, J., Marcelli-Barge, A., Ropartz, C., and Lanset, S. Definition in man of a polymorphic system of the normal colonic secretions. *Tissue Antigens, 6*:121, 1975.

21

Transplantation of Lymphoid Cells in Patients with Severe Combined Immunodeficiency (SCID)

C. Griscelli, A. Durandy,
J. J. Ballet, J. L. Virelizier,
and F. Daguillard

INTRODUCTION

There have been many successful instances of reconstitution of patients with severe combined immunodeficiencies (SCID) by transplantation of bone marrow, fetal tissue, and more recently, cultured thymic epithelium (1,2,3,4). This report will summarize 13 attempts made in our hospital to correct various types of SCID by grafts of fetal liver and/or thymic cells or by transplantation of phenotypically or genotypically identical bone marrow tissue. Special emphasis will be given to the complete correction, with compatible bone marrow cells, of SCID patients carrying B-lymphocytes. These patients, who seem to have a defect of precursor T-cells, are reconstituted by donor marrow stem cells that mature into host thymus cells and become capable of expressing cell-mediated functions and of cooperating with recipient B-cells in antibody production.

CHARACTERISTICS OF PATIENTS WITH SCID

Distribution of patients by sex was 11 males and two females. Two patients had adenosine deaminase (ADA) deficiency. Among the 11 remaining patients, the family history strongly suggested that two had an autosomic form and five a sex-linked form of the disease. In four other patients, a sex-linked transmission was probable. At the time of their transplant, all the patients were less than 8 months old.

There was considerable variability in the white blood cell counts of the 13 patients: lymphopenia in four, including the ADA-negative patients, and near normal lymphocyte counts in nine. B-lymphocyte studies were done on 12 patients. Eight of these patients had increased numbers of B-cells in their peripheral blood, but in all 13 cases, the immunoglobulin levels were markedly depressed (except for the presence of maternal IgG). In all 13 cases, the anti-A and anti-B isoagglutinins were negative, and no significant rise in antibody titers was obtained following immunization. Data on erythrocyte-rosette-forming cells (E-RFC) were available for 12 SCID patients. In all cases, the E-RFC were less than 1 percent, and skin tests were negative. Mitogenic responses were evaluated in the presence of various polyclonal mitogens and allogeneic cells. The results showed that the response to phytohemagglutin (PHA), concanavalin-A (Con-A), or pokeweed mitogen (PWM) were absent. In four cases with an increased number of B-cells, there was a significant response to allogeneic leukocytes. The cells of three patients could not stimulate normal allogeneic cells.

GRAFTING OF FETAL TISSUE

One patient received fetal thymus alone and three other patients received fetal thymus and liver cells (i.v. or i.p.). The fetuses were obtained from neighboring hospitals in Paris and varied in age from 13 to 20 weeks' gestation. In each case, cell suspensions were prepared and injected within 2 hr following hysterotomy. Cell viability was more than 90 percent by trypan blue exclusion. The characteristics of these grafts are summarized in Table 21-1. The four patients were placed in Trexler Isolators within 8 days to 6 months after birth. Bacterial decontamination by oral antibiotics was attempted in all four cases and was successful in three cases (Patients 1, 3, and 4). In patient 4, however, *Candida albicans* continued to be present. The take of the graft was thought probable in two patients (patients 2 and 4) because of partial T- and B-cell reconstitution and strong signs of a graft-versus-host disease) (GvHD) (Patient 4). These observations demonstrate that fetal tissues can induce fatal GvHD. The take was certain in patient 1, who had a mild GvHD and a rise of E-RFC bearing the HL-A markers of the fetal donor. However, this patient, who survived 11 months, had no evidence of humoral functions. In patient 3, who died from infection 2 weeks after grafting, the graft did not take. Infection was also the cause of death in the three other cases. Patient 2 died within 48 hr after oral antibiotics were stopped. His septicemia was caused by gastrointestinal residual bacteria. This was unfortunate, since the patient had, 1 month earlier, shown signs of partial reconstitution. Patient 4 died from GvHD with gastrointestinal involvement and *C. albicans* septicemia. The first patient who developed cell-mediated functions left the isolator at the age of 7 months and was receiving gammaglobulins. His cell-mediated functions declined however, and he subsequently died from *Pneumocystis carinii* pneumonia.

GRAFTING OF BONE MARROW TISSUE

Nine patients received bone marrow cells in varying amounts. The results are summarized in Table 21-2. Five bone marrow transplants were HLA-A, -B, -D-genoidentical. The other four were HLA-A, -B-phenoidentical. Mixed leukocyte reactions were negative (proliferative index less than 2) in all cases, except patient 5, in whom it was difficult to assess the response to allogeneic cells because of a high spontaneous incorporation of labeled thymidine. The appearance of severe GvHD in this case indicated that the donor and the recipient were not identical at the D locus. In six other patients who survived long enough, mild signs of GvHD were observed, characterized by transitory (15 days) or persisting eosinophilia (several months) and skin rash, but no gastrointestinal involvement. Definite signs of T- and B-cell reconstitutions were obtained in seven cases, and chimerism was demonstrated in four of them by karyotyping, Y chromosome staining with quinacrine (5) or donor HLA typing. In patient 7, the rise of lymphocyte ADA suggested that these cells were of donor origin. Patient 8 (female) was reconstituted with marrow cells from an HLA-identical sister and showed red blood cells markers of donor origin.

Two patients died before any sign of take (patients 1 and 2). Three other patients died despite evidence of reconstitution. In patient 3, death was caused by the neurological sequelae of various episodes of meningitis experienced before grafting. Upon postmortem examination, B- and T-cell compartments could be visualized in most lymphoid organs, except the lymph nodes and Peyer patches (6). Patient 4 was grafted with cells from an HLA-A, -B, -D-identical grandmother. Before grafting, she had received twice-irradiated packed blood cells from her father. Despite the fact that these cells had been irradiated with 5,000 rad, they were

TABLE 21-1 Results of Fetal Tissue Transplantation in SCID *Groupe des enfants malades Paris*, Aug. 1977

PATIENT[a]	FETAL TISSUE[b]	NUMBER OF CELLS (10^7/kg)	GvH	RECONSTITUTION T	RECONSTITUTION B	POST-GRAFT SURVIVAL (Months)	FOLLOW-UP
1(1)	TH(20)	2.5	±	++	0	11	Pneumocytosis
2(7)	TH + L(13)	NA[c]	0	+	+	2.9	Septicemia
3(5)	TH + L(14)	NA[c]	0	0	0	0.5	Septicemia
4(3)	TH + L(14)	2.5 + 80	+++	±	±	2	Septicemia + GvHD

[a]Numbers in parentheses represent age (months).
[b]TH, Thymus: L, Liver; numbers in parentheses represent fetal age (weeks).
[c]NA, not available.

TABLE 21-2   Results of Bone Marrow Transplantation in SCID *Groupe des enfants malades Paris,* Aug. 1977

| PATIENT[a] | DONOR | TYPING COMPATI-BILITY | | NUMBER OF CELLS (10^7/kg) | RECONSTI-TUTION | | | POST-GRAFT SURVIVAL (MONTHS) | FOLLOW-UP |
		ABO	HLA-D (MLC)		GvHD	T	B		
1(8)	Father	+	+	92	?	0	0	0.3	Pneumocystosis
2(7)	Sister	+	+	35	?	?	?	0.1	Pneumocystosis
3(9)	Sister	+	+	100	±	+++	+++	2[a]	Sequelae of meningitis
4(3)	Grand-mother	+	+	6.5	+	++	+	2[a]	Graft versus graft
5(3)[c]	Grand-aunt	+	?	30	++	++	±	2	GvHD?
6(2)	Brother	+	+	49	±	+++	+++	63[a]	Healthy
7(2)[c]	Brother	+	+	27	+	+++	++	26[a]	Healthy
8(4)	Sister	+	+	40	±	+++	+++	15[a]	Healthy
9(1.5)	Aunt	+	+	30	+	+++	++	9[a]	Healthy

[a]Numbers in parentheses represent age (months).
[b]Proven chimerism.
[c]ADA negative.

present 2 months later, by the time of the marrow transplant and could be detected by HLA typing and radiation-induced chromosomal abnormalities. This special situation resulted in a fetal graft-ver-sus-host graft reaction. Patient 5 had a fatal GvHD. In the four patients surviving in good health, complete and persisting reconsition of humoral- and cell-mediated immune functions were obtained.

IMMUNOLOGICAL DYSREGULATION ASSOCIATED WITH MILD GvHD

Even under conditions of complete HLA: A, B, D identity, some GvHD reaction was observed after bone marrow transplantation. This produces a complex immunological situation, with alternative phases of correction and disturbance of various immunological parameters. Figure 21-1 shows the immunological follow-up after a HLA-A, -B, and -D compatible bone marrow transplant in patient 7. Fifteen days after transplantation, the number of circulating lymphocytes increased abruptly and this was followed shortly by an intense eosinophilia, which peaked at about 30,000/mm^3 on day 40 after the transplantation. A characteristic skin rash was also observed at that time, further suggesting the development of a GvHD reaction. This period was characterized by the appearance of T-cell markers in the peripheral blood, with a concomitant rise in lymphocyte reactivity to mitogens. The IgM titer in the serum rose to 170 mg %, an exaggerated level at that age. However, after this period of intense

immunological activity, a progressive decrease in lymphocyte counts and E-RFC and serum IgM levels was observed. Three months after transplantation, the three later immunological markers had returned to levels comparable to those observed before transplantation. In contrast, at that stage, proliferation induced by PHA and Con-A was very intense. This, together with the finding of a subnormal level of ADA activity in the patient's peripheral lymphocytes, indicated that immunological reconstitution had taken place and persisted. It was thus decided not to perform any new bone marrow transplantation. In an attempt to stimulate the immunological system, five injections of *Corynebacterium parvum* were administered intramuscularly. This was followed shortly by another rise in the number of peripheral lymphocytes and eosinophils, with a concomitant rise in E-RFC and in serum IgM. In contrast, *in vitro* reactivity to mitogens diminished to almost undetectable levels. This second period of intense immunological activity was again followed by a progressive decrease in E-RFC and in serum IgM levels. A stabilization of immunological activity was again followed by a progressive decrease in E-RFC and in serum IgM levels. A stabilization of immunological markers was eventually obtained, and the patient was successfully taken out of the sterile environment. He is now fully reconstituted, 26 months after his bone marrow transplant. The study of patients 6 and 8 showed similar variations in immunological parameters during the months following transplantation. Thus, signs of immunological dysregulation appear to be frequent after such transplantation.

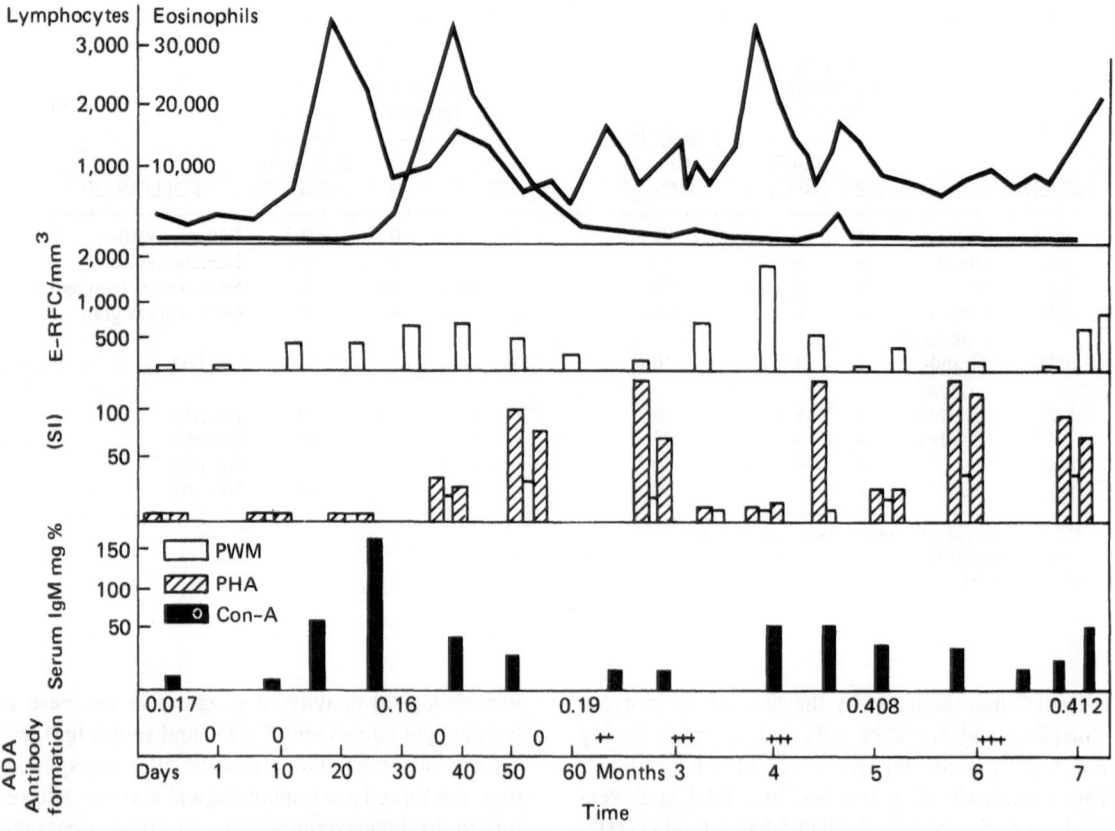

FIGURE 21-1. Immunological follow-up of patient 7.

DEFECT OF PRECURSOR T-CELLS IN SCID WITH B-LYMPHOCYTES

Recently, SCID patients have been described with peripheral B-lymphocytes that increase in number but fail to mature into antibody-producing cells (7–12). Eight such patients were present in this series. In two cases (patients 6 and 9), a phenotypically or a genotypically identical marrow transplantation allowed the development of a thymus that became visible on X-rays, strongly suggesting that the primary defect was a selective abnormality of bone marrow precursor T-cells, at a prethymic level (13).

Almost all the lymphocytes (95 to 100 percent) of these two patients carried surface Ig determinants, with a large percentage having both μ and δ chains. These cells had also receptors for complement and were stained by a fluorescent antiserum specific for B-cells. *In vitro* studies of the maturation of the cells of patient 9 in the presence of PWM and normal T-cells suggested that the B-cells remained blocked at the level of μ and δ chain expression, perhaps due to an absence of T-cell helper functions. Indeed, in the presence of PWM and normal T-cells, the B-cells of this patient matured into Ig-containing cells (13). Following graft-

ing, both patients proved to carry T-cells of donor origin by karyotyping or fluorescent staining of Y chromosome with quinacrine. Lymphocytes with surface or intracytoplasmic immunoglobulins were from the recipients. Allotype studies by Dr. L. Rivat of anti-A isohemaglutinins present in the serum of patient 9 after repeated immunization with blood group substances, revealed the presence of a Gm factor not present on the immunoglobulins of the donor. Taken together, these studies demonstrated that host B-cells were able to cooperate with donor T-cells in immunoglobulin and antibody production. Why the donor's B-cells, which were presumably present in the marrow graft, were never identified in the recipient is a matter of speculation. The peripheral monocytes and granulocytes also carried the recipient's markers after reconstitution, only the missing lineage being readily replaced.

CONCLUSION

Our studies indicate that, even in the group of patients with B-cells, bone marrow transplantation is the best way to reconstitute cellular and humoral functions. The slow maturation in host thymus of T-cells capable of cooperation with the recipient's

B-cells could take several months. In several cases, it was accompanied by a mild GvHD, causing alternative increases and decreases of various immune functions.

The benefits of bone marrow transplantation were awaited by keeping the patients in a Trexler isolator until immune competence was obtained, to avoid reinjection of new marrow cells. Unfortunately, a histocompatible bone marrow donor is not always available. In patients with a prethymic deficiency, the logical alternative seems to be the injection of precursor cells from a fetus. In fact, several fetal liver or thymus grafts have been used to reconstitute partially SCID patients (14–18). A thymus graft can, on one hand, correct epithelial function to allow the maturation of recipient stem cells. On the other hand, thymocytes present in the graft can directly reconstitute the T-cell compartment, with the risk of GvHD. The recent report stating that various cases of SCID have benefited from injections of cultured epithelium from normal thymus offers a new therapeutic approach that does not entail the risks of GvHD (4). It remains to be seen whether this procedure will be effective in patients who do not seem to have an intrathymic defect.

Severe combined immunodeficiency is really a heterogeneous disease and every effort should be made to assess the level of the defect. Preliminary studies indicate that *in vitro* culture of patient's stem cells in the presence of normal thymus epithelium (19) or determination of serum thymic factor (20) may be helpful in this respect.

SUMMARY

The effects of bone marrow and fetal lymphoid organ transplants into patients with severe combined immunodeficiency disease is reported. The efficacy of various transplant regimes are discussed. The studies indicate that even in the group of patients with B-cells, bone marrow transplants may be the best procedures for the reconstitution of cellular and humoral functions.

ACKNOWLEDGMENT

This work was supported by INSERM grants (A.T.P. 7-74-28 and 8-74-29) and Fondation pour la Recherche Médicale Française.

REFERENCES

1. Ackeret, C., Plüss, H.J., and Hitzig, W.H. Hereditary severe combined immunodeficiency and adenosine deaminase deficiency. *Pediat. Res.*, *10*:67, 1976.

2. Amman, A.J., Wara, D.W., Salmon, S., and Perkins, H. Thymus transplantation. Permanent reconstitution of cellular immunity in a patient with sex-linked combined immunodeficiency. *New Eng. J. Med.*, *5*:5, 1973.

3. Buckley, R.H. Reconstitution: grafting of bone marrow and thymus. In D. B. Amos, ed., *Progress In Immunology*. New York: Academic Press, 1971, p. 1061.

4. Buckley, R.H., Whismant, J.K., Schiff, R.I., Gilbertsen, R.B., Huang, A.T., and Platt, M.S. Correction of severe combined immunodeficiency by fetal liver cells. *New Eng. J. Med.*, *294*:1076, 1976.

5. Capersson, T., Zech, L., Johansson, C., and Modest, E.J. Identification of human chromosomes by DNA binding fluorescent agents. *Chromosoma (Berlin)*, *30*:215, 1970.

6. De Fazio, S., Criswell, B.S., Kimzey, S.L., South, M.A., and Montgomery, J.R. A paraprotein in severe combined immunodeficiency disease detected by immunoelectrophoresis analysis of plasma. *Clin. Exp. Immunol.*, *19*:563, 1975.

7. Dooren, L.J., Kamphuis, R.P., De Koning, J., and Vossen, J.M. Bone marrow transplantation in children. *Sem. Hematol.*, *11*:369, 1974.

8. Geha, R.S., Schneeberger, E., Gatien, J., and Rosen, F.S. Synthesis of an M-component by circulating B lymphocytes in severe combined immunodeficiency. *New Eng. J. Med.*, *290*:726, 1974.

9. Good, R.A., and Bach, F.H. Bone marrow and thymus transplants: Cellular engineering to correct primary immunodeficiency. In F. H. Bach and R. A. Good, eds., *Clinical Immunobiology, Vol. 2*. New York: Academic Press, 1974, p. 63.

10. Griscelli, C. T and B markers in immunodeficiencies. In D. Bergsma, ed., *Birth Defects: Immunodeficiency in Man and Animals, Vol. 11*. New York: The National Foundation March of Dimes, 1975, p. 45.

11. Griscelli, C. Bone marrow transplantation in SCID. In D. Bergsma, ed., *Birth Defects: Immunodeficiency in Man and Animals, Vol. 11*. New York: The National Foundation March of Dimes, 1975, p. 426.

12. Griscelli, C., Durandy, A., Ballet, J.J., Virelizier, J.L., and Daguillard, F. Selective defect of precursor T cells in severe combined immunodeficiency with B lymphocytes. *J. Pediat.* (in press).

13. Hong, R., Santosham, M., Schulte-Wissermann, H., Horowitz, S., Hsu, S.F., and Winkelstein, J.A. Reconstitution of B and T lymphocyte functions in severe combined immunodeficiency disease after transplantation with thymic epithelium. *Lancet, ii*:1270, 1976.

14. Incefy, G.S., Dardenne, M., Pahwa, S., Grimes, E., Pahwa, R., Smithwick, E., O'Reilly, R., and Good, R.A. Thymic activity in severe combined immunodeficiency disease. *Proc. Nat. Sci. (USA)* (in press).

15. Keightley, R.G., Lawton, A.R., Cooper, M.D., and Ynis, E.J. Successful fetal liver transplantation in a child with severe combined immunodeficiency. *Lancet ii*:850, 1975.

16. Preud'homme, J.L., Griscelli, C., and Seligmann, M. Immunoglobulins on the surface of lymphocytes in

fifty patients with primary immunodeficiency diseases. *Clin. Immunol. Immunopathol., 1*:241, 1973.

17. Pyke, K.W., Dosch, H.M., Ipp, M.M., and Gelfand, E.W. Demonstration of an intrathymic defect in a case of severe combined immunodeficiency disease. *New Eng. J. Med., 293*:424, 1975.

18. Rachelefsky, G.S., Stiehm, E.R., Amman, A.J., Cederbaum, S.D., Opelz, G., and Terasaki, P.I. T-cell reconstitution by thymus transplantation and transfer factor in severe combined immunodeficiency. *Pediat., 55*:114, 1975.

19. Seeger, R.C., Robins, R.A., Stevens, R.H., Klein, R.B., Waldman, D.J., Zeltzer, P.M., and Kessler, S.W. Severe combined immunodeficiency with B lymphocytes: *In vitro* correction of defective Ig production by addition of normal T lymphocytes. *Clin. Exp. Immunol., 26*:1, 1976.

20. Seligmann, M., Griscelli, C., Preud'homme, J.L., Sasportes, M., Herzog, C., and Brouet, J.C. A variant of severe combined immunodeficiency with normal *in vitro* response to allogeneic cells and an increase in circulating B lymphocyte persisting several months after successful bone marrow. *Clin. Exp. Immunol., 17*:245, 1974.

PART V
Treatment of Leukemia

George W. Santos

22

Experience with Syngeneic Marrow Transplantation in Brown-Norway and Wistar-Furth Rat Models of Acute Myelogenous Leukemia

George W. Santos
and Saul J. Sharkis

INTRODUCTION

Acute myelogenous leukemia (AML) continues to be a highly fatal disease in man. The major approach to this disease has been that of intensive chemotherapy. Under current chemotherapy protocols, remission rates of 70 percent can be achieved but the median survival (3 to 19 months) of responders is relatively poor (6,7,15,18,19). Immunotherapy added to intensive chemotherapy has shown only limited benefit (14). Allogeneic bone marrow transplantation has shown modest success in end-stage patients (17) but still remains in the experimental stages of development. Unfortunately, the failure to completely eradicate residual AML and a high recurrence rate following bone marrow transplantation remains a serious clinical problem.

Recently two animal models of AML have been described in inbred rat strains. Arguments have been made both for the Brown-Norway (BN) (1, 3–5,11) and Wistar-Furth (WF) (8–10) rat AML as being pertinent models for investigating new therapeutic approaches applicable to the disease in humans.

The purpose of this communication is to report our initial findings regarding the therapeutic effects of lethal doses of cyclophosphamide (CY), busulfan (BU), and total-body irradiation (TBI) followed by syngeneic bone marrow transplantation in these two rat models.

MATERIALS AND METHODS
Animals

Female BN (AgB3) and WF (AgB2) rats, 10 to 12 weeks of age, were obtained from Microbiological Associates (Bethesda, Md.). The animals were housed in polycarbonate cages, four to a cage. They were provided with tapwater and Purina chow *ad libitum*.

Drugs and TBI

Cyclophosphamide, generously supplied by Dr. Walter A. Zygmunt (Mead Johnson, Evansville, Indiana), was prepared in saline. Busulfan, generously supplied by Dr. George Hitchings (Burroughs Wellcome and Company, Research Triangle Park, North Carolina), was prepared in 2.5 percent carboxymethylcellulose in water. All drug injections were given intraperitoneally a few minutes after preparation of the drug in a volume of 10 ml/kg of body weight. A dual-source, ^{137}Cs small-animal irradiator delivering 136 rad/min was used for TBI.

Acute Myelogenous Tumors

The transplantable BN AML tumor was induced in female BN rats with 7, 12 dimethylbenzanthracine, which was kindly supplied to us by Dr. A. Hagenbeek and Dr. D. W. vanBekkum. The transplantable WF AML tumor was induced in female WF rats with methylcholanthracene (12), which was kindly supplied to us by Dr. J. S. Greenberger and Dr. W. C. Moloney.

Preparation of Cell Inocula

Under light ether anesthesia, normal donor animals were killed by cervical dislocation. Marrow from the femur, tibia, and humerus was collected in cold RPMI 1640 solution, cell counts were made, and viability was estimated with trypan blue as described previously (16). Marrow was given intravenously (i.v.) in a constant 1 ml volume at the desired cell concentration. Marrow preparations showed 90 to 95 percent viability as estimated by trypan blue exclusion.

Spleens were taken from leukemic rats (who had received 10^6 leukemic cells 3 to 4 weeks previously), minced, and gently teased in Petri dishes containing cold RPMI 1640 solution. Clumps of cells were gently dispersed with the aid of a small syringe. Cell suspensions were counted and adjusted to the proper concentration as noted above. Viability as judged by trypan blue exclusion was 85 to 90 percent. Cells were injected (i.v.) into rats in a constant volume of 1 ml.

Hematological Examination

Blood for peripheral white blood cell counts, hematocrits and the preparation of blood films were taken from the tail veins. The hematocrit values and white cell and differential counts were determined by routine hematological methods.

RESULTS

Tumor Dose—Survival Studies

Groups of female BN rats received an i.v. injection of spleen cells from a tumor-bearing rat at the following cell doses: 10^6, 10×10^6, 25×10^6, and 100×10^6. The median survivals were 25, 24, 16, and 15 days, respectively. All animals showed evidence of anemia and leukemia blasts in the peripheral blood 5 to 6 days before death. The white cell count at death was around 50×10^6/ml in the majority of animals (normal values 10×10^6). Hematocrit values were low, in the range of 15 to 25 percent at death (normal values 45%).

Experiments were performed and similar results obtained in the WF model. The WF animals given 10^6 tumor cells (i.v.) had a median survival of 24 days.

Therapeutic Response to CY

Brown-Norway female rats were given 10^6 leukemic cells (i.v.). Two weeks later, when all rats

showed 10 to 30 percent AML cells in the peripheral blood, they were divided into groups of five animals each and received no further treatment or CY (25, 50, 100, or 150 mg/kg). Animals were examined daily for survival and evidence of leukemia. The results are shown in Table 22-1. The data indicate a therapeutic effect of CY up to 100 mg/kg. Animals given 150 mg/kg died early, with evidence of profound marrow aplasia and no evidence of leukemia. Clearly, the optimal single dose is around 100 mg/kg. Doses higher than this are too toxic for the leukemia-bearing animal to survive.

Therapeutic Response to Lethal Doses of CY, BU, and TBI Followed by Syngeneic Marrow

Groups of BN or WF female rats were given 10^6 leukemia cells (of the appropriate origin) i.v. Two weeks later, when all animals showed 10 to 30 percent AML cells in the peripheral blood, they were divided into various treatment groups: no further treatment (controls); CY, 200 mg/kg; BU, 30 mg/kg; or TBI, 1,000 rad (all supralethal doses) followed in 24 hr by the i.v. infusion of 64×10^6 nucleated syngeneic marrow cells. Animals were observed daily for survival and evidence of leukemia. The results of the experiments for the BN and WF systems are shown in Table 22-2.

The data for the BN tumor system indicate that lethal doses of BU and TBI only modestly prolong survival but that lethal doses of CY followed by marrow transplantation are curative. In the WF tumor system, similar findings of only modest prolongation of survival with BU and TBI are noted. In this system, CY treatment gives the greatest therapeutic effect and, although not curative, more than doubles survival time.

One striking difference between the leukemia models was that about 60 to 70 percent of WF animals that relapsed following treatment with CY and BU did so with evidence of hindlimb paralysis.

TABLE 22-1 Therapeutic Effect of Cyclophosphamide (Cy) on Brown-Norway (BN) Acute Myeloid Leukemia (AML)

GROUP	DOSE CY[a] mg/kg	SURVIVAL IN (DAYS) FOLLOWING TUMOR TRANSFER
1	0	18, 18, 18, 27, 28
2	25	27, 29, 29, 30, 33
3	50	25, 37, 38, 39, 40
4	100	45, 45, 48, 50, 51
5[b]	150	19, 20, 20, 20, 22

[a]Animals were given CY 14 days after the i.v. injection of 10^6 AML cells.
[b]These animals died in aplasia without evidence of leukemia. All other animals died with evidence of leukemia.

TABLE 22-2 Therapeutic Effect of Cyclophosphamide (CY), Busulfan (BU), and Total-Body Irradiation (TBI) followed by Syngeneic Marrow on BN and WF Acute Myelogenous Leukemia (AML) Survival

GROUP	NUMBER OF ANIMALS	CYTOTOXIC THERAPY[a]	MEDIAN SURVIVAL (DAYS) AFTER TUMOR TRANSFER[b]	RANGE OF SURVIVAL (DAYS) AFTER TUMOR TRANSFER
BN				
1	20	None	24	16-29
2	15	CY, 200 mg/kg	>183	N.A.
3	10	BU, 30 mg/kg	34	23-35
4	15	TBI, 1,000 rad	35	27-40
WF				
1	13	None	22	21-26
2	10	CY, 200 mg/kg	50	33-60
3	10	BU, 30 mg/kg	32	27-34
4	10	TBI, 1,000 rad	28	24-41

[a]Animals were given CY, BU, or TBI 14 days after the i.v. injection of 10^6 AML cells; 64×10^6 syngeneic marrow cells were given 1 day later.
[b]All animals that died showed evidence of florid leukemia except for two BN animals given CY that died at 18 and 21 days without evidence of tumor. Thirteen animals of BN group 2 remained free of disease at 183 days.

Histological examination of three of such animals indicated involvement of the spinal cord and meninges. Although the BN rats were not examined in the same way, only an occasional rat showed similar clinical neurological signs.

DISCUSSION

"Predictive" animal models for the therapy of malignancy are, at best, never complete satisfactory. Indeed, they seem to be very unsatisfactory for human AML (8,13,15,17). Recent AML models in WF and BN rats have been described that show some promise of improving this situation (1,8,10, 14). Evidence has been presented that both of these tumors have the morphological and cytochemical characteristics, growth patterns and kinetics, histological patterns of dissemination, physiological effects, and effects on normal hemopoiesis that are found in the human disease (8,15). Preliminary studies with chemotherapy suggest that these tumors respond to agents active in human AML (3, 4,8,11). In addition, sensitive bioassays for residual clonogenic tumor cells and normal hemopoietic stem cells may be performed simultaneously in the BN model (2). Furthermore, in the WF, AML meningeal leukemia has been seen in 8 percent of rats autopsied following the systemic injection of leukemic cells and in 93 percent of relapses postchemotherapy (8). This latter feature, a valuable characteristic for an animal model, takes into account the problem of sanctuary relapses in clinical cases.

The present studies indicate that lethal doses of BU and TBI followed by the infusion of syngeneic marrow are relatively ineffective in prolonging sur-

vival in leukemic WF and BN rats. Nonlethal doses of CY are able to prolong markedly survival, but are not curative in the BN AML. Doses of CY (200 mg/kg) that are supralethal to the tumor-bearing BN rat appear to be curative when syngeneic marrow is given after CY.

The therapeutic effects of nonlethal doses of CY in the WF model were not investigated in this study. Others, however, have noted that, although nonlethal doses of CY may have marked anti-leukemic effects in this model, it is not curative (8). Supralethal doses of CY (200 mg/kg) followed by syngeneic marrow transplantation, although not curative as in the BN model, more than doubled the survival time as compared to controls.

The present findings, as well as the published results with both of these AML tumors, suggest that they will be valuable models for investigating various possible therapeutic approaches to AML in man, including the role of autologous, syngeneic, and allogeneic bone marrow transplantation.

SUMMARY

The therapeutic effects of supralethal doses of cyclophosphamide (CY), busulfan (BU) and total-body irradiation (TBI) followed by syngeneic marrow infusion was studied in transplantable carcinogen-induced acute myelogenous leukemia tumors in the Brown-Norway and Wistar-Furth rat strains. If 10^6 syngeneic leukemic blasts are injected into BN rats and, 2 weeks later, groups of rats receive supralethal doses of CY, BU, or TBI followed 24 hr later by a syngeneic marrow graft, only animals that receive Cy conditioning survive (survival over 183 days) following transplantation. The BU- and

TBI-treated groups show only minimal prolongation of survival and all die of florid leukemia. Nonlethal doses of CY prolong survival but are not curative. Similar studies in the WF rat demonstrate almost identical results regarding the therapeutic effects of BU and TBI. Supralethal doses of CY, followed by syngeneic marrow infusion, although not curative, as in the BN model, more than doubles the survival of the animals. The present results as well as published data regarding the therapeutic effects of more conventional chemotherapeutic treatments suggest that the BN and WF AML tumors may be useful models for investigating new treatment protocols in human AML.

ACKNOWLEDGMENT

This work was supported by Public Health Service research grants CA06973 and CA15396 from the National Cancer Institute, National Institutes of Health, Bethesda, Maryland.

REFERENCES

1. vanBekkum, D.W., and Hagenbeek, A. The relevance of the BN Leukemia as a model for human acute myelocytic leukemia blood cells. (In press) 1977.

2. vanBekkum, D.W., vanOosterram, P., and Dieke, K.A. *In vitro* colony formation of transplantable rat leukemias in comparison with human acute myeloid leukemia. *Cancer Res., 36*:941, 1976.

3. Colly, L.P., and Hagenbeek, A. Experimental chemotherapy in a rat model for human leukemia. *Exp. Hematol., 4* (Suppl.):196, 1976.

4. Colly, L.P., and Hagenbeek, A. Experimental chemotherapy: A rat model for human acute myeloid leukemia. In S. J. Baum, and G. P. Ledney, eds., *Experimental Hematology Today*. New York: Springer-Verlag, 1977, p. 211.

5. Colly, L.P., Tons, A., and Hagenbeek, A. Experimental chemotherapy in a rat model for acute myelocytic leukemia. *Exp. Hematol., 5* (Suppl.):50, 1977.

6. Crowther, D., Bateman, C.J.T., Vartan, C.P., Whitehouse, J.M.A., Malpas, J.S., Hamilton, F.G., and Scott, R.B. Combination chemotherapy using L-asparaginase, daunorubicin and cytosine arabinosine in adults with acute myelogenous leukemia. *Br. Med. J., IV*:513, 1970.

7. Graun, V., Erickson, R., Flannery, J., Finch, S., and Clarkson, B. The therapy of acute granulocytic leukemia in patients more than fifty years old. *Ann. Intern. Med., 80*:15, 1974.

8. Greenberger, J.S., Bocaccino, C.A., Szot, S.J., and Moloney, W.C. Chemotherapeutic remissions in Wistar-Furth rat acute myelogenous leukemia: A model for human AML. *Acta Haemat., 57*:233, 1977.

9. Greenberger, J.S., Aaronson, S.A., Rosenthal, D.S., and Moloney, W.C. Continuous production of peroxidase, esterase, alkaline phosphatase and lysozyme by clones of promyelocytes. *Nature (Lond.), 257*:143, 1975.

10. Greenberger, J.S., Rosenthal, D.S., Aaronson, S.A., and Moloney, W.C. Acute myelogenous leukemia of the Wistar/Furth rat: Establishment of a continuous tissue culture line producing lysozyme *in vitro*. *Blood, 46*:27, 1975.

11. Hagenbeek, A., and vanBekkum, D.W. (eds.) Proceedings of a workshop on "Comparative evaluation of the L5222 and BNML rat leukemia models and their relevance for human acute leukemia." *Leukemia Res.,* (in press) 1977.

12. Kim, V., Clifoton, K.H., and Furth, J. A highly inbred line of Wistar rats yielding spontaneous mammo-somatrophic pituitary and other tumors. *J. Nat. Cancer Inst., 24*:1031, 1966.

13. Moloney, W.C. Primary granulocytic leukemia in the rat. *Cancer Res., 34*:3049, 1974.

14. Powles, R. Immunotherapy for acute myelogenous leukemia using irradiated and unirradiated leukemia cells. *Cancer, 34*:1558, 1974.

15. Rosenthal, D.S., and Moloney, W.C. The treatment of acute granulocytic leukemia in adults. *New Eng. J. Med., 286*:1176, 1972.

16. Santos, G.W., and Owens, A.H., Jr. Syngeneic and allogeneic marrow transplants in the cyclophosphamide pretreated rat. In J. Dausset, J. Hamburger, and G. Mathé, eds., *Advances in Transplantation*. Copenhagen: Munksgaard, 1968, pp. 432–436.

17. Thomas, E.D., Buckner, C.D., Banaji, M., Clift, R.A., Fefer, A., Flournoy, N., Goodell, B.W., Hickman, R.O., Lerner, K.G., Neiman, P.E., Sale, G.E., Sanders, J.E., Singer, J., Stevens, M., Storb, R., and Weiden, P.L. One hundred patients with acute leukemia treated by chemotherapy, total body irradiation and allogeneic marrow transplantation. *Blood, 49*:511, 1977.

18. Walters, T.R., Aur, R.J.A., Hernandez, K., Vietti, T., and Pinkel, D. 6-Azouridine in combination chemotherapy of childhood acute myelocytic leukemia. *Cancer, 29*:1057, 1972.

19. Wiernick, P.H., and Serpick, A.A. A randomized clinical trial of daunorubicin and combination of prednisone, vincristine, 6-mercaptopurine and methotrexate in adult acute nonlymphocytic leukemia. *Cancer Res., 32*:2023, 1972.

23

Marrow Transplantation for Acute Leukemia

C. D. Buckner, R. A. Clift,
A. Fefer, P. E. Neiman,
J. E. Sanders, R. Storb,
P. L. Weiden, E. D. Thomas,
and the Seattle Marrow
Transplant Team

Marrow transplantation has been used with increasing frequency over the past several years for the treatment of refractory leukemia. In theory, marrow transplantation should allow a greater cell kill to be achieved with irradiation and chemotherapy, since marrow toxicity can be obviated. In addition, the transplantation of immunologically competent allogeneic marrow could theoretically be of benefit in eradicating residual leukemic cells immunologically (1,7).

Patients with acute leukemia who have a normal identical twin offer a unique opportunity to evaluate transplantation regimens uncomplicated by rejection or graft-versus-host disease (GvHD). Table 23-1 presents the long-term results in the initial group of 29 patients. Ninety percent of patients with terminal acute leukemia with an identical twin donor will achieve a complete remission. However, 70 percent will relapse in the first 1½ years and the remaining 30 percent represent "cures," since no relapses have occurred beyond 2 years, with observation up to 6½ years. These patients did not receive maintenance therapy. The data clearly indicate that patients with identical twin donors should be transplanted early in the course of their disease before they develop drug resistance.

Extrapolation of the information derived from syngeneic transplantation to the allogeneic situation would indicate that if all transplantation biological problems were solved and the current treatment regimen utilized, the relapse rate would be 70 percent in patients with refractory leukemia. On the other hand, patients with aplastic anemia successfully engrafted should give information about projected optimal survival if recurrent leukemia could be eliminated by earlier transplantation or more effective transplant regimens. Patients with aplastic anemia successfully engrafted from an HLA-identical sibling have an overall survival of 65 percent (90 percent with sex-matched donors and 35 percent with sex-mismatched donors). This would presumably be the current survival rate if leukemic relapse did not occur. This type of information, derived from twin transplants and alloge-

TABLE 23-1 Identical Twin Transplants

	ACUTE LYMPHOCYTIC LEUKEMIA	ACUTE MYELOCYTIC LEUKEMIA
Number of patients	15	14
Infectious deaths	2	3
Recurrent leukemia	8	8
Patients surviving in remission	5	3
Length of survival (months)	34–84	52–67

neic transplants in aplastic anemia patients, is important in designing therapeutic strategies in patients with leukemia transplanted from allogeneic-matched siblings, in that it clearly delineates the problems remaining in transplantation biology and intensive treatment of leukemia.

Patients with acute leukemia in the end stage of the disease after failure of combination chemotherapy have been transplanted in Seattle since 1969. Total-body irradiation (TBI) has been utilized in all studies, since it is an effective antileukemic modality and can eradicate leukemic cells in such "privileged" sites as the central nervous system and the testicle (6). Total-body irradiation alone was used to prepare for engraftment in the initial series of 10 Seattle patients with end-stage acute leukemia treated by marrow transplantation from a HLA-identical sibling (7). A "supralethal" exposure of 1,000 rad midpoint tissue dose from opposing ^{60}Co sources was employed to kill leukemic cells and to suppress the immune response of the recipient in order to permit engraftment and prevent rejection. The four patients with acute myelogenous leukemia (AML) died early. Five of the six patients with acute lymphocytic leukemia (ALL) showed recurrent leukemia. In two of these five cases, the recurrent leukemia was in donor-type cells (7). One male patient is now 7 years post-engraftment without evidence of leukemia and with normal marrow function of donor (female) origin.

Due to the high incidence of recurrent leukemia, the basic regimen was changed late in 1971 to include a high dose of cyclophosphamide (CY) and/or other antileukemic drugs before the 1,000 rad of TBI. The results of 100 consecutive cases of acute leukemia prepared for engraftment in this manner (as of September 1976) have been published (2). One hundred patients, 54 with AML and 46 with ALL, considered to be in the end stages of their disease after combination chemotherapy, were treated by marrow transplantation. All patients were given a marrow graft from an HLA-identical sibling after receiving 1,000 rad TBI. One group of 43 patients was given CY, 60 mg/kg on each of two days, five and four days before TBI. In a second group of 31 patients, additional chemotherapy was given before CY and TBI. In a third group of 19 patients, BCNU was given before CY and TBI. A fourth group of seven patients received other chemotherapy regimens before TBI. Six patients died 3 to 17 days after marrow infusion without evidence of engraftment. Ninety-four patients were engrafted, and only one patient rejected the graft. Thirteen patients are alive with a marrow graft, without maintenance antileukemic therapy and without recurrent leukemia 1 to 4½ years after transplantation. Three have chronic graft-versus-

host disease (GvHD). Four patients are alive 1½ to 3½ years after grafting, but have had a relapse of their leukemia. Of 93 evaluable patients, 19 did not develop GvHD, and 24 developed very mild GvHD. Fifty patients developed moderate to severe GvHD, and 40 of these were treated with anti-thymocyte globulin (ATG). Interstitial pneumonia occurred in 54 patients and was the primary cause of death in 34. Interstitial pneumonia often occurred in association with GvHD, and the most common etiological agent was cytomegalovirus. A total of 31 patients have had a relapse of leukemia. There was no definite correlation between relapse of leukemia and the presence or absence of GvHD. The relapse rate appeared to be relatively constant over the first 2 years and was extremely low after that time. Neither survival nor leukemic relapse appear to be influenced by the type of leukemia nor by the preparative chemotherapy regimen before TBI. Patients in fair clinical condition at the time of transplantation showed significantly longer survival times than patients in poor condition ($p = 0.001$). This observation, coupled with the observation that some patients may be cured of their disease, indicates that marrow transplantation should now be undertaken earlier in the management of patients with acute leukemia who have an HLA-identical sibling marrow donor. As of September 1977, no additional leukemic relapses have occurred. The longest disease-free survivor treated with CY and TBI who had ALL is now 5 years and 7 months post-grafting; for AML, 5 years and 5 months.

An actuarial analysis of the survival curve and the leukemic relapse curve of these patients can be summarized as follows (5): One hundred and ten patients with end-stage acute leukemia received a marrow transplant from an HLA-identical sibling after chemotherapy and supralethal TBI. A plot on a logarithmic scale of the survival of these patients discloses a decreasing death rate after transplantation. The initial rapid loss of patients in the first 130 days is due primarily to infection and GvHD. The rate of leukemic relapse is relatively constant for the first 2 years, after which the relapse rate becomes very low. There have been no deaths between 2 and 6 years. The flat slope of the long tail of the survival curve and the corresponding low relapse rate constitute an operational definition of cure of leukemia for the long-term survivors.

The survival of the long-term disease-free patients without maintenance chemotherapy is encouraging, since no other therapeutic approach in end-stage patients has resulted in such survival. However, recurrent leukemia is still a major problem.

In an attempt to prevent the recurrence of leukemia (3,4), we gave more intensive chemotherapy

TABLE 23-2 Recent Allogeneic Transplants Following CY and TBI

DISEASE	NUMBER OF PATIENTS	NUMBER OF RELAPSES	NUMBER DEAD	DAYS TO DEATH MEDIAN (RANGE)	NUMBER OF SURVIVORS	DURATION OF SURVIVAL (DAYS)
Acute lymphatic leukemia	20	11	17	98 (10–271)	3	464, 190, 141
Acute myelocytic leukemia	11	4	11	73 (14–234)	0	—

before CY and TBI in studies conducted in late 1975 and early 1976. Nineteen patients were given 8 to 14 mg of BCNU/kg prior to CY and TBI. Three of these patients relapsed and only one is a long-term survivor. Seven patients were given a 5-day continuous infusion of cytosine arabinoside, 600 mg/m^2/daily before CY and TBI. Only one of seven patients survived more than 100 days, and he relapsed 17 months after transplantation.

Since that time, many additional patients with acute leukemia have been transplanted, with several protocols employed. By early 1976, it was apparent that giving additional chemotherapy before CY and TBI resulted in increased toxicity, and there was no clear evidence of a decrease in the rate of leukemic relapse. Accordingly, we decided to return to the basic regimen of CY and TBI in a series of patients while proceeding with the studies of protective environments, prophylactic granulocyte transfusions, and interstitial pneumonia. The results in patients transplanted between April 1976 and June of 1977 are described in Table 23-2.

The results have been rather discouraging. The number of relapses of leukemia following transplantation were expected and are not different from the initial 110 patients described in Table 23-2. However, the overall long-term survival has, if anything, been worse. Although this series of patients has not yet been analyzed by a multifactorial regression analysis, it appears that these patients have had much more intensive chemotherapy, particularly with anthracyclines and cytosine arabinoside, before coming to marrow transplantation, and many have been critically ill upon arrival on

the transplantation ward. Accordingly, additional approaches have been undertaken.

In a pilot study, six patients with ALL in relapse (two in third relapse, one in second relapse, and three drug-resistant cases in first relapse) were given 5 mg of dimethyl myleran/kg, followed by CY, TBI, and the marrow transplant. All the transplants were successful. One patient died of interstitial pneumonia 81 days post-transplant. The others are now outpatients 2 to 4 months post-transplant.

Another approach is based upon the fact that marrow transplanting has produced some very long unmaintained remissions in end-stage patients. Therefore, it is logical to undertake marrow transplanting earlier in the course of the disease, i.e., in patients with a poor prognosis but who are in remission. In theory, marrow transplanting in remission offers the advantage of (a) treatment when the body burden of leukemic cells is minimal, (b) treatment before the leukemic cells become refractory to therapeutic agents, and (c) treatment while the patient is in optimal clinical condition. In January of 1976, the Human Subjects Review Committees of the University of Washington and the Fred Hutchinson Cancer Research Center approved a protocol for transplantation of selected patients in remission. The current status of this ongoing study is summarized in Table 23-3.

The patients with ALL were in the second to fourth remission, whereas the patients with AML were in the first to third remission. The median survival has not been reached in either group. Of particular note, in contrast to previous experience

TABLE 23-3 Allogeneic Marrow Transplants Performed during Remission

DISEASE	NUMBER OF PATIENTS	NUMBER OF RELAPSES FOLLOWING TRANSPLANTATION	NUMBER LIVING IN REMISSION	SURVIVAL RANGE (MONTHS)
Acute lymphatic leukemia	18	5	8	3–14
Acute myelocytic leukemia	11	0	6	3–19

in end-stage patients, was the fact that all patients were able to leave the hospital after transplantation. The earliest death, from interstitial pneumonia, was on day 59. It is already apparent that a significant number of patients with ALL have resistant disease, and this strategy will not be wholly successful without changes in the basic transplant regimen. There have been no relapses yet in patients with AML, but a minimum follow-up of 2 years is necessary in order to evaluate this study in regard to the long-term disease-free survival.

SUMMARY

Marrow transplantation has been used to treat refractory patients with acute myelogenous leukemia (AML) and acute lymphoblastic leukemia (ALL) who have identical twin or HLA-identical sibling donors. The basic treatment regimen consisted of total-body irradiation (TBI) preceded in most instances by high doses of cyclophosphamide (CY) with or without additional chemotherapeutic agents.

Six of 16 patients with hematological malignancy refractory to conventional therapy transplanted from identical twin donors are presently in complete remission 46 to 78 months after transplantation. All six are leading normal lives without maintenance therapy.

One of six patients with ALL transplanted from an HLA-identical sibling following TBI alone is alive 6½ years after transplantation. Relapse of leukemia occurred in the other five patients and in two of these the relapse was in donor cells.

Thirteen of 100 patients transplanted following CY and TBI are alive 1½ to 5½ years after transplantation, off all treatment since day 100 posttransplant. Four additional patients are alive after grafting but have had a relapse of their leukemia.

These results were sufficiently encouraging to warrant transplantation earlier in the course of the patient's disease, whenever an identical twin or HLA-identical sibling was available as a donor.

Results of early transplantation in patients with AML and ALL are presented as well as results of varying pre-transplant regimens on relapse rate.

ACKNOWLEDGMENT

This investigation was supported by Grant Numbers CA 18029, CA 18579 and CA 15704, awarded by the National Cancer Institute, DHEW.

REFERENCES

1. Fefer, A., Einstein, A.B., Jr., and Cheever, M.A. Adoptive chemoimmunotherapy of cancer in animals: A review of results, principles and problems. *Ann. NY. Acad. Sci., 277:*492, 1976.
2. Thomas, E.D., Buckner, C.D., Banaji, M., Clift, R.A., Fefer, A., Flournoy, N., Goodell, B.W., Hickman, R.O., Lerner, K.G., Neiman, P.E., Sale, G.E., Sanders, J.E., Singer, J., Stevens, M., Storb, R., and Weiden, P.L. One hundred patients with acute leukemia treated by chemotherapy, total body irradiation, and allogeneic marrow transplantation. *Blood, 49:*511, 1977.
3. Thomas, E.D., Buckner, C.D., Fefer, A., Neiman, P.E., and Storb, R. Marrow transplantation in the treatment of acute leukemia. *Adv. Cancer Res.* (in press).
4. Thomas, E.D., Buckner, C.D., Fefer, A., Sanders, J.E., and Storb, R. Efforts to prevent recurrence of leukemia in marrow graft recipients. *Transplant. Proc.* (in press).
5. Thomas, E.D., Flournoy, N., Buckner, C.D., Clift, R.A., Fefer, A., Neiman, P.E., and Storb, R. Cure of leukemia by marrow transplantation. *Leukemia Res., 1:*67, 1977.
6. Thomas, E.D., Storb, R., and Buckner, C.D. Total-body irradiation in preparation for marrow engraftment. *Transplant. Proc., 8:*591, 1976.
7. Thomas, E.D., Storb, R., Clift, R.A., Fefer, A., Johnson, F.L., Neiman, P.E., Lerner, K.G., Glucksberg, H., and Buckner, C.D. Bone-marrow transplantation. *N. Eng. J. Med., 292:*832, 895, 1975.

24

Use of Decontamination and a Protected Environment to Prevent Secondary Disease Following Adoptive Immunotherapy of Acute Leukemia in Mice

Robert L. Truitt

INTRODUCTION

Bone marrow transplantation offers exciting possibilities for the treatment of leukemia and related malignancies; yet, its potential has not been fully realized. Despite the encouraging report by Thomas et al. (14,15), several major problems with this form of therapy still exist including recurrent leukemia and secondary disease, manifested by graft-versus-host disease (GvHD) and infections. In patients undergoing marrow transplantation for the treatment of leukemia, the risk of fatal infection is substantial because of the immunosuppressive effects of the malignancy, of the GvH reaction, and of the treatments given pre-transplant to assure engraftment and post-transplant to avoid GvHD.

Germfree technology can be of particular value in conditions associated with depressed immune response in that patients can be protected from endogenous or exogenous sources of infection. The role of bacteria in secondary disease mortality has been demonstrated in several studies of marrow transplantation in germfree mice and rats (1,6,8–11,18,19). In addition, procedures for rendering animals free of bacteria by antibiotic treatment were developed (5,21–24) and applied to marrow transplantation in mice and monkeys (4,17,19,20). In each instance, a significant reduction in mortality from secondary disease was reported. Although the GvH reaction was not entirely eliminated in germfree or antibiotic-decontaminated chimeras, a germfree state permitted the transplanted animals to live longer and, in some animal systems, survive free of clinical evidence of secondary disease.

Germfree environments have been used successfully to prevent secondary disease in the treatment of acute leukemia in AKR mice by allogeneic bone marrow transplantation (16,17). In these studies, antileukemic treatment was not initiated until *after* the appearance of leukemia; however, the AKR recipients were germfree *before* the disease appeared, either by nature of birth (germfree mice) or through decontamination initiated 4 to 5 months prior to the median age at which leukemia developed. The results encouraged us to develop a model for the application of germfree techniques to the treatment of end-stage leukemia that could be applied to man.

In the treatment model reported here, we have attempted to simulate the clinical situation in which a patient with leukemia, while being considered for marrow transplantation, is placed on a remission-induction (RI) chemotherapy regimen and either started on a decontamination protocol or maintained in a conventional environment. Our objectives were (a) to develop procedures for decontamination of mice with acute leukemia after the disease is diagnosed, (b) to employ decontamination together with a protected environment (axenic isolation) to prevent secondary disease mortality, and (c) to demonstrate that adoptive immunotherapy can cure acute leukemia that is refractory to other therapeutic strategies.

MATERIALS AND METHODS

Mice. Female AKR/J (H-2^k) retired breeders and young AKR and SJL/J (H-2^s) mice were obtained from The Jackson Laboratory, Bar Harbor, Me. The AKR retired breeders were used when they developed leukemia; AKR and SJL/J mice were used as donors of bone marrow and lymph node cells between 11 and 20 weeks of age.

AKR Spontaneous Leukemia

Acute lymphoblastic leukemia develops spontaneously in AKR mice at a median age of 9 months (12). AKR leukemia resembles several human malignancies, but most closely simulates human T-cell leukemia (2). Both the human and murine T-cell leukemias share numerous features, including a mediastinal mass, internal organomegaly, and refractoriness to cure with chemotherapy. The disease can be diagnosed accurately in mice by physical examination; the spleen plus inguinal and axillary or brachial lymph nodes are all more than three times normal size. For these experiments, a colony of approximately 900 conventional AKR mice from 6 to 11 months of age was palpated once a week for evidence of leukemia. The colony was restocked weekly, and mice that had not developed leukemia by 12 months of age were removed. Our diagnostic accuracy over the last 4 years has exceeded 99 percent as evidenced by death from leukemia within 60 days after diagnosis in untreated mice.

Housing

Mice kept in our regular animal facility are referred to as conventional mice. Conventional mice were housed (four in a cage) in polycarbonate Isosystem® cages with filter lids (Lab Products, Inc., Garfield, N.J.). The mice were given autoclaved mouse chow and acidified, chlorinated water (but no antibiotics) *ad libitum*. All cage materials, including the bedding, were autoclaved before use.

Mice undergoing antibiotics decontamination were housed in sterile polycarbonate cages in sterile horizontal laminar airflow (LAF) hoods with HEPA filters (Integrated Air Systems, Burbank, Ca.). All cage materials and food were double-wrapped in Kraft wrapping paper, autoclaved, and

transferred into the LAF hoods aseptically by removing the outer layer and passing the inner package to a technician wearing ethylene oxide-sterilized, elbow-length plastic gloves. Cages, bedding, food, water bottles, and cage tops were changed every 3 or 4 days during the first 2 weeks of antibiotic treatment, and weekly thereafter. No more than five mice were housed in one cage; the usual number was four.

Antibiotic Decontamination

Mice were rendered germfree by the oral administration of nonabsorbable broad-spectrum antibiotics. In the early experiments, the mice were given a single, filter-sterilized, aqueous solution of streptomycin (5 mg/ml), neomycin (4 mg/kg), and bacitracin (60 u/ml); in later experiments, four antibiotic combinations were alternated weekly—bacitracin plus either streptomycin, kanamycin (5 mg/ml), gentamicin (6 mg/ml), or neomycin. Mice that had been on antibiotics for 90 days or more were given antibiotics at one-tenth the original concentration. Antibiotics were gifts from The Upjohn Co., Kalamazoo, Mi. (bacitracin), Bristol Laboratories, Syracuse, N.Y. (kanamycin), Schering Corp., Bloomfield, N.J. (gentamicin), and E.R. Squibb & Sons, Princeton, N.J. (neomycin). Streptomycin was purchased from Calbiochem (La Jolla, Ca.).

Fecal Cultures

Fresh feces were collected for culturing in fluid thioglycollate medium (Difco, Detroit, Mi). Culture tubes were incubated at 37°C and examined after 24 and 120 hours. Absence of microbial growth in fecal cultures prior to transplantation was used as a definition of decontamination in these experiments; however, the animals were not strictly equivalent to germfree mice.

Drugs and Irradiation

Cyclophosphamide (CY), methyl cyclo-hexylnitrosourea (MeC), and PalmO cystosine arabinoside (PalmO) were provided by the Drug Synthesis and Chemistry Branch, Division of Cancer Treatment, National Cancer Institute (Bethesda, Md.) and injected i.p. Total-body radiation (TBR) was administered as gamma-rays using a Gammacell-40 Irradiator (Atomic Energy of Canada, Ltd., Quebec) with twin ^{137}Cs sources at a rate of 120 R/min. Decontaminated mice were transported and irradiated inside sterile screwcap plastic bottles in order to prevent exposure to exogenous microorganisms. A breathing port was provided by removing the bottom of the transport vessel and replacing it with fiberglass-filter-covered wire screen.

Statistical Analysis and Survival Data

Statistical comparisons were made using X^2 analysis. Median survival times (MST) and survival rates for all the mice entered into an experiment are calculated from the last day of treatment to avoid inflation of the data by time expended during treatment and exclusion of mice that did not survive treatment.

RESULTS

Decontamination of Leukemic Mice: Technical Considerations

The initial parts of this study were undertaken to determine whether procedures that had been used to decontaminate nonleukemic mice could be applied to mice after leukemia was diagnosed. Conventional AKR mice diagnosed as having advanced leukemia on the basis of palpation were randomized and entered into experimental or control groups. To eliminate endogenous sources of infection, leukemic mice were provided with antibiotics in their drinking water; the rationale for this treatment has been discussed in detail elsewhere (5,21–24). In order to prevent exposure of mice undergoing decontamination to exogenous microorganisms, the mice were housed in an axenic environment in LAF hoods.

Mice must consume sufficient antibiotics to achieve and maintain bactericidal concentrations in their intestinal tract for decontamination to be successful (22). Figure 24-1 shows data on the quantity of antibiotic solution consumed by leukemic AKR mice that were given or not given chemotherapy for leukemia remission-induction. When untreated, the leukemia progressed and led to a deteriorated clinical state which, in turn, led to reduced antibiotic intake and retarded or prevented decontamination. In contrast, with chemotherapy, the clinical state of the mice improved and consumption of the antibiotic solution accelerated rapidly; 1 week after a single dose of a chemotherapeutic agent, antibiotic consumption increased to approximately 9 ml/mouse daily (Figure 24-1). Using the RI chemotherapy regimen described later in this paper, we eliminated the intestinal bacteria in 89 percent (323/346) of the leukemic animals within 13 to 20 days. No single species of bacteria was consistently isolated from mice that had been on antibiotics for 2 weeks or more, and no attempt was made to determine the antibiotic-sensitivity pattern of persistent microorganisms.

Experimental Treatment Model

Conventional AKR mice, diagnosed as bearing advanced spontaneous leukemia, were randomized and entered into control or experimental groups

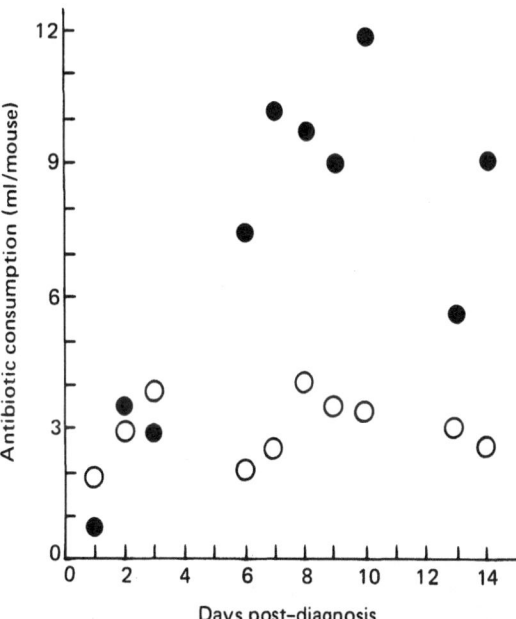

FIGURE 24-1. Average daily consumption of antibiotic solution by leukemic AKR mice given remission-induction chemotherapy (●) or no treatment (o). Each point represents nine to twelve mice.

with or without antibiotic decontamination and axenic isolation. The treatment model is diagrammed in Figure 24-2. Results of recent experiments are shown in Table 24-1. One group of leukemic mice was entered into the standard decontamination protocol and placed in axenic isolation but given no further treatment. Their MST was 17 days post-diagnosis, and all mice were dead of leukemia within 48 days (Group 2, Table 24-1). The survival data for 237 conventional leukemic mice, which were randomized and set aside as untreated controls is also shown in Table 24-1. There was no significant difference in the survival data of conventional or decontaminated untreated AKR mice.

Remission-induction chemotherapy was initiated immediately after diagnosis in order to simulate the clinical situation and to keep a significant number of leukemic AKR mice alive until decontamination was accomplished. Fecal specimens from leukemic mice undergoing decontamination and RI chemotherapy were cultured 13 to 20 days after the start of antibiotic treatment. Mice with negative fecal cultures were continued in the experiments; mice with positive fecal cultures were excluded.

Twelve RI chemotherapy regimens of various lengths were tested in leukemic AKR mice undergoing decontamination as well as in conventional mice. The most successful regimen consisted of CY (100 mg/kg) and MeC (10 mg/kg) given on the day of leukemia diagnosis (day 0) and 1 week later (day 7), followed by four injections of PalmO

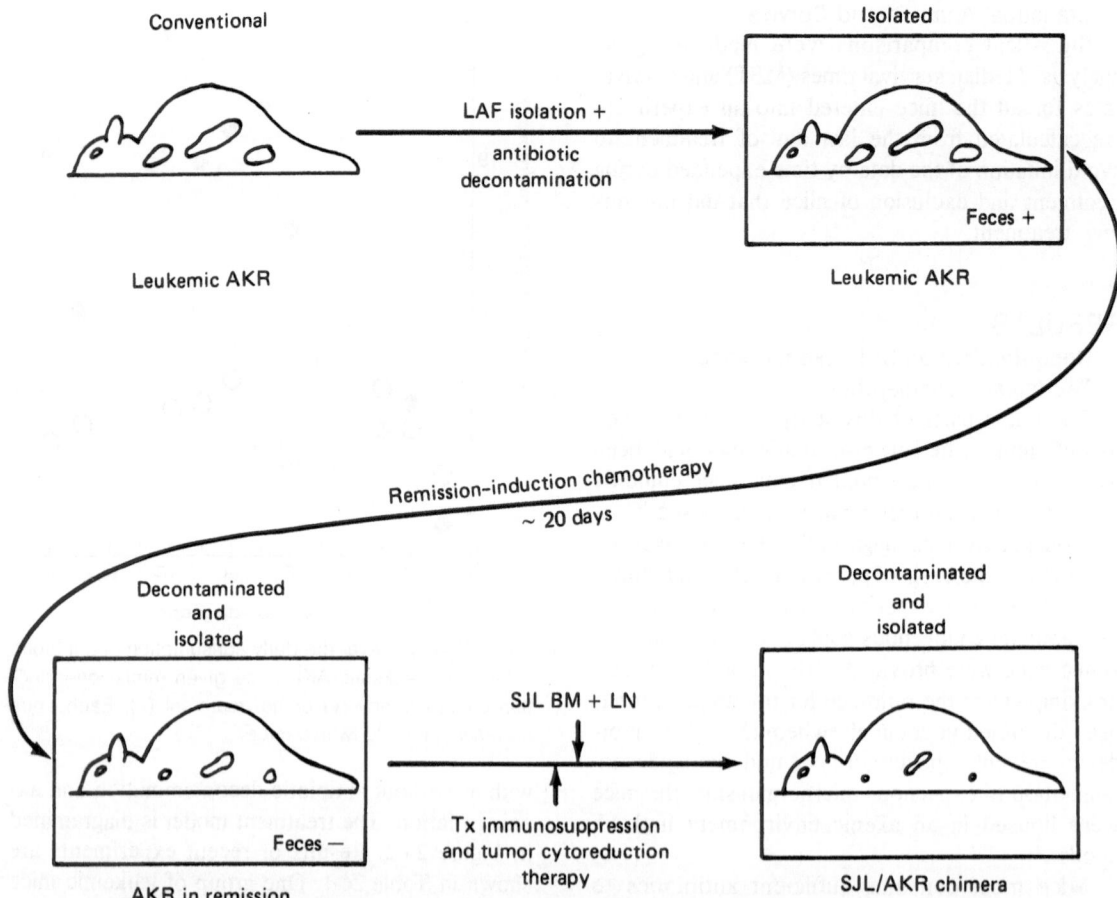

FIGURE 24-2. Experimental design to test adoptive immunotherapy in combination with chemo-radiotherapy of spontaneous T-cell leukemia in AKR mice undergoing bacterial decontamination and isolation in a protected environment. Conventional AKR mice diagnosed as having advanced leukemia were placed in sterile laminar airflow hoods (Isolation) and given high doses of antibiotics in their drinking water (Antibiotic Decontamination). Remission-induction chemotherapy was instituted to keep a significant proportion of the leukemic mice alive during antibiotic decontamination. Decontaminated AKR mice (Feces -) in leukemia remission underwent more intense chemoradiotherapy for transplant immunosuppression and tumor cytoreduction, followed 24 hr later by injection of H-2 incompatible SJL bone marrow and lymph node cells for adoptive immunotherapy. Antibiotic treatment and axenic isolation were continued post-transplant.

(10 mg/kg) on days 13, 15, 17, and 20. Mortality during RI was 14 percent in the conventional mice (Group 3) and 16 percent in the decontaminated mice (Group 4). Conventional mice given RI therapy alone had an MST of 39 days; decontaminated mice 51 days. However, only 9 percent (4/46) of the mice in these two groups survived to 120 days post-treatment. Leukemia relapse accounted for 59 percent and 83 percent of the deaths between the end of treatment (day 20) and 140 days in Groups 3 and 4, respectively.

Leukemic AKR mice in remission underwent more intense chemoradiotherapy in order to (a) reduce the tumor burden and (b) immunosuppress the recipient to prevent rejection of the subsequent transplant of allogeneic cells. The transplant immunosuppressive and tumor cytoreductive therapy

(TITC) consisted of 150 mg/kg CY and 700 R TBR, and if marrow was not transplanted, resulted in the death of 96 percent (25/26) of the mice within 20 days (Groups 5 and 6).

Drug and radiation toxicity from TITC therapy could be overcome if hemopoietic support was provided in the form of syngeneic bone marrow (Groups 7 and 8). Syngeneic cells, however, provided no adoptive immunotherapeutic benefit, and 68 percent of the decontaminated mice (Group 8) that completed treatment died of recurrent leukemia within 120 days. Conventional mice (Group 7) died primarily from infectious complications before leukemia recurred. The survival data for leukemic AKR mice that were decontaminated (Group 8) was not significantly better than their conventional counterparts (Group 7).

TABLE 24-1 Survival Data from Experiments To Test the Effect of Antibiotic Decontamination and Axenic Isolation on Bone Marrow Transplantation in Combination with Chemoradiotherapy for Acute T-Cell Leukemia in AKR Mice

		TREATMENT			NUMBER OF AKR MICE		MEAN SURVIVAL TIME (DAYS)[d]	PERCENT 120-DAY SURVIVAL (PERCENT)[d]	LEUKEMIC DEATHS PERCENT[e]
GROUP	MICROBIAL STATUS	RI[a]	TITC[b]	Donor BM + LN[c]	Entered	Completing treatment percent			
1	CV	−	−	−	237	NA[f]	20	0	100
2	DC	−	−	−	77	NA	17	0	100
3	CV	+	−	−	21	18(86)	39	14	59
4	DC	+	−	−	25	21(84)	51	4	83
5	CV	+	+	−	10	9(90)	14	0	0
6	DC	+	+	−	16	9(56)	12	6	0
7	CV	+	+	AKR	21	17(81)	16	5	38
8	DC	+	+	AKR	52	37(71)	33	10	68
9	CV	+	+	SJL	31	27(87)	20	6	4
10	DC	+	+	SJL	30	23(77)	119	50	10

[a]Remission-induction (RI), 100 mg CY/kg plus 10 mg MeC/kg (day 0 and 7); 10 mg PalmO/kg (day 13, 15, 17, and 20).
[b]Tumor cytoreductive therapy (TITC), 150 mg CY/kg plus 700 R TBR (day 22).
[c]Bone marrow (BM), 2×10^7 viable cells; lymph node (LN), 5×10^5 viable cells.
[d]Survival data calculated from the last day of treatment for all mice entered.
[e]$\dfrac{\text{Number dead with leukemia in 120 days}}{\text{Number completing treatment—nonevaluable deaths}} \times 100.$
[f]Not applicable.

For adoptive immunotherapy, conventional control (Group 9) and decontaminated experimental (Group 10) mice were given 2×10^7 bone marrow and 5×10^5 lymph node cells from H-2 incompatible SJL donors. Conventional animals developed acute secondary disease; their MST was 20 days, and only one of twenty-five deaths post-treatment was attributed to leukemia. In contrast, when a protected environment and decontamination was provided (Group 10), secondary disease was not a major problem. The MST and 120-day survival rate for decontaminated experimental mice (Group 10) were significantly better than all other groups ($p < 0.01$). In addition, transplanted H-2 incompatible cells had an adoptive immunotherapeutic effect as evidenced by a significant ($p < 0.01$) decrease in the rate of leukemia relapse as compared with cells from syngeneic donors (68 percent for Group 8 versus 10 percent for Group 10).

CONCLUSIONS

The results of these experiments indicate that it is possible to apply germfree technology to end-stage leukemia in mice. Antibiotic decontamination and isolation in a protected environment permitted the use of a therapeutic strategy that could not be used in conventional animals.

The benefits of decontamination/isolation were not apparent in the survival data for untreated leukemic animals (Group 1 versus 2, Table 24-1) or in animals given RI chemotherapy alone (Group 3 versus 4). Use of decontamination/isolation in combi-

nation with more aggressive chemoradiotherapy (RI + TITC) and syngeneic bone marrow transplantation resulted in a slight but not significant increase in the MST, as compared to conventionally housed animals (Group 7 versus 8). Despite the additional chemoradiotherapy and protection from infection, recurrent leukemia accounted for a large portion of the deaths in decontaminated mice given syngeneic marrow (Group 8).

Of the evaluable deaths in Group 8 that occurred within 125 days after completion of treatment, 52 percent (13/25) were directly attributable to leukemia that recurred more than 60 days' post-treatment; 79 percent (11/14) of the mice at risk 90 days' post-treatment died with leukemia within the next 35 days. According to the cytokinetic characteristics of spontaneous AKR leukemia, a single viable (clonogenic) leukemia cell would be expected to multiply to a lethal number in approximately 60 days (13). Thus, a survival of 60 days' post-treatment can be considered a "cell cure" (i.e., elimination of the original disease), assuming no perturbation of the cytokinetic growth characteristics during treatment (13). It seems unlikely (though possible) that the late recurrence of leukemia was due to the persistence of leukemic host cells. Rather, leukemic deaths more than 60 days' post-treatment probably represented induction of a new disease episode in susceptible cells. We have previously reported that the etiological agent of AKR leukemia, endogenous (Gross-type) murine leukemia virus, was not affected by adoptive immunotherapy (16). Thus, it appears likely that in Group 8 the susceptibility of lymphocytes in the transplanted AKR marrow and lymph node inoculum to Gross virus leukemogenesis accounted for the late recurrence of leukemia.

It is important to note that the problem of late recurrent leukemia was evident only when a large number of mice were still alive 60 days beyond the end of the treatment. That is to say, in transplanted conventional animals (Groups 7 and 9) or animals undergoing lethal chemoradiotherapy without hemopoietic support (Groups 5 and 6), most mice died before leukemia could reappear. Therefore, analysis of leukemia relapse rates were made using decontaminated mice reconstituted with marrow; use of other mice could lead to false conclusions.

Adoptive immunotherapy in the form of marrow and lymph node cells from H-2 incompatible SJL/J mice was necessary to significantly improve the problem of recurrent leukemia. The H-2 compatible donors were not used because cells from such donors provided no detectable antileukemic effect (3,7). Although histoincompatible cells eliminated the leukemia, they also caused severe secondary disease in conventional mice (Group 9). Much of the mortality in Group 9 was due to the susceptibility of these animals to endogenous and exogenous sources of infection. When decontamination/isolation was incorporated into the same treatment regimen, there was a significant ($p < 0.01$) improvement in survival (Group 9 versus 10). Furthermore, leukemia recurred within 120 days in only two (10 percent) decontaminated mice completing treatment in Group 10.

Several important conclusions can be drawn from this study: First, procedures for rendering mice bacteria-free can be applied successfully to mice bearing a spontaneous, end-stage acute leukemia after diagnosis of the disease. Second, although decontamination and axenic isolation had a slight beneficial effect when syngeneic cells were transplanted following aggressive chemoradiotherapy, recurrent leukemia posed a major problem in this experimental system. Finally, adoptive immunotherapy in the form of H-2 incompatible bone marrow and lymph node cell transplantation, in decontaminated leukemic AKR mice in remission, resulted in significant ($p < 0.01$) long-term leukemia-free survival without apparent secondary disease.

SUMMARY

Procedures for elimination of bacteria from the gastrointestinal tract were successfuly applied to AKR ($H-2^k$) mice after diagnosis of spontaneous acute T-cell leukemia. Leukemic mice were rendered bacteria-free (decontaminated) by continuous treatment with poorly absorbed, high dose antibiotics;

a sterile laminar airflow environment (axenic isolation) was used to protect decontaminated mice from exogenous microorganisms. Leukemia remission-induction (RI) with chemotherapy was necessary to keep a significant proportion of leukemic mice alive during decontamination.

An experimental model for the application of adoptive immunotherapy to the treatment of end-stage leukemia was developed using decontamination/isolation to prevent secondary disease (GvHD and infections). Conventional AKR mice, diagnosed as bearing advanced spontaneous leukemia, were randomized and entered into control or experimental groups with or without antibiotic decontamination and axenic isolation. Control and experimental groups included conventional and decontaminated mice that were (a) untreated, (b) undergoing RI chemotherapy only, (c) undergoing RI therapy plus transplant immunosuppression and tumor cytoreduction (TITC) chemoradiotherapy but no transplant, (d) undergoing RI plus TITC therapy followed by bone marrow and lymph node cell injection from nonleukemic AKR mice, and (e) undergoing RI plus TITC therapy followed by bone marrow and lymph node cell injection from histoincompatible SJL ($H-2^s$) mice for adoptive immunotherapy.

Decontamination/isolation had no significant effect on survival of untreated leukemic mice or animals undergoing RI chemotherapy only. Remission-induction plus TITC chemoradiotherapy had a lethally toxic effect on the hemopoietic system, but this could be overcome with marrow transplantation. Fatal relapse or recurrence of leukemia occurred in 68 percent of the decontaminated mice reconstituted with marrow from syngeneic donors within 120 days' post-treatment. Adoptive immunotherapy using histoincompatible SJL cells resulted in severe secondary disease in conventional leukemic AKR mice. However, when decontamination and axenic isolation were used with adoptive immunotherapy, there was a significant improvement in survival over the conventional controls (mean survival time = 119 versus 20 days; 120-day survival rates = 50 versus 6 percent) and all other treatment groups ($p < 0.01$). In addition, transplantation of H-2 incompatible cells resulted in a significant decrease in the rate of leukemia recurrence as compared with mice given cells from syngeneic donors (10 versus 68 percent; $p < 0.01$).

ACKNOWLEDGMENTS

This investigation was supported by United States Public Health Service Research Grant CA-18440 from the National Cancer Institute, the Allen-Bradley Foundation and the Board of Trustees, Mount

Sinai Medical Center, Milwaukee. Dr. Truitt is a Special Fellow of the Leukemia Society of America, Inc.

The author is grateful for the technical assistance of Kathleen M. Kohut, Evangeline R. Reynolds, and Barbara J. Stephan.

REFERENCES

1. Bealmear, P.M., Loughman, B.E., Nordin, A.A., and Wilson, R. Evidence for graft vs. host reaction in the germfree allogeneic radiation chimera. In J. B. Heneghan, ed., *Germfree Research: Biological Effects of Gnotobiotic Environment*. New York: Academic Press, 1973, p. 471.

2. Bortin, M.M., and Truitt, R.L. AKR T cell acute lymphoblastic leukemia: A model for human T cell leukemia. *Biomedicine, 26*:315, 1977.

3. Bortin, M.M., Truitt, R.L., and Rimm, A.A. Nonspecific adoptive immunotherapy of T cell acute lymphoblastic leukemia in AKR mice: A model for the treatment of T cell leukemia in man. In H. Waters, ed., *The Handbook of Cancer Immunology:* Immunotherapy, Vol. 5, New York: Garland (In press) 1978.

4. Heit, H., Wilson, R., Fliedner, T.M., and Kohne, E. Mortality of secondary disease in antibiotic-treated mouse radiation chimeras. In J. B. Heneghan, ed., *Germfree Research: Biological Effects of Gnotobiotic Environment*. New York, Academic Press, 1973, p. 477.

5. Hendricks, W.D.H., van der Waaij, D., Korthals-Altes, C., Berghuis, J.M., Lekkerkerk, J.E.C., and de Vast, J. Elimination of the gastrointestinal microflora in monkeys with nonabsorbable antibiotics. In J. Klastersky, ed., *Clinical Use of Combinations of Antibiotics*. New York: Wiley, 1975, p. 135.

6. Jones, J.M., Wilson, R., and Bealmear, P.M. Mortality and gross pathology of secondary disease in germfree mouse radiation chimeras. *Radiat. Res., 45*:577, 1971.

7. LeFeber, W.P., Truitt, R.L., Rose, W.C., and Bortin, M.M. Graft versus leukemia. VII. Donor selection for adoptive immunotherapy in mice. In S. J. Baum, and G. D. Ledney, eds., *Experimental Hematology Today*. New York: Springer-Verlag, 1977, p. 239.

8. Pollard, M., Chang, C.F., and Srivastava, K.K. The role of microflora in development of graft-versus-host disease. *Trans. Proc., 8*:533, 1976.

9. Pollard, M., and Truitt, R.L. Allogeneic bone marrow chimerism in germfree mice. I. Prevention of spontaneous leukemia in AKR mice. *Proc. Soc. Exp. Biol. Med., 144*:659, 1973.

10. Pollard, M., and Truitt, R.L. Allogeneic bone marrow chimerism in germfree mice. II. Prevention of reticulum cell sarcomas in SJL/J mice. *Proc. Soc. Exp. Biol. Med., 145*:488, 1974.

11. Pollard, M., and Truitt, R.L. Applications of germfree technology to experimental problems of clinical significance. In T. Hasegawa, ed., *Proceedings of the First Intersectional Congress of IAMS, Vol. 3*. Tokyo: Science Council of Japan, p. 250, 1975.

12. Schabel, Jr., F.M., Skipper, H.E., Trader, M.W., Laster, Jr., W.R., and Simpson-Herren, L. Spontaneous leukemia (lymphoma) as a model system. *Cancer Chemo. Rept., 53*:329, 1969.

13. Skipper, H.E., Schabel, Jr., F.M., Trader, M.W., Laster, Jr., W.R., Simpson-Herren, L., and Lloyd, H. Basic and therapeutic trial results obtained in the spontaneous AK leukemia (Lymphoma) model—End of 1971. *Cancer Chemo. Rept., 56*:273, 1972.

14. Thomas, E.D., Buckner, C.D., Banaji, M., Clift, R.A., Fefer, A., Flournoy, N., Goodell, B.W., Hickman, R.O., Lerner, K.G., Neiman, P.E., Sale, G.E., Sanders, J.E., Singer, J., Stevens, M., Storb, R., and Weiden, P.L. One hundred patients with acute leukemia treated by chemotherapy, total body irradiation, and allogeneic marrow transplantation. *Blood, 49*:511, 1977.

15. Thomas, E.D., Flournoy, N., Buckner, C.D., Clift, R.A., Fefer, A., Neiman, P.E., and Storb, R. Cure of leukemia by marrow transplantation. *Leukemia Res., 1*:67, 1977.

16. Truitt, R.L. Application of germfree techniques to the treatment of leukemia in AKR mice by allogeneic bone marrow transplantation. In H. Waters, ed., *The Handbook of Cancer Immunology: Immunotherapy, Vol. 5*. New York: Garland, (in press) 1978.

17. Truitt, R.L., Pollard, M., and Srivastava, K.K. Allogeneic bone marrow chimerism in germfree mice. III. Therapy of leukemic AKR mice. *Proc. Soc. Exp. Biol. Med., 146*:153, 1974.

18. van Bekkum, D.W., de Vries, M.J., and van der Waaij, D. Lesions characteristic of secondary disease in germfree heterologous radiation chimeras. *J. Nat. Cancer Inst., 38*:223, 1967.

19. van Bekkum, D.W., Roodenburg, J., Heidt, P.J., and van der Waaij, D. Mitigation of secondary disease of allogeneic mouse radiation chimeras by modification of the intestinal microflora. *J. Nat. Cancer Inst., 52*:401, 1974.

20. van Bekkum, D.W., and van der Waaij, D. Total body irradiation and bone marrow transplantation in monkeys. In H. Balner, and W. I. Beveridge, eds., *Infections and Immunosuppression in Subhuman Primates*. Copenhagen: Munksgaard, 1970, p. 225.

21. van der Waaij, D., de Vries, J.M., and Lekkerkerk, J.E. Eliminating bacteria from monkeys with antibiotics. In H. Balner, and W. I. Beveridge, eds., *Infections and Immunosuppression in Subhuman Primates*. Copenhagen: Munksgaard, 1970, p. 21.

22. van der Waaij, D., and Sturm, C.A. Antibiotic decontamination of the digestive tract of mice: Technical procedures. *Lab. Animal Care, 18*:1, 1968.

23. van der Waaij, D., and Sturm, C.A. The production of "bacteria-free" mice. Relationship between fecal flora and bacterial population of the skin. *Antonie van Leeuwenhoek, 37*:139, 1971.

24. van der Waaij, D., and Vossen, J.M. Antibiotic decontamination in animals and in human patients. In T. Hasegawa, ed., *Proceedings of the First Intersectional Congress of IAMS, Vol. 3*. Tokyo: Science Concil of Japan, 1975, p. 233.

25

Procurement of Proliferative and Quiescent Acute Leukemia Cells by Centrifugal Elutriation

H. D. Preisler, G. Hausner,
I. Walczak, S. Bruno,
V. Von Fliedner, G. Christoff,
Y. M. Rustum, and J. Renick

INTRODUCTION

Leukemic cells do not represent homogeneous populations; rather, within the bone marrow and peripheral blood of individual patients there are subpopulations of leukemic cells that differ in size, morphology, and reproductive kinetic characteristics. Presumably, biochemical differences parallel these more obvious differences and, hence, the various subpopulations of cells may differ with respect to their sensitivity to antineoplastic agents as well. A previous communication from this laboratory has reported that centrifugal elutriation can be used to obtain subpopulations of leukemic cells that differ in [³H]TdR labeling index by more than 1 log (1). This communication represents a continuation of our elutriator studies and demonstrates that this method can be used to obtain large numbers of leukemic cells that differ morphologically. We have also refined our analytical methods to include the use of flow cytofluorometry so that rapid analysis of the subpopulations obtained can be accomplished.

METHODS

Bone marrow and peripheral blood specimens were obtained and separated by centrifugal elutriation as previously described (4). Labeling indices were also determined as previously described (4).

For determination of cell DNA content, the propidium iodide method of Krishan was employed (3). Two million cells were suspended in a solution containing 0.05 mg/ml of propidium iodide in 0.1 percent sodium citrate. The cells were pelleted by centrifugation at 1,500 rpm, and the pellet was resuspended in 3 ml of the propidium iodide solution. The tube containing this cell suspension was placed on ice for 10 min. The nuclei were pelleted once more and washed three times in Earle's balanced salt solution (EBSS). The pellet was resuspended in 1 ml of normal saline.

Propidium iodide-stained nuclei were analyzed by a Becton-Dickinson Fluorescence Activated Cell Sorter (FACS-II) using 0.5 W of laser power at 488 nm. Filters for 520 nm and 620 nm were used to prevent direct laser radiation from entering the photomultiplier tube. Sample flow rate was 800 cells/sec., and 10^5 cells were interrogated for each sample.

To estimate the number of cells in G_0-G_1 and S-G_2-M, the G_1 was identified and the number of cells in the left one-half of the G_1 peak was determined by counting the cells in the interval between the beginning of the G_1 peak to the modal peak. This number (i.e., the cells in the left one-half of the G_1 curve) was doubled to give an estimate of the number of cells in G_0 and G_1. The number of cells in S-G_2-M were estimated by determining the total number of cells present and subtracting the number of cells in G_0-G_1.

Light scatter analysis was done in an identical manner, except the cells to be analyzed were not exposed to propidium iodide but were simply suspended in EBSS at a concentration of 5×10^6 cells/ml and analyzed by the FACS-II.

Coulter Counter Analysis

A Coulter channelizer model H4 with a Coulter Counter ZBI was used. The instrument was calibrated with 10 μ polystyrene spheres and the scale factor determined. The sample was adjusted to give a 10 percent concentration index. Proper amplification and aperture current settings were determined, and the analysis used 2,000 as the maximum counts per channel. Results for population volume were then printed out and converted to diameter in microns (μ).

RESULTS

Acquisition of Cells with Extremely High Proliferative Fraction

Ms. M.N., a 36-year-old black female, had a peripheral blood white cell count of 508,000/μl, which consisted of blast cells. Her marrow was hypercellular, with 97 percent blasts. The morphological classification of her leukemia was unclear, but the presence of PAS-positive granules in the cytoplasm of the leukemic cells suggested acute lymphocytic leukemia. Both the peripheral blood cells and the bone marrow cells from this patient were subjected to centrifugal elutriation. The blood and marrow separations each produced three significant subpopulations of cells. Table 25-1 gives the size, labeling index, and flow cytofluorography estimate of S, G_2, and M of these subpopulations. Figure 25-1 gives the cell sorter DNA histograms of the unseparated cells and subpopulations 1, 2, and 3.

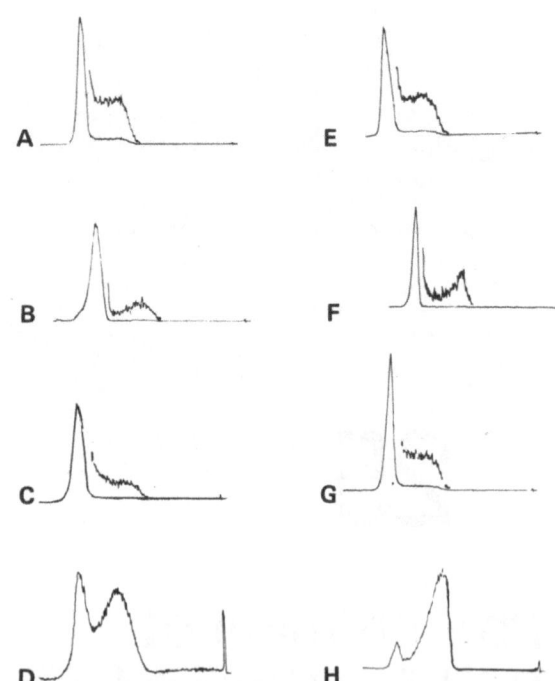

FIGURE 25-1. DNA histograms of leukemic cells from patient M. N. The ordinate represents cell number, the abscissa fluorescence. The major peak represents G_0-G_1 cells with G_2-M cells being represented by the peak to the right. In graphs A-C and E-G, the partial curve to the right of the G_0-G_1 peak and above the continuous curve represents a scale expansion of the curve below. Calculations for G_0-G_1, S, and G_2-M distribution of cells are given in Table 25-1. Marrow: (A) unseparated cells, (B), (C), and (D), subpopulation 1, 2, and 3, respectively. Peripheral blood: (E) unseparated cells; (F), (G), and (H) subpopulation 1, 2, and 3, respectively.

From the marrow, two subpopulations of leukemic cells were obtained, with an estimated S-G_2-M fraction of 3.6 and 56 percent, respectively. From the peripheral blood, two subpopulations were obtained, with estimated S-G_2-M fractions of 2.2 and 85 percent, respectively, and labeling indexes of 1 and 74 percent. This study demonstrates

TABLE 25-1 Separation of Leukemia Cells of Patient M.N.

CELLS	BONE MARROW			PERIPHERAL BLOOD		
	Size[a]	LI[b]	S-G_2-M[c]	Size[a]	LI[b]	S-G_2-M[c]
Unseparated	14.1 ± .4	25	22	16.4 ± .4	22	14
Subpop. 1	12.2 ± .2	0	3.6	13.6 ± .3	1	2.2
Subpop. 2	ND[d]	15	12	ND[d]	ND[d]	14
Subpop. 3	14.6 ± .6	ND	56	17.2 ± .3	74	87

[a]Mean ± standard error (μ) as determined by means of an image-splitting device (1).
[b]LI, labeling indexes (percent).
[c]S-G_2-M, synthesis-gap 2-mitosis (percent).
[d]ND, Not done.

TABLE 25-2 Separation of Peripheral Blood Leukemic Cells from Patient E. VD.

	BLASTS+PRO[a]	MODAL SIZE (μ)[b]	LI[c]	S-G_2-M[d]	CELL NUMBER
Unseparated	72	16.48	2%	10%	2.7×10^9
Subpop. 1	67	15.0	0	23	9.9×10^8
Subpop. 2	79	15.8	1.5	7	6.5×10^8
Subpop. 3	78	17.7	3	6	7.3×10^8
Subpop. 4	87	22.7	28	23	0.6×10^8

[a]Myeloblasts and promyelocytes (percent).
[b]Determined by Coulter Counter analysis.
[c]LI, labeling indexes (percent).
[d]S-G_2-M, synthesis-gap 2-mitosis (percent).

Acquisition of Leukemic Cells that Differ in Both Morphological Appearance and Proliferative Characteristics

Mr. E. VD., a 59-year-old white male, with acute promyelocytic leukemia had a peripheral white blood cell count of 6,500/μl, with 31 percent myeloblasts and 11 percent promyelocytes. The patient was leukapheresed, and 2.7×10^9 peripheral blood leukemic cells were subjected to centrifugal elutriation. Four subpopulations of cells were obtained. Table 25-2 gives modal size, [³H]TdR labeling index, and estimated S-G_2-M by flow cytofluorography. Figure 25-2 gives Coulter volume curves, light scatter profile, and DNA histograms of these populations. By Coulter volume analysis, there are four distinct populations of cells in the unseparated populations, with modal diameters of 16.5 μ, 12.5 μ, 7.9 μ, and 6.1 μ. The light scatter

that subpopulations of leukemic cells with extremely different proliferative characteristics can be obtained by centrifugal elutriation.

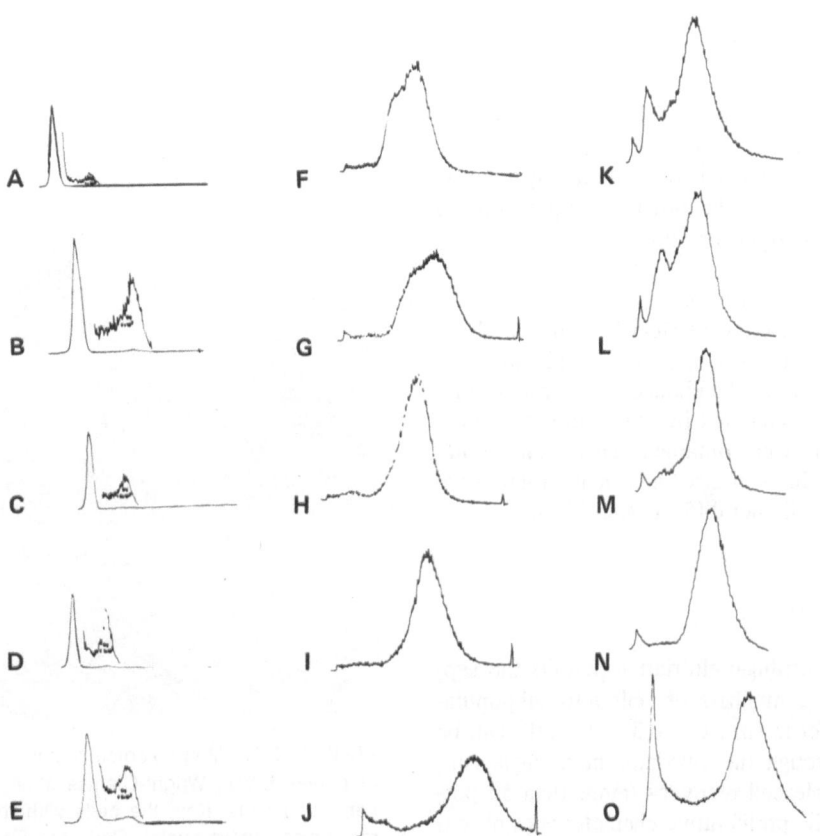

FIGURE 25-2. The DNA histograms of leukemic cells from patient E.VD. are as described in legend for Figure 25-1. The distribution of cells in the various phases of the cell cycle and in terms of cell size are given in Table 25-2. (A), (F), and (K), unseparated cells; (B), (G), and (J) subpopulation 1; (C), (H), and (M) subpopulation 2; (D), (I), and (N) subpopulation 3; (E), (J), and (O) subpopulation 4.

analysis of population size also demonstrated four subpopulations. Analysis of subpopulations 1-4 by light scatter and Coulter analysis were also similar. Figure 25-3 illustrates the morphological differences between the cells of subpopulations 1 and 4, with the former consisting of cells with round nuclei and high nuclear/cytoplasmic ratio, and the latter consisting of cells with monocytoid nuclei and a much lower nuclear/cytoplasmic ratio.

Separation of Previously Unrecognized Subpopulation of Cells

Mr. F.J., a 39-year-old white male, with acute lymphocytic leukemia was shown to have, on karyotypic analysis, a Philadelphia chromosome (Ph'). Bone marrow examination revealed 86 percent lymphoblasts and 0.6 percent promyelocytes; the remaining cells were lymphocytes. The marrow cells were subjected to elutriation, and three significant subpopulations were obtained. Table 25-3 gives the percentage blasts, modal size, labeling index, and estimated S-G$_2$-M; in Fig. 25-4 the Coulter Counter and light scatter profiles of the populations are compared. Coulter Counter analysis demonstrated three distinct populations in the unseparated specimen and in subpopulation 1 and 2. Fraction 3 contained four population peaks. The light scatter profile agreed with this analysis, except that, in subpopulation 3, only three population peaks were discerned. Figure 25-5 shows that the cells in subpopulation 3 were primarily myeloblasts and promyelocytes. The other subpopulations of cells were lymphoid cells only.

Nonseparation of Cells

J.R., a 34-year-old white female, had acute lymphocytic leukemia, which, on karyotypic analysis, was found to contain the Philadelphia chromosome. Her cells were separated by elutriation, and three subpopulations were obtained. These subpopulations did not differ in size, and their proliferative characteristics did not differ (Table 25-4).

DISCUSSION

The use of centrifugal elutriation permits the separation of large numbers of cells into subpopulations that differ in size. Up to 3×10^9 cells can be processed through the elutriator in a single run, with reasonable cell recovery (more than 50 percent). Since the proliferative characteristics of leukemic cells vary with cell size, the subpopulations of cells obtained can differ by more than 30-fold with respect to the percent of cells in S phase. In one-half of the marrow specimens separated, we obtained subpopulations of leukemic cells of the

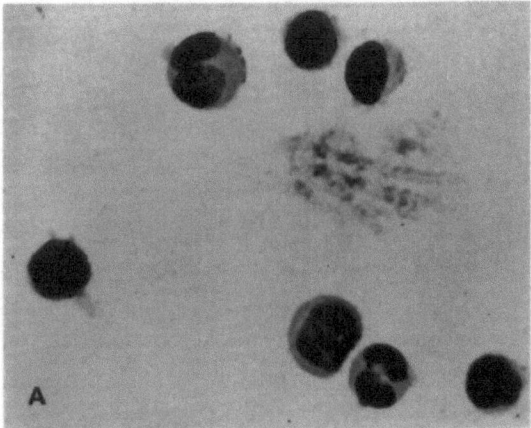
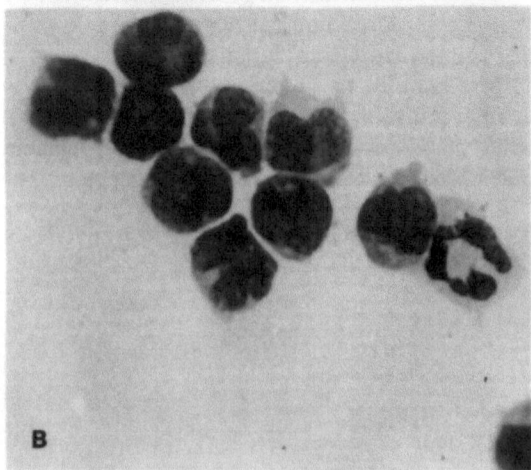
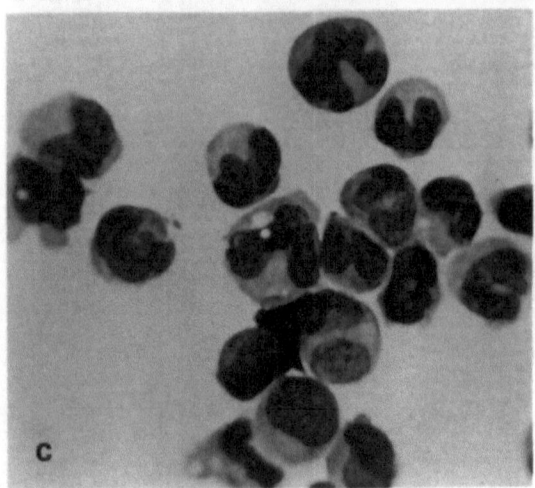

FIGURE 25-3. Morphological features of leukemic cells of patient E.VD.; Wright-Giemsa stain, X 900. (A) Unseparated cells. Note the cells with round nuclei and myelomonocytoid nuclei. Cells are small, with a high nuclear:cytoplasmic ratio. (B) Cells of subpopulation 1. The cells have a very high nuclear/cytoplasmic ratio. Nuclei are irregularly kidney shaped. (C). Cells of subpopulation 4. Large cells with a moderate nuclear/cytoplasmic ratio. Nuclei are markedly monocytoid in nature.

TABLE 25-3 Separation of Bone Marrow Leukemic Cells of Patient F.J.

	BLASTS[a]	MODAL SIZE (μ)[c]	LI[d]	S-G$_2$-M[e]	CELL NUMBER
Unseparated	80	12.0	6.5	7.6	1.5×10^9
Subpop. 1	71	10.2	1	5.4	1.9×10^8
Subpop. 2	94	13.9	4.5	4	6.7×10^8
Subpop. 3	94[b]	21.5	38	22	0.092×10^8

[a]Lymphoblasts (percent).
[b]Fifty percent myeloblasts and 16 percent promyelocytes.
[c]Determined by Coulter Counter analysis.
[d]LI, labeling indexes (percent).
[e]S-G$_2$-M, synthesis-gap 2-mitosis (percent).

peripheral blood, and only occasionally a subpopulation with a high labeling index (4). The separation of the peripheral blood of patient M.N., however, demonstrates that, in some cases, subpopulations of cells can be obtained from peripheral blood with extremely high labeling indexes (74 percent). Our inability to separate the cells of patient J.R. into different-sized cells correlated with the lack of proliferative differences between the fractions obtained. Hence, the fractionation of leukemic cells into subpopulations with different proliferative characteristics depends upon a separation of the original cells into subpopulations of different-sized cells.

With some marrow or blood specimens, elutriator separation of leukemic cells has provided subpopulations of cells that not only differ in size and proliferative characteristics, but differ in morphological appearance and in cell lineage. For example, the separation of the peripheral blood leukemic cells of patient E. VD. produced subpopulations, one of which appeared to be of the conventional myeloblastic type, whereas another was myelomonocytic. Presumably, these subpopulations are both members of the same cell series. On the other hand, the separation of marrow cells obtained from patient F.J. resulted in the acquisition of a subpopulation of cells of at least 50 percent myelob-

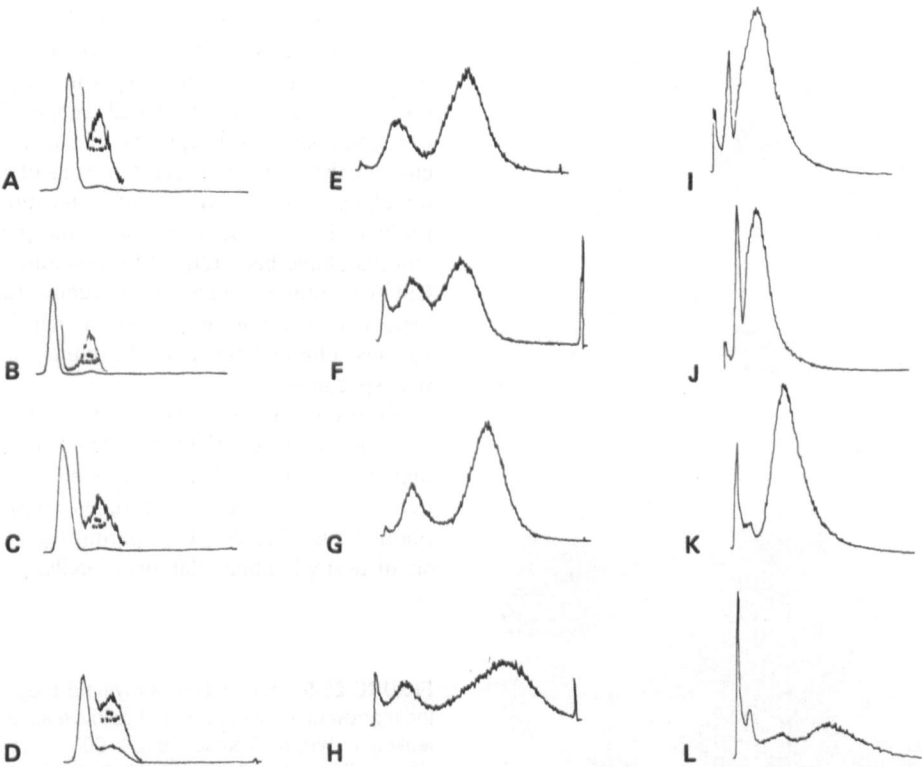

FIGURE 25-4. The DNA histograms of leukemic cells from patient F. J. are as described in the legend for Figure 25-1. The distribution of cells in the various phases of the cell cycle and in terms of cell size are given in Table 25-3, (A), (E), and (I) unseparated cells; (B), (F), and (J) subpopulation 1; (C), (G), and (K), subpopulation 2; (D), (H), and (I) subpopulation 3.

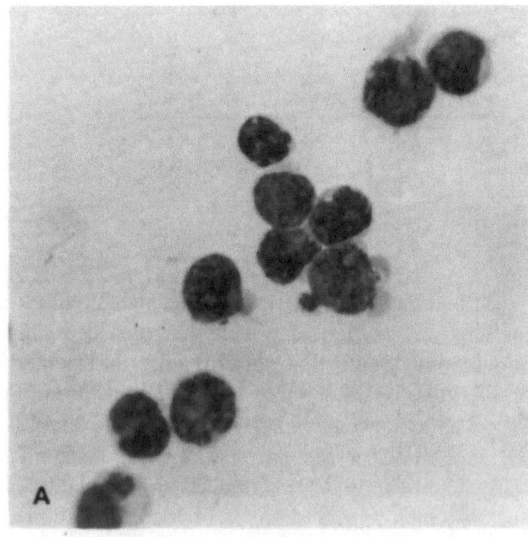

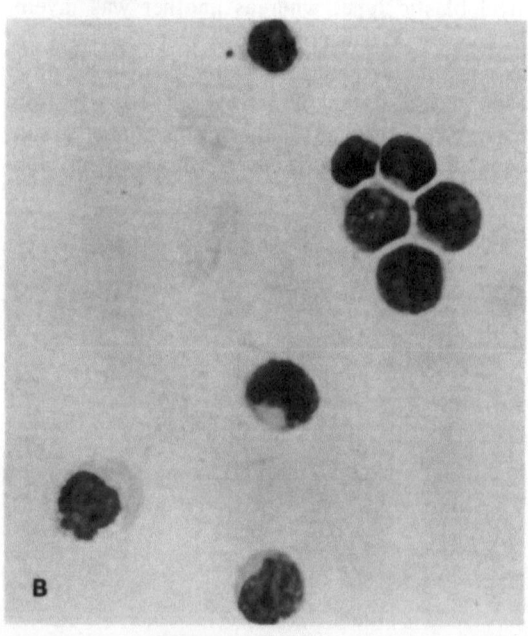

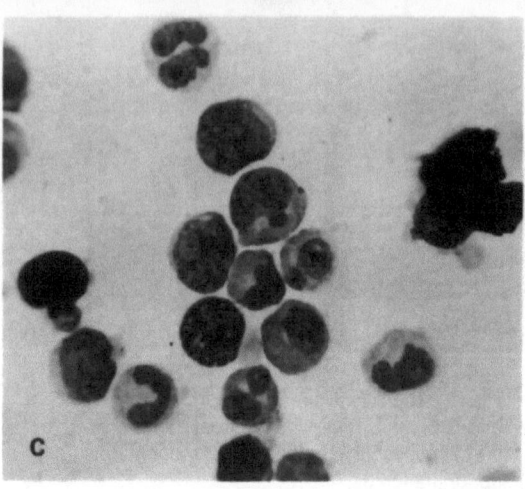

lasts and promyelocytes, cells that were hardly noticed in the unseparated specimen. Since this patient had acute lymphocytic leukemia with a Philadelphia chromosome, it is not clear at the present time whether the immature myeloid cells were normal cells or chronic myelocytic leukemic cells. Studies to delineate the leukemic or nonleukemic nature of the myeloblasts are in progress.

The sequential use of the elutriator and cytofluorography and the Coulter Counter channelizer has permitted rapid evaluation of the size and proliferative characteristics of the subpopulations of cells obtained. Hence, within 1 hr of separation, we can identify the subpopulations, which consist of the smallest and largest cells and obtain an estimate of the proliferative characteristics of the cells. Our analysis of the DNA histograms is not as accurate as we would like. Perhaps the most glaring example is the analysis of subpopulation 1 of patient E. VD. In this case, the cell sorter estimate of S-G_2-M was 23 percent, whereas the labeling index of this subpopulation was zero. This inaccuracy occurred because the G_1 peak in the DNA histogram was skewed to the left. This greatly magnified the error intrinsic to our method of estimation. We hope that the use of computer analysis will obviate this problem (1,2).

A second problem with which we are faced related to the determination of cell size. Although there is good general agreement between light scatter analysis and Coulter sizing with respect to the number of different-sized cell subpopulations, if one computes (by integrating the area under the curves) the number of cells in each of these subpopulations, one arrives at different estimates. This problem is also under review at the present time and may have been related to a malfunction of the Coulter Counter, since more recent studies have yielded comparable estimates of cell numbers in various subpopulations of cells in unseparated marrow specimens.

In any event, these studies have demonstrated the utility of the elutriator as a device for separating human leukemic cells into subpopulations. A flow cytofluorograph greatly facilitates analysis of the populations obtained, thus permitting rapid analysis of desired subpopulations of cells.

FIGURE 25-5. Separation of myeloid progenitors from the marrow of a patient with Ph¹ positive acute lymphatic leukemia; Wright-Giemsa stain, X 900. (A) The unseparated cells are lymphoblasts. (B) Cells of subpopulation 1 are small lymphoblasts and lymphocytes. (C) Cells of subpopulation 3 are of the granulocytic series. Myeloblasts and promyelocytes, as well as more mature granulocytic elements.

TABLE 25-4 Separation of Peripheral Blood Leukemic Cells of Patient J.R.

	BLASTS[a]	MODAL SIZE (μ)	LI[b]	S-G$_2$-M[c]	CELL NUMBER
Unseparated	81.5	12.3	13.5	11	6.8×10^8
Subpop. 1	62	12.3	19	15	2.9×10^8
Subpop. 2	63	12.3	ND[d]	ND[d]	1.2×10^6
Subpop. 3	83	12.3	17	12	0.2×10^8

[a]Lymphoblasts (percent).
[b]LI, labeling indexes (percent).
[c]S-G$_2$-M, synthesis-gap 2-mitosis (percent).
[d]ND, not done.

SUMMARY

Centrifugal elutriation can be used to separate human leukemic cells into subpopulations that differ in cell size. This separation technique has made it possible to obtain subpopulations of cells whose labeling index differ by more than 1 log. Leukemic cells from the bone marrow or peripheral blood are suitable for separation, and up to 3×10^9 cells can be processed in 1 hr, thus permitting the acquisition of a suitable number of cells for biochemical and tissue culture studies. Subpopulations of cells that differ morphologically can also be obtained. For example, two subpopulations of cells, one almost exclusively myeloblasts and the other monocytoid, were obtained from a myelomonocytic leukemia patient. To facilitate analysis of cell subpopulations, we have employed a FACS-II cell sorter to generate DNA histograms and light scatter profiles of subpopulations of cells. The DNA histograms were in good agreement with the labeling index studies, and the distribution of cells as determined by light scatter was in good agreement with cell volume distribution as determined by a Coulter Channelizer. Hence, elutriation and flow cytofluorography can be used for the rapid separation and evaluation of subpopulations of large numbers of human leukemic cells.

ACKNOWLEDGMENT

The authors would especially like to thank the family of Mrs. Perry without whose help the FACS-II would not have been available for these studies.

The authors would also like to thank Dr. Donald Higby for performing the leukapheresis and Mrs. Irene Rakowski for the photographs of her cells.

REFERENCES

1. Fried, J. Method for the Quantitative Evaluation of Data from Flow Microfluorometry. *Computers Biomed. Res., 9*:263, 1976.
2. Jett, J.H. Analysis of DNA Histograms from Asynchronous and Synchronous Cell Populations. *Third International Symposium on Pulse Cytophotometry. March 30–April 1, 1977*, Vienna, Austria, p. 12.
3. Krishan, A. Rapid Flow Cytofluorometric Analysis of Mammalian Cell Cycles by Propidium Iodide Staining. *J. Cell Biol., 66*:188, 1975.
4. Preisler, H.D., Walczak, I., Renick, J., and Rustum, Y.M. Separation of Leukemic Marrow into Proliferative and Nonproliferative Subpopulations by Centrifugal Elutriation. *In press, Cancer Res.*, 1977.

PART VI

Advances in Experimental Hematology

D. H. Pluznik

26

Cellular Requirements for Induction of Colony-Stimulating Factor in Lymphoid Cell Cultures Stimulated by T- and B-Cell Mitogens

Ron N. Apte, Eliahu Heller,
Chanita F. Hertogs, and
Dov H. Pluznik

INTRODUCTION

The clonal growth of hemopoietic cells *in vitro* into colonies of granulocytes and/or macrophages requires the presence of a stimulatory substance designated colony-stimulating factor (CSF) (7,21). Data from various clinical and animal experiments suggest that CSF is, or is related to, a physiological regulator of granulopoiesis (review, references 3,12,22,26,27). In the mouse assay, this substance is produced and/or released *in vitro* by cell suspensions from many organs, including hemopoietic tissue, and by cell lines; it is also present in the serum and urine (12,22,26,27). However, in man, the distribution of this humoral factor is limited mainly to mature leukocytes (12,22,26,27). These observations on the production of CSF by leukocytes led to hypotheses concerning feedback control, whereby mature leukocytes regulate their own production (12,22,26,27).

Increased numbers of granulocytes and macrophages are found in many immunological reactions (5). It is also shown that elevated levels of serum CSF are associated with a wide variety of immunological situations, such as stimulation by antigen, by allograft and tumor rejection, and by bacterial and viral infections; they are also seen in some lymphoproliferative diseases (review, reference 18). In addition, lymphoid cells that are undergoing a mixed lymphocyte reaction, or stimulated by mitogens and antigens, secrete large amounts of CSF (2,18,19,23,24).

Macrophages and granulocytes are cardinal cellular elements in host resistance and in inflammation, and they are essential in the elimination of invading microorganisms. Bacterial lipopolysaccharides (LPS), which are produced by gram-negative bacteria, have pronounced effects on the granulopoietic system. After an injection of LPS into mice, a rise in tissue and serum CSF, as well as a rise in the number of granulocyte/macrophage progenitor cells in the marrow and spleen, is observed (review, reference 3). Analogous results were obtained when staphylococcal enterotoxin B (SEB), an exotoxin derived from a gram-positive bacteria, *Staphylococcus aureus,* was injected into mice (14). In addition, both bacterial products are potent lymphocyte mitogens, LPS being a B-cell mitogen (11,20) and SEB a T-cell mitogen (1). Both mitogens also induce CSF production in lymphoid cell suspensions (14,15,19).

This study was undertaken to elucidate the cellular requirements of CSF production in lymphoid cell suspensions, as induced by SEB and LPS.

MATERIALS AND METHODS
Mice
Our C₃H/eB, C₃H/HeJ, Balb/c, and ICR, as well
as C₃H/eB congenitally athymic (nude) male mice
were purchased from the Weizmann Institute of
Science, Rehovot, Israel, and CBA/N mice generously provided by Dr. I. Scher (Naval Medical Research Institute, Bethesda, Md.) were subsequently
bred and raised in our facilities. All mice were used
at the age of 9 to 12 weeks.

Mitogens
The LPS-W from *Escherichia coli* 055:B5 was
purchased from Difco Laboratories, Detroit, Mi.
Staphylococcal enterotoxin B (SEB) was kindly
provided by Dr. I. Hertman of the Israel Institute
for Biological Research, Ness-Ziona, Israel.

Lymphoid Cell Cultures
Spleens or thymi were removed aseptically from
mice, dispersed by passage through a stainless steel
screen into phosphate-buffered saline, and washed
twice. Nucleated viable cells (2 × 10⁶/ml) were incubated in RPMI-1640 medium supplemented with
5 percent heat-inactivated horse serum (Bio-Lab,
Jerusalem, Israel), in 1 ml aliquots at 37°C in a humidified atmosphere of 5 percent CO₂, usually for 4
days. At the end of the incubation, the cells were
removed by centrifugation and the cell-free supernatant was stored at −20°C until used.

Enriched Cell Populations
(a) *Peritoneal macrophages* were obtained from
unstimulated mice or mice that had been injected
3 days before with 1.5 ml of 10 percent proteose-
peptone or thioglycolate solution (Difco Laboratories, Detroit, Mi.). The mice were sacrificed, and
peritoneal cells were collected from the peritoneal
cavity after the injection of 5 ml of PBS. After being
aspirated from the peritoneal cavity, these cells
were washed and cultured under conditions similar
to those described above. The cells were transferred to Petri dishes, incubated for 2 hr, and then
washed vigorously three times to remove non-
phagocytic loosely attached cells; they were subsequently cultured. (b) *Splenic or thymic lymphocytes* were suspended, and the suspensions were
depleted of adherent cells by passage through two
columns of Pyrex wool (Pyrex Wool Owens-Corn-
ing Fiberglass Corp., Corning, N.Y.). Adherence
columns were constructed in plastic syringes, and
each contained 1 g Pyrex wool, which had previously been washed, autoclaved, and rinsed with
warm PBS supplemented with 5 percent horse
serum. Approximately 1 × 10⁸ cells in a volume of
2 ml were loaded onto each column. The cells were
first incubated for 20 min at the upper part of the

column and then were washed into the lower part
for an additional incubation period of 25 min. All
separation procedures were performed at 37 °C. At
the end of the incubation time, each column was
washed with 30 ml of PBS, and the cells were collected, washed, and cultured as described above.

Assay for CSF
Colony-stimulating factor from the culture fluids
of the different cell cultures was assayed by its ability to stimulate colony formation from murine bone
marrow cells in a soft agar system. Bone marrow
cells (10⁵) were cloned in soft agar medium on a
harder agar base (0.5 percent) supplemented with
either 15 or 30 percent conditioned medium and 25
percent horse serum. Cultures were incubated for
7 days, and cell colonies growing in soft agar were
counted under the microscope.

The results shown in each table (below) are average results obtained from five experiments.

RESULTS
Interactions between Macrophages and Lymphocytes in CSF Production Induced by LPS
Spleen cell cultures, when stimulated by an op-
timal dose of LPS (50 μg/ml) and cultured for 4
days, secrete large amounts of CSF. When such
cell suspensions are passed through two adherence
columns to deplete adherent macrophages, the ability of the nonadherent effluent cells to secrete CSF
in response to LPS gradually decreases. However,
when such nonadherent cell suspensions are supplemented with peritoneal macrophages (at a final
concentration of 10 percent) large amounts of CSF
are produced. Macrophages in the amount added
to lymphocyte cultures secrete only minute
amounts of CSF (Table 26-1).

In order to elucidate the nature of the interac-
tion(s) between macrophages and lymphocytes in
the process of secretion of this mediator, macrophage supernatants were prepared by incubating 2
× 10⁶ macrophages/ml in culture medium with or
without 50 μg of LPS/ml for 24 or 48 hr. Supernatants were diluted 1:1 in culture medium and
added to nonadherent lymphocytes, which were
subsequently cultured for 4 days with or without
50 μg of LPS/ml. The culture fluids from such lymphocyte suspensions were incorporated into the
agar as a source of CSF. The results shown in Table
26-2 indicate that active macrophage supernatants
are produced when LPS is added to the macrophage cultures and that such supernatants can activate lymphocytes to secrete CSF, even when the
lymphocyte cultures are not stimulated by the mitogen. Supernatants prepared from macrophage

TABLE 26-1 Comparison of the Ability of Purified Lymphoid Cell Cultures To Secrete CSF in Response to LPS

CELL POPULATION	NUMBER OF COLONIES/10^5 BONE MARROW CELLS[a]
Unfractionated spleen cells[b]	175.6
Effluent cells from the adherence columns (nonadherent lymphocytes)	38.8
Nonadherent lymphocytes + macrophages[c]	190.0
Macrophages[d]	18.8

[a]Agar cultures were supplemented with 15 percent of the CSF from the culture supernatants of the different cell populations.

[b]Lymphoid cell populations were incubated (2×10^6 cells/ml) for 4 days with 50 μg LPS/ml.

[c]Nonadherent lymphocytes (2×10^6) were supplemented with 2×10^5 proteose-peptone stimulated peritoneal macrophages and incubated as described (b).

[d]Proteose-peptone–stimulated peritoneal macrophages (2×10^5) were incubated as described (b).

cultures not stimulated by LPS contain only small amounts of lymphocyte-activating factors. Only small amounts of CSF were produced when the supernatants of LPS-activated lymphocytes were added to cultures of peritoneal macrophages (Apte et al., unpublished observations), indicating that the events that lead to CSF secretion by LPS activation occur via the transfer of a soluble factor generated by macrophages to lymphocytes, which, in turn, secrete CSF, and not vice versa.

In the next series of experiments, we studied the specificity of CSF production and the mode of action of such macrophage supernatants. The LPS-induced macrophage supernatants were obtained from peritoneal macrophages from C_3H/eB and C_3H/HeJ mice, the latter being known as low re-

sponders to the differential effects induced by LPS (summary, reference 3). The activity of each type of macrophage supernatant was tested by incubating it with lymphocytes from both strains of mice. The results of these experiments are shown in Table 26-3. Macrophages obtained from low responder mice (C_3H/HeJ) do not elaborate lymphocyte-activating factors. In contrast, such lymphocyte-activating factors generated by macrophages from high responder mice (C_3H/eB), can activate lymphocytes from low responder mice (C_3H/HeJ) to secrete CSF. Thus, it seems that a macrophage-derived factor activates lymphocytes to secrete CSF, not the residual LPS that is present in the macrophage supernatants, since such macrophage supernatants also activate lymphocytes from C_3H/HeJ mice that could not be directly activated by LPS (2).

The population of lymphocytes responding to such LPS-induced macrophage supernatants by secreting high levels of CSF may be B-cells. This was evidenced by showing that nonadherent lymphocytes from congenitally athymic (nude) mice respond to such macrophage factors in the same manner as lymphocytes from conventional mice (Table 26-4). The B-cell population that participates with the macrophages in the secretion of CSF, is distinct from that population that undergoes mitogenesis and blast transformation. Spleen cells from CBA/N mice, carrying an x-linked defect that alters their mitogenic and immunogenic responses to T-independent antigens including LPS (25), failed to incorporate significant amounts of [^{3}H-]thymidine into DNA when stimulated by LPS, although large amounts of CSF were generated in LPS-stimulated spleen cell cultures from such mice. These results have been reported (2) and are summarized in Table 26-5.

TABLE 26-2 Macrophage-replacing Activity of Macrophage Culture Supernatants (Mϕ-Sup) in the CSF Production by Splenic Purified Lymphocytes

ADDITION TO LYMPHOCYTES[b]	NUMBER OF COLONIES/10^5 BONE MARROW CELLS[a]	
	– LPS[c]	+ LPS[c]
– LPS	11.4	155.2
+ LPS	63.0	142.2

[a]Agar cultures were supplemented with 15 percent lymphocyte CSF from the culture supernatant from the different cell populations.

[b]Macrophage supernatants (diluted 1:1) were added to purified lymphocytes (2×10^6/ml) and incubated for 4 days. Lymphocyte cultures were either not stimulated or stimulated with 50 μg LPS/ml.

[c]Macrophage supernatants were obtained from proteose-peptone-stimulated peritoneal macrophages (2×10^6/ml) incubated for 24 or 48 hr with or without 50 μg LPS/ml.

TABLE 26-3 Generation and Specificity of Action of Macrophage Supernatants (Mφ-Sup) from LPS High (C₃H/eB)- and Low (C₃H/HeJ)-Responding Mice

SOURCE OF LYMPHOCYTES[b]	NUMBER OF COLONIES/10⁵ BONE MARROW CELLS[a]	
	C₃H/eB[c]	C₃H/HeJ[c]
C₃H/eB	155.0	54.4
C₃H/HeJ	164.2	46.6

[a]Agar cultures were supplemented with 15 percent lymphocyte CSF from the culture supernatants from the two strains.

[b]Macrophage supernatants (diluted 1:1) were added to 2×10^6 purified lymphocytes and incubated for 4 days.

[c]Macrophage supernatants were obtained from proteose-peptone–stimulated peritoneal macrophages (2×10^6/ml) incubated for 24 or 48 hr in RPMI-1640 + 5 percent horse serum, and supplemented with 50 μg LPS/ml.

Interactions between Macrophages and Lymphocytes in CSF Production Induced by SEB

In a previous study, it was shown that spleen cell suspensions, although containing less mature T-cells than thymic suspensions, secrete larger amounts of CSF when stimulated by SEB (15). These results were interpreted as indicating that macrophages and T-cells must interact to produce this type of CSF, since the spleen is richer in macrophages than the thymus. In order to test this hypothesis, similar experiments to those described in the previous section were designed.

When unseparated thymus cells were incubated for 4 days with an optimal dose of SEB (10 μg/ml), a significant amount of CSF was secreted. However, when the thymocytes were filtered through Pyrex wool adherence columns, the effluent nonadherent cells largely lost their potential to secrete CSF. Supplementation of nonadherent thymocytes

with 5 percent peritoneal macrophages from unstimulated mice restored the ability of these lymphoid cultures to secrete CSF. Activated macrophages harvested from mice stimulated with thioglycolate only slightly increased CSF secretion in cultures of nonadherent thymocytes.

Supernatants obtained from cultures of unstimulated peritoneal macrophages incubated for 24 hr in culture medium without the mitogen and added to nonadherent thymocytes together with SEB restored the ability of T-cells to secrete CSF. Such supernatants, when prepared from thioglycolate-activated macrophages, had only a slight effect on T-cell CSF secretion. It should be noted that macrophage supernatants were always dialyzed before use in order to remove dialyzable inhibitors that depress lymphocyte proliferation (28). The results of these experiments are shown in Table 26-6.

In an additional experiment, the thymic cell pop-

TABLE 26-4 Effects of Macrophage Supernatants (Mφ-Sup) on CSF Production by Splenic Lymphocytes from Congenitally Athymic (nude) Mice

CELL POPULATION[a]	NUMBER OF COLONIES/10⁵ BONE MARROW CELLS[b]	
	Conventional Mice	Nude Mice
Unfractionated spleen cells	189.4	91.6
Splenic lymphocytes	25.4	29.0
Splenic lymphocytes + Mφ-Sup	171.0	168.2

[a]Unfractionated spleen cells or splenic purified lymphocytes were incubated for 4 days in culture medium supplemented with 50 μg LPS/ml. Splenic lymphocytes were incubated with LPS-stimulated macrophage supernatants (diluted 1:1) for 4 days.

[b]Agar cultures were supplemented with 15 percent CSF from the culture supernatant from the different cell populations.

TABLE 26-5 Comparison of the Ability of Spleen Cell Suspensions from Different Strains of Mice To Undergo Mitogenesis and To Secrete CSF in Response to LPS

SOURCE OF SPLEEN CELLS	cpm/2×10^6 SPLEEN CELLS[a]	NUMBER OF COLONIES/10^5 BONE MARROW CELLS[b]
C_3H/eB	65,320	194.4
CBA/N	10,412	250.4
C_3H/HeJ	11,584	24.8

[a]Spleen cell cultures were incubated at a concentration of 2×10^6 cells/ml, in culture medium supplemented with 50 μg LPS/ml. Cultures were assessed for uptake of [^{3}H-]thymidine by cells after 2 days (2) and for assay of CSF in the culture supernatants after 4 days.

[b]Agar cultures were supplemented with 15 percent CSF from the culture supernatant from the different spleen cell populations.

ulation that secretes CSF in response to SEB was identified as a mature immunocompetent population of hydrocortisone-resistant cells (Table 26-7).

DISCUSSION

Lymphocytes stimulated *in vitro* by specific antigens or nonspecific mitogens produce soluble factors (lymphokines) that have diverse biological activities (review, reference 6). Initial reports pointed to thymus-dependent lymphocytes (T-lymphocytes) as the cell type responsible for the production of these mediators, and their secretion was considered to be related to delayed hypersensitivity (6). However, subsequent reports have shown that both B- and T-cells can produce lymphokines (17). It was observed that some lymphokines may affect different macrophage functions, such as adherence, phagocytosis, spreading, motility, inhibition of bacterial growth, as well as enhancement of nonspecific destruction of tumor cells (9). In addition, it was shown that lymphokines may affect the proliferative state of cells of the macrophage granulocyte series. Lymphokines may trigger proliferation of

quiescent pluripotent stem cells capable of forming colonies in the spleen of irradiated mice (8). Hadden et al. (13) have shown that lymphocyte products are secreted in response to an antigen-induced proliferation of mature peritoneal macrophages. Data are also accumulating that supernatants from antigen- or mitogen-stimulated lymphocyte cultures are rich sources of CSF (2,4,15,18,19,23,24).

This study indicates that the secretion of CSF by B- or T-cells requires some type of interaction with macrophages; however, these processes are regulated by different mechanisms. It is now widely accepted that T-cell proliferation, as well as the production of lymphokines by T-cells, is macrophage dependent (17). Such macrophage dependency has been shown in human peripheral lymphocytes for the production of interferon induced by PHA and in the rat for induction of lymphotoxin induced by antigens (10,29). But most studies indicate that B-cell mitogens activate such lymphocytes directly without auxiliary signals provided by macrophages (summary, reference 17). Moreover, it has been reported that the addition of macrophages to lymphoid cultures stimulated by B-cell mitogens suppress the proliferative response

TABLE 26-6 Comparison of the Ability of Lymphoid-cell Populations to Generate CSF in Response to SEB

CELL POPULATION	NUMBER OF COLONIES 10^5 BONE MARROW CELLS[a]
Unseparated spleen cells[b]	403.8
Unseparated thymus cells	140.2
Nonadherent thymocytes + macrophages[c]	130.4
Nonadherent thymocytes + supernatant of macrophages[d]	94.4
Nonadherent thymocytes + activated macrophages[e]	40.2
Macrophages (nonactivated or activated)[f]	4.8

[a]Agar cultures were supplemented with 30 percent of the culture supernatants from the different cell populations.
[b]Lymphoid cell populations (2×10^6/ml) were incubated for 4 days with 10 μg SEB/ml.
[c]Nonadherent thymocytes (2×10^6) were supplemented with 10^5 unstimulated macrophages, as described in (b).
[d]Thymus nonadherent enriched cells were suspended in macrophage supernatants obtained from 10^5 unstimulated macrophages incubated for 24 hr and subsequently cultured as described in (b).
[e]Nonadherent thymocytes (2×10^6) were supplemented with 10^5 thioglycolate-stimulated macrophages and cultured as described in (b).
[f]Unstimulated or thioglycolate-stimulated macrophages (10^5) were incubated as described in (b).

TABLE 26-7 Identification of the Thymus-cell Population that Secretes CSF in Response to SEB

CELL POPULATION	NUMBER OF COLONIES/10⁵ BONE MARROW CELLS[a]
Unseparated thymus cells[b]	150.4
Hydrocortisone-resistant thymus cells[c]	200.0
Hydrocortisone-resistant nonadherent thymocytes[d]	42.6

[a]Agar cultures were supplemented with 30 percent of the culture supernatants from different cell populations.
[b]Thymus cells (2×10^6/ml) were incubated for 4 days with 10 μg SEB/ml.
[c]Hydrocortisone-resistant thymus cells were obtained from mice injected intraperitoneally 48 hr before with 2.5 mg hydrocortisone acetate/mouse (Frederiksberj Chemical Laboratories, Denmark).
[d]Thymus cell suspensions were filtered through Pyrex wool adherence columns and cultured as described in (b).

(16,30). Since the concept that B-lymphocytes secrete lymphokines is novel, we could not find any report concerning its macrophage dependency.

The experiments described here show that macrophage supernatants can replace intact macrophages in CSF secretion by LPS and SEB stimulation. When macrophage supernatants prepared from unstimulated peritoneal macrophages are added with SEB to enriched nonadherent thymocytes, large amounts of CSF are secreted. It seems that the mitogen binds to the T-lymphocyte through a specific receptor to produce one activating signal. A second activating signal is given by the macrophage-soluble factor. The two signals are probably produced simultaneously. On the other hand, activation of B-lymphocytes by LPS seems to occur via two subsequent signals. Macrophages incubated for 24 hr with LPS secrete soluble factors, which, in turn, activate B-cells to secrete CSF, without addition of the mitogen to lymphocytes. The fact that lymphocytes from low-responder mice (C_3H/HeJ) that have an altered response to LPS respond to such macrophage-derived factors to secrete CSF indicates that soluble factors, rather than the residual LPS in the macrophage supernatants, activate the B-cell. However, it has not been shown that such macrophage supernatants also contain active LPS, which has been handled and processed by macrophages, so that its recognition sites have been changed and it activates B-cells from C_3H/HeJ mice directly. In a subsequent study we have shown that the macrophage-derived factors required for CSF secretion by LPS stimulation cannot be replaced by the synthetic compound 2-mercaptoethanol, further indicating that macrophage factor(s) may act specifically and not only to promote lymphocyte viability or reactivity (Apte et al., unpublished observations). Our study also indicates that the degree of macrophage activation also plays an important role in the capacity of these cells to secrete lymphostimulatory molecules. The mechanisms proposed for the secretion of CSF in response to LPS and SEB are schematically shown in Figure 26-1.

Different soluble macrophage-derived lymphostimulatory molecules have recently been described by others (review, references 17,28). Such molecules were found to enhance the proliferation of thymocytes in response to mitogens and to cause the rapid maturation of immature thymocytes into competent mature cells. In addition, such factors also increase the helper activity in hapten-protein assays and enhance the differentiation of memory B-lymphocytes to antibody-secreting cells (17,28).

In summary, this study emphasized the mutual relationship between macrophages and lymphocytes in generating granulopoietic factors by B- and T-cell mitogens of bacterial origin; however, there are different sequences of events for each of these processes.

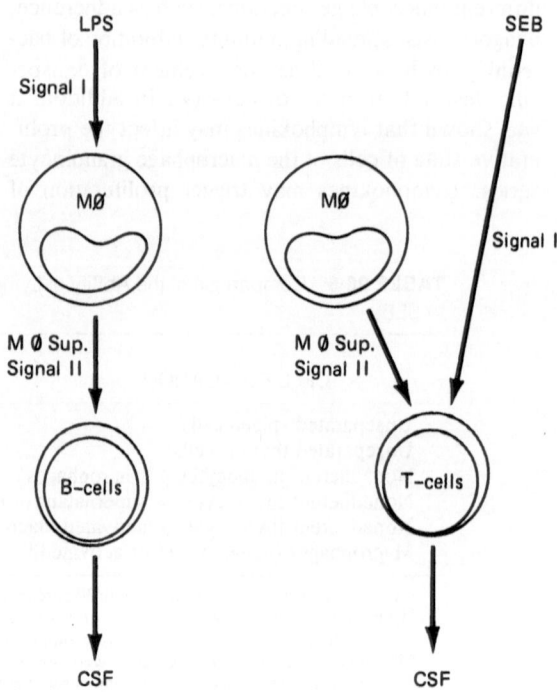

FIGURE 26-1. Mechanism proposed for the secretion of CSF stimulated by LPS and SEB.

SUMMARY

Murine lymphoid cells incubated with a B-cell mitogen, a bacterial lipopolysaccharide (LPS), or a T-cell mitogen staphylococcal enterotoxin B (SEB) elaborate factors with granulopoietic stimulatory activity. Such factors enhance the growth of macrophage/granulocyte progenitor cells *in vitro* in soft agar cultures and are designated colony-stimulating factors (CSF). Lymphocytes and macrophages must interact in order to produce these factors. Macrophage supernatants can replace intact macrophages in these interactions with lymphocytes. Different reactions between macrophages and lymphocytes mediate the secretion of CSF by T- and B-cells. For T-cell secretion of CSF, it seems that two signals occurring simultaneously in culture are needed. The T-cell mitogen, SEB, binds to the T-cell through a specific receptor to produce one activating signal, whereas the second signal is produced by a macrophage-soluble factor. As for activation of B-cells by LPS to secrete CSF, two sequential signals are needed. First, LPS interacts with the macrophage, which secretes helper factor that, in turn, activates the lymphocyte; the mitogen does not need to interact directly with the B-cell. The B-cell subpopulation that secretes CSF in response to LPS is distinct from the B-cells that undergo mitogenesis and cell proliferation; the T-cell reactive to SEB is a mature immunocompetent cell, resistant to the cytotoxic action of hydrocortisone.

ACKNOWLEDGMENTS

This study was partially supported by a grant from the Hematology Research Foundation, Chicago, Ill., and a grant from the Research Committee, Bar-Ilan University, Ramat-Gan, Israel.

REFERENCES

1. Anderson, J., Möller, G., and Sjöberg, O. Selective induction of DNA synthesis in T and B lymphocytes. *Cell. Immunol., 4*:381, 1972.
2. Apte, R.N., Hertogs, C.F., and Pluznik, D.H. Regulation of lipopolysaccharide-induced granulopoiesis and macrophage formation by spleen cells. I. Relationship between colony-stimulating factor release and lymphocyte activation *in vitro*. *J. Immunol., 118*:1435, 1977.
3. Apte, R.N., and Pluznik, D.H. Genetic control of lipopolysaccharide-induced generation of serum colony stimulating factor and proliferation of splenic granulocyte/macrophage precursor cells. *J. Cell. Physiol., 89*:313, 1976.
4. Apte, R.N., and Pluznik, D.H. Control mechanisms of endotoxin and particulate material stimulation of hemopoietic colony forming cell differentiation. *Exp. Hematol., 4*:10, 1976.
5. Bechler, E.L., and Hensen, P.M. *In vitro* studies of immunologically induced secretion of mediators from cells and related phenomena. *Adv. Immunol., 17*:93, 1973.
6. Bloom, B.R. *In vitro* approaches to the mechanism of cell mediated immune reactions. *Adv. Immunol., 13*:101, 1971.
7. Bradley, T.R., and Metcalf, D. The growth of mouse marrow cells *in vitro*. *Austral. J. Exp. Biol. Med. Sci., 44*:287, 1966.
8. Cerny, J. Stimulation of bone marrow haemopoietic stem cells by a factor from activated T cells. *Nature (Lond.), 249*:63, 1974.
9. David, J.R. Macrophage activation by lymphocyte mediators. *Fed. Proc., 34*:1730, 1975.
10. Epstein, L.B., Cline, M.J., and Merigan, T.C. The interaction of human macrophages and lymphocytes in the PHA-stimulated production of interferon. *J. Nat. Cancer Inst., 50*:744, 1971.
11. Gery, I., Krüger, K., and Spiesel, S.Z. Stimulation of B lymphocytes by endotoxin: Reactions of thymus-deprived mice and karyotypic analysis of dividing cells in mice bearing T_6T_6 thymus grafts. *J. Immunol., 108*:1088, 1972.
12. Golde, D.W., and Cline, M.J. Regulation of granulopoiesis. *New Eng. J. Med., 291*:1388, 1974.
13. Hadden, J.W., Sadlik, J.R., and Hadden, E.M. Macrophage proliferation induced *in vitro* by a lymphocyte factor. *Nature (Lond.), 257*:483, 1975.
14. Heller, E., Hertman, I., and Pluznik, D.H. Staphylococcal enterotoxin B: a bacterial exotoxin affecting the granulopoietic system in mice. *Exp. Hematol. 3* (Suppl):12, 1975.
15. Heller, E., and Pluznik, D.H. Interaction between macrophages and thymocytes in generating a mediator which stimulates proliferation of macrophage/granulocyte progenitor cells. *Israel J. Med. Sci.,* (in press), 1977.
16. Nelson, D.N. Production by stimulated macrophages of factors depressing lymphocyte transformation. *Nature (Lond.), 246*:306, 1973.
17. Oppenheim, J.J., and Rosenstreich, D.L. Signals regulating *in vitro* activation of lymphocytes. *Prof. Aller., 20*:65, 1976.
18. Parker, J.W., and Metcalf, D. Production of colony-stimulating factor in mixed leukocyte cultures. *Immunology, 26*:1039, 1974.
19. Parker, J.W., and Metcalf, D. Production of colony-stimulating factor in mitogen-stimulated lymphocyte cultures. *J. Immunol., 112*:502, 1974.
20. Peavy, D.L., Adler, W.H., and Smith, R.T. The mitogenic effect of endotoxin and staphylococcal enterotoxin B on mouse spleen cells and human peripheral lymphocytes. *J. Immunol., 105*:1453, 1970.
21. Pluznik, D.H., and Sachs, L. The cloning of normal "mast" cells in tissue culture. *J. Cell. Comp. Physiol., 66*:319, 1965.
22. Robinson, W.A., and Mangalik, A. The kinetics and regulation of granulopoiesis. *Sem. Hematol., 12*:7, 1975.

23. Ruscetti, F.W., and Chervenick, P.A. Regulation of the release of colony-stimulating activity from mitogen-stimulated lymphocytes. *J. Immunol., 114*:1513, 1975.

24. Ruscetti, F.W., Cypess, R.H., and Chervenick, P.A. Specific release of neutrophilic- and eosinophilic-stimulating factors from sensitized lymphocytes. *Blood, 47*:757, 1976.

25. Scher, I., Ahmed, A., Strong, D.M., Steinberg, A.D., and Paul, W.E. X-linked and B-lymphocyte immune defect in CBA/HN mice. I. Studies of the function and composition of spleen cells. *J. Exp. Med., 141*:788, 1975.

26. Stanley, E.R., Hansen, G., Woodcock, J., and Metcalf, D. Colony-stimulating factor and the regulation of granulopoiesis and macrophage production. *Fed. Proc., 34*:2272, 1975.

27. Till, J.E., Price, G.B., Mak, T.W., and McCulloch, E.A. Regulation of blood cell differentiation. *Fed. Proc., 34*:2279, 1975.

28. Unanue, E.R., Beller, D.I., Calderon, J., Kiely, J.M., and Stadecker, M.J. Regulation of immunity and inflammation by mediators from macrophages. *Amer. J. Pathol., 85*:465, 1977.

29. Yoshinaga, M., and Waksman, B.H. Regulation of lymphocyte responses *in vitro*. IV. Role of macrophages in rat lymphocyte responses and their inhibition by cytochalasin B. *Ann. Inst. Pasteur, 124C*:97, 1973.

30. Yoshinaga, M., Yoshinaga, A., and Waksman, B.H. Regulation of lymphocyte responses *in vitro*. I. Regulatory effect of macrophages and thymus-dependent (T) cells on the response of thymus-independent (B) lymphocytes to endotoxin. *J. Exp. Med., 136*:956, 1972.

27

Serum Blocking Factor-Mediated Decrease of T-Cell Immunity Induced by Soluble Histocompatibility Antigens (SHA); Similar Effects of Antibody-free SHA-raised and Hemagglutinating-enhancing Sera

O. Halle-Pannenko,
L. Berumen, N. Abuaf,
and G. Mathe

INTRODUCTION

Alloantigen-extract–mediated graft enhancement has been widely studied. Many investigators have observed a decreased host-versus-graft (HvG) (10,16,17,19,22,30) or graft-versus-host (GvH) (11,13,23,29) reaction after the injection of crude or soluble tissue extracts, although contradictory results indicating an accelerated graft rejection have also been reported (7,8). We have previously noted (11) that several factors may influence the response to alloantigen-extract administration. These include the organ used for the extraction; the "density" of the H-2 D and K antigens; the physical form of the extract; the dosage, frequency, and route of administration; the interval between time of administration and subsequent challenge; and the amount of an as yet undefined substance in the extract. The variability of these factors, plus the variety of species used and the donor-recipient genetic disparities involved, along with the inevitable variation in microbiological and environmental conditions influencing the immunoreactivity of experimental animals, provides, at least in part, an explanation for many of the discrepancies among the results reported by different laboratories.

A number of immunological mechanisms have been invoked to explain specific immunosuppression of transplantation reactions (tolerance, enhancing antibodies, blocking effect of antigen-antibody complexes, suppressor cells). Although the discrepancy between the phenomena that were observed may be only apparent and due to the fact that different observations reflect various stages of one complex process, the interrelationships of all the factors involved in specific immunosuppression remain undefined.

In earlier communications we have reported that in mice repeatedly injected with soluble alloantigen extract, there is a definite prolongation of skin allograft survival (11,12). In the study reported here, we have analyzed the cellular and humoral immune state of treated animals to try to determine the nature of the immune mechanism involved in the observed immunodepression. Simultaneously, since the success of bone marrow transplantation is one of our main objectives, we have tried to determine whether the treatment that reduced the HvG reaction could also reduce the GvH reaction.

MATERIAL AND METHODS

Animals

In all experiments, 2.5- to 3-month-old $C_{57}Bl/6$ (B6), DBA/2 (D2), and $(C_{57}Bl/6 \times DBA/2)F_1$ (B_6D2F_1) male mice were used. Experimental mice were purchased from La Source CNRS Laboratories, Orléans, France.

Extraction of Soluble Histocompatibility Antigens (SHA)

These antigens were obtained from D_2 and B_6 mouse liver. The extraction of SHA by exposure of cells to a series of hypotonic shocks and their solubilization by autodigestion in Tris-HCl buffered medium was performed according to the technique described by Nathenson (21). The biological activity of the SHA was determined by inhibition of hemagglutinating antibodies.

Treatment with SHA

B_6 mice were treated with D_2 (experimental) or B_6 (control) SHA, 10 μg/i.p. injection, administered five times a week over a 2-week period unless otherwise indicated. On the day of the last injection, the serum from the mice was collected and stored at $-20°C$. The animals were then sacrificed and the reactivity of their lymph node and spleen cells was tested in GvH and mixed lymphocyte reaction (MLR) tests.

Preparation of Hemagglutinating-enhancing Serum

Donor anti-host hemagglutinating enhancing serum was produced according to the method of Voisin (28), by injecting B_6 mice with D_2 lyophilized spleen cells. The serum contained hemagglutinating, non-cytotoxic antibodies. The enhancing properties of the serum were confirmed by its capacity to reduce the mortality due to secondary disease under experimental conditions previously described (13).

GvH Reaction Test

The GvH reaction was induced in 3-month-old $B_6D_2F_1$ hybrid mice by the i.v. injection of 10^7 B_6 lymph node cells or by the i.v. or i.p. injection of 10^7 or 5×10^7 spleen cells, respectively. Mortality after the injection of mice with 10^7 spleen cells was somewhat low. Therefore, in subsequent experiments the higher dose of cells was employed. These mice were injected i.p. to avoid death from emboli. About 5 hr before grafting, the F_1 hosts were irradiated with 300 rad (85 rad/min) with a Gravatom R^x 33/55 M Cesium source (Gravatom Industries Ltd., Gosport, Hampshire, England). The intensity of the GvH reaction was measured by mortality and splenomegaly; in the latter case, relative spleen weights and spleen indexes were calculated 9 days after grafting according to the method described by Simonsen (24). In one experiment, we studied the effect of serum from SHA-treated mice on the GvH reaction by incubating 10^7 B_6 donor cells in 0.5 ml of 199 culture medium with 0.5 ml of undiluted serum at 37°C in a humidified 5 percent CO_2 at-

mosphere for 30 min. After incubation, the cells were washed twice with serum-free medium before injection into $B_6D_2F_1$ mice.

Skin Graft Test

Skin grafts were performed as previously described (9). The donors of a tail skin graft were adult D_2 mice and the hosts were 10-week-old B_6 mice.

Mixed Lymphocyte Reaction (MLR) Test

This test was performed as previously described (12). Briefly, cultures were set up using Microtest II tissue culture plates and lids (Falcon Plastics, Oxnard, Ca, USA). Each well contained 5×10^5 responding cells and 5×10^5 irradiated stimulating cells in 0.2 ml of RPMI 1640 medium. When MLR experiments were carried out in the presence of sera from treated mice, the serum was added at the dilutions tested, and the total volume in each well was 0.25 ml. In experiments in which we studied the effect of serum from SHA-treated animals on responder cells, 10^7 cells in 1 ml of culture medium and 1 ml of serum were incubated prior to MLR in a humidified 5 percent CO_2 atmosphere at 37°C for 30 min; then the cells were washed twice with serum-free medium and subsequently tested in MLR. Cultures were incubated 72 hr at 37°C in a humidified 5 percent CO_2 atmosphere. [^{3}H]Thymidine (1 μC in 0.05 ml NaCl) was then added to each well, and the cultures were incubated for an additional 16 hr. [^{3}H]Thymidine incorporation was measured in a Packard Tri-Carb Model 3320 liquid scintillation counter. [^{3}H]Thymidine uptake is expressed as mean counts per minute (cpm) of six cultures.

RESULTS

Cellular Immunity of Mice Treated with SHA

GvH mortality test The results are shown in Table 27-1. As compared to the mortality induced by grafting the control (isogeneic SHA-treated) B_6 donor cells, the mortality induced by grafting cells from experimental (D_2 SHA-treated) donors was delayed and reduced in all cases when lymph node cells were grafted, and in two out of three cases when spleen cells were grafted.

GvH splenomegaly test In the same three experiments, when the intensity of the GvH reaction developed by grafting experimental donor cells was measured by GvH splenomegaly (Table 27-2), the spleen indexes decreased in only one instance (experiment 3) when spleen cells were employed.

TABLE 27-1 Cellular Immunity of Mice Treated with Soluble Histocompatibility Antigens, as Measured by the GvH-Mortality Test

| EXPERIMENT NUMBER | NUMBER OF CELLS GRAFTED | MEDIAN SURVIVAL IN DAYS (Range of Mortality) | | SURVIVAL (PERCENT) | | | | SIGNIFICANCE[c] |
| | | | | EXPERIMENTAL[a] | | CONTROL[b] | | |
		Experimental[a]	Control[b]	Day 40	Day 100	Day 40	Day 100	
	Lymph node							
1	10^7, i.v.	40 (21–80)	23 (19–33)	50	20	12	12	$p < 0.01$
2	10^7, i.v.	27 (17–31)	17 (14–24)	25	25	0	0	$p < 0.01$
3	10^7, i.v.	42 (19–100)	20 (16–50)	50	33	25	8	$p < 0.02$
	Spleen							
1	10^7, i.v.	No mortality	44 (26–75)	100	100	60	50	$p < 0.02$
2	5×10^7, i.p.	24 (17–32)	19 (14–41)	16	16	24	16	N.S.
3	5×10^7, i.p.	>100 (27–45)	27 (12–32)	67	56	0	0	$p < 0.02$

[a]The 3-month-old $B_6D_2F_1$ hosts were irradiated with 300 rad and grafted on day 0 with B_6 cells from donor treated with allogeneic (D_2) soluble histocompatibility antigens.

[b]As in (a) except that the donors were treated with syngeneic (B_6) soluble histocompatibility antigens.

[c]The significance of the difference between experimental and control groups was determined by Wilcoxon's non-parametric test; N.S., not significance ($p > 0.05$).

MLR test The response to D_2 stimulating cells was found to be significantly lower in the cultures with B_6 responding cells from D_2 SHA-treated experimental mice than in the cultures with B_6 responding cells from syngeneic SHA-treated control mice (Table 27-3).

When the reactivity of spleen cells from the same SHA-treated animals was compared in the three tests above, a better correlation was found between MLR and GvH mortality than between MLR and GvH splenomegaly tests; in fact, the reactivity of the same spleen cell population was

TABLE 27-2 Cellular Immunity of Mice Treated with Soluble Histocompatibility Antigens, as Measured by the GvH Splenomegaly Test

| EXPERIMENT NUMBER | NUMBER OF GRAFTED CELLS | MEAN RELATIVE SPLEEN WEIGHT ± S.E.[a] | | SIGNIFICANCE[d] |
		Experimental[b]	Control[c]	
	Lymph node			
1	10^7, i.v.	5.85 ± 0.32	5.02 ± 0.58	N.S.
2	10^7, i.v.	4.47 ± 0.25	5.03 ± 0.26	N.S.
3	10^7, i.v.	6.63 ± 0.27	5.96 ± 0.66	N.S.
	Spleen			
1	10^7, i.v.	4.68 ± 0.47	5.48 ± 0.61	N.S.
2	5×10^7, i.p.	6.28 ± 0.31	6.70 ± 0.59	N.S.
3	5×10^7, i.p.	3.51 ± 0.26	4.68 ± 0.31	$p < 0.001$

[a]S.E., standard error of the mean.

[b]The 3-month-old $B_6D_2F_1$ hosts were irradiated with 300 rad and grafted with B_6 cells from donors treated with allogeneic (D_2) soluble histocompatibility antigens.

[c]As in (b) except that the donors were treated with syngeneic (B_6) soluble histocompatibility antigens.

[d]Significance of the difference between experimental and control groups by the Student-Fisher t test; N.S., not significant ($p > 0.05$).

TABLE 27-3 Cellular Immunity of Mice Treated with Soluble Histocompatibility Antigens, Comparison of MLR, GvH Mortality, and GvH Splenomegaly Tests

CONDITONS OF TEST	TEST	MEASUREMENT	RESULTS Experimental[a]	Control[b]	SIGNIFICANCE[c]
B_6 responding cells; D_2 irradiated stimulating cells	MLR	[^{3}H]thymidine uptake; mean cpm ± S.E.[d]	29,459 ± 2007	51,689 ± 3633	$p < 0.001$[e]
3-Month-old $B_6D_2F_1$ mice irradiated with 300 rad and grafted (i.v.) with 10^7 B_6 spleen cells	GvH mortality	Median survival in days (mortality range)	No mortality	44 (26–75)	$p < 0.02$[f]
	GvH splenomegaly	Mean relative spleen weight ± C.L.[g]	4.68 ± 0.95	5.48 ± 1.22	N.S.[e]

[a]Responder (MLR) or donor (GvH) spleen cells from B_6 mice treated *in vivo* with D_2 soluble histocompatibility antigens.
[b]Responder (MLR) or donor (GvH) spleen cells from B_6 mice treated *in vivo* with B_6 soluble histocompatibility antigens.
[c]Significance of the difference between the experimental and control groups.
[d]Counts per minute ± standard error of the mean.
[e]Student-Fisher *t* Test.
[f]Wilcoxon's Non-parametric Test; N.S., not significant ($p > 0.05$).
[g]Confidence limit calculated at $p = 0.05$.

significantly decreased in the experimental group, as measured by MLR and mortality tests but not as measured by the splenomegaly test (Table 27-3).

HUMORAL IMMUNITY OF MICE TREATED WITH SHA

Effect of SHA-raised Serum on Skin Graft Survival

The serum from B_6 mice treated with D_2-SHA contains neither hemagglutinating antibodies nor the hemagglutinating synergistic antibodies de-scribed in tolerant mice (27) nor complement-dependent cytotoxic antibodies. However, as shown in Table 27-4, this serum delays D_2 skin graft rejection upon injection into normal B_6 hosts. Moreover, the blocking activity of SHA-raised serum depends on the amount of SHA per injection and also on the duration of treatment. Thus, a higher dose requires a shorter treatment regimen and a lower dose requires a longer. The best survival was obtained after injection of serum from animals injected with 40 μg of SHA/injection for 1 week, with 10 μg for 2 weeks, and with 2.5 μg for 3 weeks.

TABLE 27-4 Antigen Dose and Treatment Duration-dependent Effect of Serum from Mice Treated with Soluble Histocompatibility Antigens on Skin Graft Survival

PREPARATION OF SERUM[a] Duration of treatment (weeks)	Antigen dose (μg)/ injection	EFFECT OF SERUM ON SKIN GRAFT SURVIVAL[b] MEAN SURVIVAL IN DAYS Experimental Serum[c] (B_6-anti-D_2)	Control Serum[c] (normal B_6)
1	2.5	15.2 ($p < 0.05$)[d]	
	10	15.5 ($p < 0.05$)	
	40	17.9 ($p < 0.01$)	
2	2.5	15.5 ($p < 0.05$)	13.1
	10	18.9 ($p < 0.01$)	
	40	16.1 ($p < 0.03$)	
3	2.5	16.3 ($p < 0.03$)	
	10	14.6 (N.S.)	
	40	14.6 (N.S.)	

[a]The B_6 mice were treated with D_2 soluble histocompatibility antigens, (given in 5 i.p. injections/week).
[b]D_2 skin graft on B_6 hosts.
[c]Normal hosts received 0.5 ml/i.v. injection of either experimental or control serum on day 0, +2, +5, and +8 after skin grafting.
[d]Statistical significance of difference between experimental and control serum groups by the Student-Fisher *t* Test; N.S., not significant ($p > 0.05$).

224

Effect of SHA-raised Serum on MLR

For this MLR Test and for the experiments that follow, SHA-raised sera were collected from the B_6 mice injected with 10 μg of D_2-SHA (test serum) or B_6-SHA (control serum) for 2 weeks. Because of the nonspecific inhibitory activity of mouse serum on the MLR, the effect of the test serum was always compared to the effect of the control serum. The results obtained in the MLR of normal B_6 responder and D_2 stimulating cells indicate that the effect of the tested sera is concentration dependent. Figure 27-1(A) shows that for serum number 1 stored for 1 month inhibition was observed with high and low serum concentrations. Stimulation was obtained with an intermediate concentration of the stored serum. However, the effect of the serum may be modified by storage: Figure 27-1(B) shows

that the same serum tested 1 month later was highly inhibitory at a concentration (1:64) that previously had had statistically a non-significant stimulatory effect. Moreover, as shown in Figure 27-2, another batch of serum (No. 2) stored for a longer period (3 months) was not inhibitory at high concentrations but only stimulation followed by inhibition at lower concentrations was observed with this serum.

To determine whether the decrease in MLR reactivity observed in the presence of SHA-raised serum is related to the effect of the serum on the responding cells, serum No. 1 (after 13 weeks' storage) was incubated with the responding cells prior to MLR. Results in Figure 27-3 show that, under these experimental conditions, a highly significant inhibition occurred at the three serum concentrations tested.

Effect of SHA-raised Serum on GvH Reaction

The GvH reaction was induced in $B_6D_2F_1$ hosts by grafting parental strain B_6 lymph node cells; we compared the mortality induced by untreated donor cells with that induced by donor cells preincubated with either control serum (from B_6 mice treated with B_6-SHA) or test serum (from B_6 mice treated with D_2-SHA). Results in Figure 27-4 show that *in*

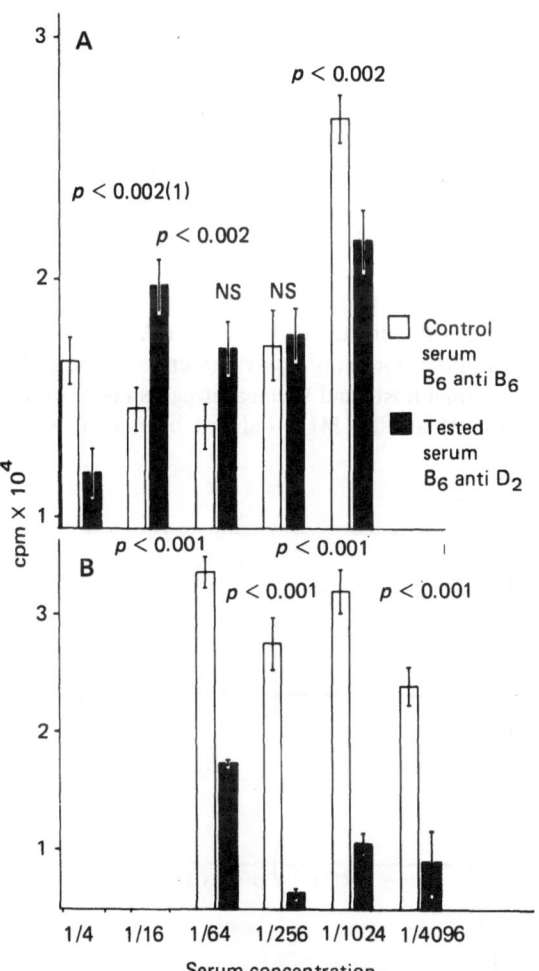

FIGURE 27-1. Concentration and storage-dependent effects on MLR of (antibody-free) serum from mice treated with soluble histocompatibility antigens (serum No. 1). (A) Sera stored 1 month; (B) sera stored 2 months, both at −20°C. Statistical significance of difference between test and control sera by the Student-Fisher t Test; NS, not significant ($p > 0.05$).

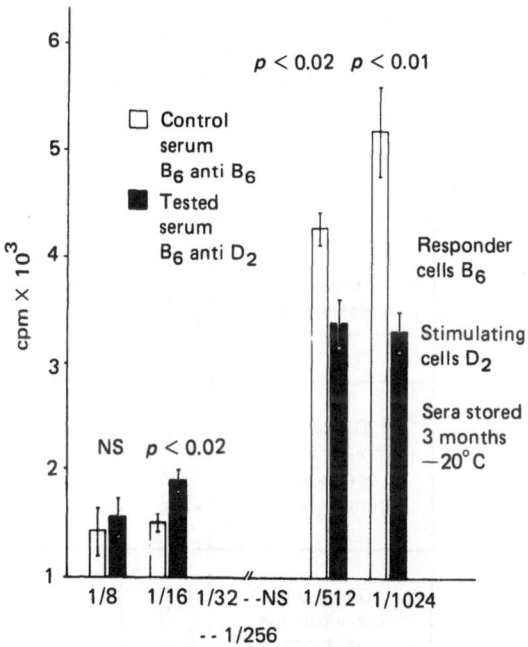

FIGURE 27-2. Concentration-dependent effect on MLR of (antibody-free) serum from mice treated with soluble histocompatibility antigens. (Serum No. 2). Statistical significance of difference between test and control sera by the Student-Fisher t Test; NS, not significant ($p > 0.05$).

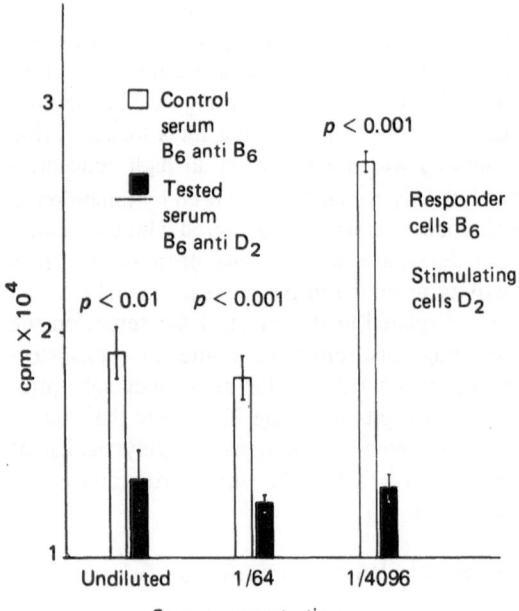

FIGURE 27-3. Decreased reactivity in MLR of normal responder cells incubated with antibody-free serum from mice treated with soluble histocompatibility antigens. Statistical significance of difference between test and control sera by the Student-Fisher *t* Test.

vitro preincubation of donor cells with D_2-SHA–raised serum delayed the GvH mortality as compared to the mortality induced either by untreated cells or by cells incubated with control serum.

Effect of Hemagglutinating-enhancing Serum on MLR

In the last series of experiments, the effect on MLR of the classical enhancing serum, raised by injection of lyophilized spleen cells, was compared to the effect of the SHA-raised serum. Figure 27-5 shows that the results obtained in MLR performed in the presence of the hemagglutinating-enhancing serum (stored for 4 months) are similar to those (Figure 27-2) obtained with antibody-free, SHA-raised serum (stored for 3 months). A concentration-dependent effect of hemagglutinating serum (stimulation with high concentrations, inhibition with low concentrations) on MLR was observed.

DISCUSSION

In this study, our aim was to determine whether the repeated administration of SHA, which reduces the HvG reaction (11,12), can also reduce the GvH reaction and whether its action is mediated by cellular and/or humoral immunity. Thus, the cellular immunity of the SHA-treated animals was studied by (a) determining the capacity of the lymph node and spleen cells to develop a GvH reaction and (b) measuring [³H]thymidine incorporation by spleen cells in MLR. The presence of humoral factors capable of modifying the immune reaction was tested by (a) examining the capacity of SHA-raised serum to prolong skin graft survival after passive transfer to normal hosts and by measuring its effect on the (b) GvH and (c) MLR induced by normal lymph

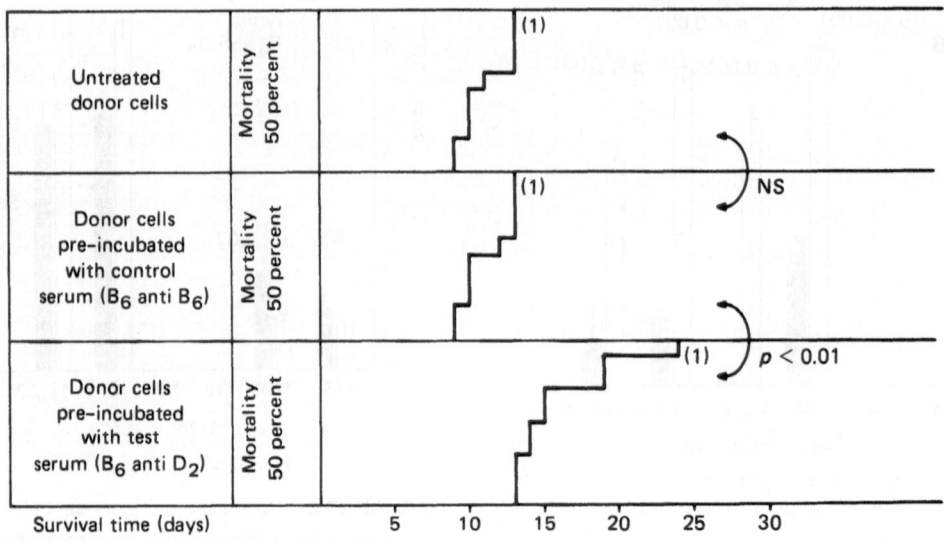

FIGURE 27-4. Decreased reactivity in the GvH mortality test of donor cells incubated with antibody-free serum from mice treated with soluble histocompatibility antigens. Three-month-old $B_6D_2F_1$ mice irradiated with 300 rad and grafted i.v. with 10^7 B_6 lymph node cells. Statistical significance of difference by the Wilcoxon's Non-parametric Test; NS, not significant ($p > 0.05$).

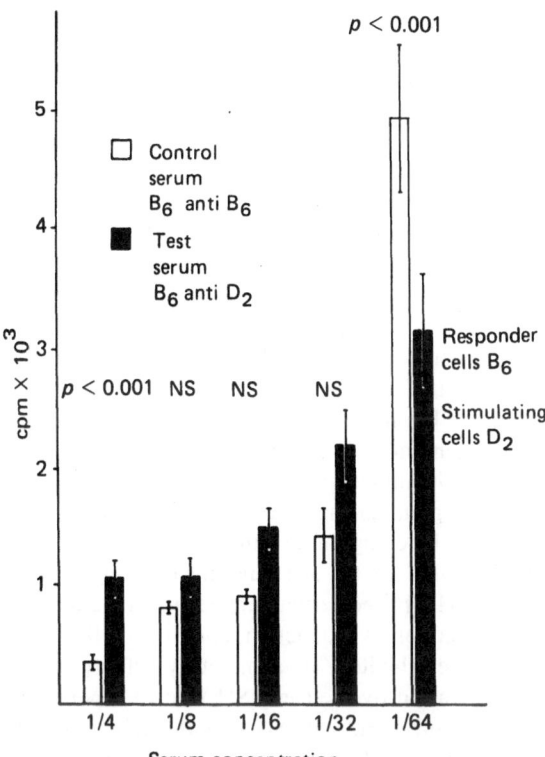

FIGURE 27-5. The concentration-dependent effect on MLR of hemagglutinating serum from mice treated with lyophilized spleen cells. Statistical significance of difference between tested and control sera by the Student-Fisher t Test; NS, not significant ($p > 0.05$).

node and spleen cells. The same genetic system was used in all experiments, and the reactivity of B_6 strain against D_2-strain histocompatibility antigens was studied.

The results show that GvH mortality subsequent to the graft of lymph node or spleen cells from experimental SHA-treated donors, was delayed and reduced, in five cases out of six, as compared to the mortality subsequent to the graft of cells from control SHA-treated donors. No correlation was found between GvH mortality and GvH splenomegaly tests, since the splenomegaly developed by experimental donor cells was reduced in only one of the same six cases. This indicates that SHA-induced immunosuppression is more often detectable by mortality than by splenomegaly and confirms our previous observations (1,12) that these two tests, which both measure the intensity of the GvH reaction, are not necessarily equivalent. Furthermore, a better correlation was found between MLR and GvH mortality than between MLR and GvH splenomegaly, since the reactivity of spleen cells from SHA-treated animals measured in MLR was also found to diminish. These results taken together indicate that SHA-induced immunodepression is

expressed at the level of the T-cells and that it is detectable during the induction (MLR) phase, with the consequences observed during the effector (GvH mortality) phase. The lack of correlation between MLR and the splenomegaly test, which, like MLR, is frequently considered to measure the induction phase may be due to the participation of host cells in this test (15).

The serum from SHA-treated animals contains neither hemagglutinating antibodies, the hemagglutinating "synergistic" antibodies described by Voisin (27) nor complement-dependent cytotoxic antibodies. This serum, however, delays skin graft rejection upon injection into normal hosts. These results indicate the presence in the serum of a blocking factor, which is not likely to be free antibody but more probably free antigen or an antigen-antibody complex. Serum blocking factor characterized as antigen-antibody complex has been reported in tolerant rats (2,31). In mice, a serum blocking factor has been induced by the injection of soluble alloantigens (32). Prolongation of heart allograft survival in rats (18) and liver allografts in baboons (20) have been obtained by the administration of antigen-antibody complexes produced *in vitro*. Moreover, many reports indicate that the immunosuppression induced by antigen-antibody complexes is more efficient than the immunosuppression induced by antigens or antibodies alone (18,19,20,26) In this regard, we had previously found that the immunosuppression induced by SHA is less effective than the immunosuppression induced by passive transfer of SHA-raised serum (12).

The fact that high doses of SHA require short treatment schedules and low doses require longer treatment schedules suggests that the blocking effect of the serum is dependent on the formation of an optimal level of blocking factor. This hypothesis is supported by our MLR experiments, where SHA-raised serum (serum No. 1) was inhibitory when tested at high and low concentrations and stimulatory when tested at an intermediate concentration. The concentration-dependent blocking activity of an alloantiserum interacting with target cells (3), and of the serum from tumor-bearing animals interacting with effector cells (6), has been described in works describing a cell-mediated cytolysis of tumor cells. Antagonistic concentration-dependent activity of regressor sera from tumor-bearing animals has also been observed (14,25). Potentiation or blocking of cell-mediated cytotoxicity was obtained, depending upon the serum dilution. As far as our experiments are concerned, it should be emphasized that the activity of SHA-raised serum may be modified by storage. Serum No. 1 tested 1 month later showed a high inhibiting activity [Fig. 27-1(B)] when used at the concentration that previously had an insignificant stimulating effect [Fig. 27-1(A)]. On the other hand, serum No. 2 stored for 3 months (Fig. 27-2) was unable to induce inhibition at a high concentration. Stimulation, followed by inhibition at lower concentrations, was observed in this case. The absence of a high-concentration inhibiting activity with serum No. 2 could be due to the fact that the initial concentration tested in this experiment was lower than that in the experiment with serum No. 1. This interpretation is unlikely, since the difference between the two initial concentrations was slight (1:4 versus 1:8). A more plausible explanation is provided by the fact that serum No. 2 was stored for 3 months and serum No. 1 for only 1 month prior to testing [Fig. 27-1(A)]. Indeed, the data obtained with serum No. 1 suggest that storage decreases serum activity such that, with time, the entire dose-response curve [Fig. 27-1(A)] is shifted to the left.

In the experiments in which responder cells were incubated with SHA-raised serum prior to MLR, a highly significant decrease in reactivity was observed. Moreover, GvH mortality was delayed after *in vitro* incubation of donor cells with SHA-raised serum. These results indicate that the SHA-raised serum may decrease the reactivity of normal T-cells, and that this effect is detectable during the induction phase (MLR) with the consequences observed in the effector phase (GvH mortality). The data further lead to the conclusion that the decreased T-cell–mediated immunity in SHA-treated animals is related to the presence of a serum-blocking factor.

Results obtained in the MLR with classical hemagglutinating enhancing serum raised by the administration of lyophilized spleen cells were comparable to those obtained with antibody-free SHA-raised serum. In the MLR experiments performed in the presence of the hemagglutinating-enhancing serum, the same concentration-dependent effect (stimulation with high and inhibition with low serum concentrations) was observed. Along with our previously reported experiments (13), in which enhancing serum delayed GvH mortality when incubated with donor cells prior to grafting, these results indicate that hemagglutinating-enhancing serum and antibody-free SHA-raised serum have similar actions. Both of them appear able to diminish the normal immunological reactivity by acting at the T-cell level. This suggests that the blocking activity of both sera is due to the same factor, and that, consequently, the immunosuppressive activity of the hemagglutinating serum may not entirely be due to the free antibodies it contains. This hypothesis is supported by recent reports (5) indicating that alloantisera may retain the antigen used for

the immunization in the form of antigen-antibody complexes and that the administration of donor-recipient immune complexes extracted from alloantisera can prolong allograft survival.

In conclusion, the results reported here indicate that the specific immunosuppression induced by SHA is expressed at the level of T-cells, and that it is detectable during the induction phase, with the consequences observed during the effector phase of the immune response. Further, this immunosuppression is mediated by a serum-blocking factor, which, contrary to the agent responsible for "peripheral" enhancement (4), probably is not an antibody but more probably is a soluble antigen or an antigen-antibody complex. Our data also suggest that the immunosuppressive properties of hemagglutinating-enhancing serum are, at least in part, due to a blocking factor other than free antibody. The fact that, on the one hand, the blocking activity of the SHA-raised serum depends upon the dose of SHA used for its production, and that, on the other hand, the inhibitory or stimulatory activity of the serum (raised by the same SHA dose) depends on its concentration, suggests that the cells interpret a signal as being either immunogenic or "tolerogenic" as a function of the level of the serum-blocking factor. It is likely that immunological unresponsiveness is the result of the interaction of several processes; and since the MLR dose-response curves for SHA-raised sera are similar to high- and low-zone tolerance, it appears as an attractive hypothesis that this latter might be governed largely through the dose-dependent regulatory effect of antigen-antibody complexes.

SUMMARY

Cellular and humoral immunity were studied in mice after treatment with allogeneic soluble histocompatibility antigen (SHA) known to decrease the host-versus-graft (HvG) reaction. Cellular immunity was found to decrease as measured by the mixed lymphocyte reaction (MLR) and graft-versus-host (GvH) tests, but more frequently when measured by GvH mortality than when measured by GvH splenomegaly. Serum from treated mice is antibody free; it enhances skin allograft survival if transferred to normal hosts, it diminishes the reactivity of normal cells when present during MLR or when incubated with the responding cells prior to MLR, and it delays GvH mortality when incubated with donor cells prior to grafting. The blocking activity of the serum is "dose" dependent: the *in vivo* blocking effect of the undiluted serum depends on the SHA dose used for its production, and the *in vitro* blocking effect of the serum (raised by the same SHA dose) depends on its concentration during MLR. Three different serum concentration-dependent effects were found in MLR experiments: inhibition with high and low serum concentrations and stimulation with intermediate serum concentrations. Similar results were observed with either antibody-free SHA-induced or hemagglutinating-enhancing sera. We concluded that the decreased cellular immunity of SHA-treated mice is related to a serum-blocking factor that interacts with T-cells, which is not an antibody but rather a soluble antigen or an antigen-antibody complex. Whether the T-cells will interpret the signal as being immunogenic or "tolerogenic" apparently depends on the level of the serum-blocking factor. The T-cell blocking activity of SHA-induced and classical enhancing sera seems to be due to a similar factor. Thus, the immunosuppressive activity of classical hemagglutinating-enhancing serum may not be entirely dependent on free antibodies.

ACKNOWLEDGMENT

The authors are grateful to Dr. L. L. Pritchard and Dr. N. Kiger for helpful advice and criticism during preparation of the manuscript; they thank Mrs. N. Richie and M. Gregoire for skillful technical assistance.

This work was supported by grants from the Délégation Générale à la Recherche Scientifique et Technique n° 75-7-737.

REFERENCES

1. Abuaf, N., Halle-Pannenko, O., and Mathe, G. Influence of genetic background in donor-recipient pairs incompatible for major histocompatibility complex (MHC). *Transplantation, 24*:74, 1977.
2. Bansal, S.C., Hellström, K.E., Hellström, I., and Sjögren, H.O. Cell-mediated immunity and blocking serum activity to tolerated allografts in rats. *J. Exp. Med., 137*:590, 1973.
3. Bonavida, B. Studies on the induction and expression of T cell-mediated immunity. II. Antiserum blocking of cell-mediated cytolysis. *J. Immunol., 112*:1308, 1974.
4. Davies, D.A.L., and Staines, N.A. A cardinal role for I-region antigens (Ia) in Immunological enhancement, and the clinical implications. *Transplant. Rev., 30*:18, 1976.
5. Debray-Sachs, M., Liegeois, A., and Hamburger, J. Presence of antigen-antibody complexes in antiallogeneic and antixenogeneic sera. *Transplantation, 22*:323, 1976.
6. Gorczynski, R.M., Kilburn, D.G., Knight, R.A., Norbury, C., Parker, D.C., and Smith, J.B. Nonspecific and specific immunosuppression in tumour-bearing mice by soluble immune complexes. *Nature (Lond.), 254*:141, 1975.
7. Graff, R.J., and Kandutsch, A.A. An attempt to produce low zone tolerance to tissue alloantigens. *Transplantation, 8*:162, 1969.
8. Graff, R.J., Mann, D.L., and Nathenson, S.G. Immunogenic properties of papain-solubilized H-2 alloantigens. *Transplantation, 10*:59, 1970.
9. Halle-Pannenko, O., and Martyre, M.C. Conditioning of allogeneic skin graft recipient with crude and water soluble H-2 antigens derived from BP-8 murine tumor cell-membrane. *Eur. J. Clin. Biol. Res., 15*:891, 1970.
10. Halle-Pannenko, O., Martyre, M.C., and Jolles, P. Conditioning of allogeneic mice with crude and purified H-2 extracts alone or combined with cyclophosphamide, for skin graft prolongation. *Transplant. Proc., 3*:257, 1971.
11. Halle-Pennenko, O., Martyre, M.C., and Mathe, G. Soluble H-2 antigens: Effect on graft-versus-host reaction and factors influencing its effect on host-versus-skin-graft reaction. *Transplant. Proc., 4*:517, 1972.
12. Halle-Pannenko, O., Abuaf, N., and Mathe, G. Use of soluble histocompatibility antigen(s) in the control of transplantation reactions: Possible dissociation of the roles of H-2 and non H-2 antigens in the prevention of mortality induced by graft-versus-host reaction. *Transplant. Proc., 8*:161, 1976.
13. Halle-Pannenko, O., Zalc-Gouget, C., Kuroiwa, A., Bourut, C., and Mathe, G. Prevention and treatment of secondary disease. II. Effects of four specific agents: Anti-recognition-structure serum, host and donor-directed sera and host soluble H-2 antigens. *Int. J. Rad. Oncol. Biol. Phys., 1*:927, 1976.
14. Hayami, M., Hellström, I., Hellström, K.E., and Lannin, D. Further studies on the ability of regressor sera to block cell-mediated destruction of Rous sarcomas. *Int. J. Cancer, 13*:43, 1974.
15. Hilgard, H.R. Dissociation of splenomegaly from graft-versus-host disease by host-x-irradiation. *Transplantation, 10*:396, 1970.
16. Lie, T.S., Ebata, H., Kim, W.I., and Grünn, U. Active enhancement of rat kidney allografts. *Transplantation, 21*:103, 1976.
17. Little, J., Myburgh, J.A., Austoker, J.L., and Smit, J.A. Detergent solubilization of baboon histocompatibility antigens and their use in prolonging liver allograft survival. *Transplantation, 19*:53, 1975.
18. Marquet, R.L., Heystek, G.A., Tank, B., and Van Es, A.A. Prolongation of heterotopic heart allograft survival in rats by use of antigen-antibody complexes. *Transplantation, 21*:454, 1976.
19. Medawar, P.B. The use of antigenic tissue extracts to weaken the immunological reaction against skin homografts in mice. *Transplantation, 1*:21, 1963.
20. Myburgh, J.A., and Smit, J.A. Prolongation of liver allograft survival by donor-specific soluble transplantation antigens and antigen-antibody complexes. *Transplantation, 19*:64, 1975.
21. Nathenson, S.G., and Davies, D.A.L. Solubilization and partial purification of mouse histocompatibility antigens from a membranous lipoprotein fraction. *Proc. Nat. Acad. Sc. (USA), 56*:476, 1966.

229

22. Owen, E.R. Prolonged survival in heterografted kidneys with transplantation antigen pretreatment. *Nature (Lond.), 219*:970, 1968.

23. Ranney, D.F., Gordon, R.O., Pincus, J.H., and Oppenheim, J.J. Biological effects of murine histocompatibility antigen solubilized with 3M potassium chloride. *Transplantation, 16*:558, 1973.

24. Simonsen, M., Engelbreth-Holm, J., Jensen, E., and Poulsen, H. A study of the graft-versus-host reaction in transplantation to embryos F₁ hybrids, and irradiated animals. *Ann. N.Y. Acad. Sci., 73*:834, 1958.

25. Skurzak, H.M., Klein, E., Yoshida, T.O., and Lamon, E.W. Synergistic or antagonistic effect of different antibody concentrations on *in vitro* lymphocyte cytotoxicity in the Moloney sarcoma virus system. *J. Exp. Med., 135*:997, 1972.

26. Stuart, F.P., Saitoh, T., and Fitch, F.W. Rejection of renal allografts: Specific immunologic suppression. *Science 160*:1463, 1968.

27. Voisin, G.A., Kinsky, R.G., and Maillard, J. Réactivité immunitaire et anticorps facilitants chez des animaux tolérants aux homogreffes. *Ann. Inst. Pasteur, 115*:855, 1968.

28. Voisin, J.E., Kinsky, R.G., and Voisin, G.A. Protection against secondary syndrome by facilitating immune sera. *Ann. Immunol. (Inst. Pasteur), 124C*:75, 1973.

29. Uphoff, D.E., and Draper, L.R. Functional capacity of bone marrow modified by contact with antigenic substances. *J. Nat. Cancer Inst., 47*:1233, 1971.

30. Wheeler, H.B., De Fronzo, A., and Corson, J.M. Prolonged skin allograft survival after treatment of the host with low doses of donor strain histocompatibility antigens. *Transplantation, 9*:78, 1970.

31. Wright, P.W., Hargreaves, R.E., Bansal, S.C., Bernstein, I.D., and Hellström, K.E. Allograft tolerance: Presumptive evidence that serum factors from tolerant animals that block lymphocyte-mediated immunity *in vitro* are soluble antigen-antibody complexes. *Proc. Nat. Acad. Sc. (USA), 70*:2539, 1973.

32. Zighelboim, J., Bonavida, B., Roa, S.V., and Fahey, J.L. Blocking activity induced by solubilized alloantigens. *J. Immunol., 112*:433, 1974.

28

Ultrastructural Analysis of Preserved Dog Granulocytes

J. E. French, M. P. Grissom,
J. F. Jemionek, T. J. Contreras,
and W. J. Flor

INTRODUCTION

Recent randomized, controlled clinical studies demonstrate that granulocyte transfusions in conjunction with antibiotics are efficacious against gram-negative bacterial infection in neutropenic patients without sufficient myeloid reserve (1,17). This probably will initiate requests by clinicians to utilize this blood component in adjunct therapy with antibiotics in the treatment of gram-negative septicemia. However, without suitable preservation methodology, granulocytes cannot be used effectively in transfusion therapy. Liquid preservation for short-term holding before transfusion or after freeze preservation for long-term holding is necessary to increase the granulocyte's *in vitro* half-life so that the logistical problems of donor selection and cross-matching to potential recipients can be overcome.

We are investigating liquid- and freeze-preserved techniques in dog and human granulocytes by *in vitro* ultrastructural and functional analysis so that we can determine (a) differences between dog and human granulocyte structure and function, (b) structural changes related to *in vitro* functional decrements in both species after preservation, (c) how dog and human granulocytes differ in their susceptibility to preservation damage, and (d) how this information can be used to improve human granulocyte preservation. These studies are being carried out in the dog because it has been demonstrated to be a suitable animal model for clinical conditions arising from granulocytopenia (10–13). It was the objective of this study to determine morphological changes in the dog granulocyte after various preservation regimens when the cells were collected by the most innocuous methods available for future comparison with current, clinically acceptable cell-collection procedures.

MATERIALS AND METHODS

Experimental Animals

AKC-registrable, male beagles (12 to 15 kg) were bled weekly (50 ml) by jugular venipuncture for leukocyte isolation or preparation of autologous serum (storage was at −80°C). Animals received double food rations (Purina Dog Chow, Ralston-Purina, St. Louis, Mo.) during periods of chronic bleedings. Periodic veterinary health inspection including hematological examinations confirmed good health of dogs and normal leukocyte kinetics.

Leukocyte-isolation Procedure

Leukocyte isolation utilizing defibrination or acid-citrate-dextrose (ACD, NIH Formula A) for anticoagulation was carried out as previously described (15). In addition, granulocyte isolation was also accomplished by counterflow centrifugation elutriation (CCE) using the Beckman JE-6 rotor (Beckman Instrument Co., Palo Alto, Ca.) at 2,000 rpm and variable flow rate from 7 ml/min (start) to 13 ml/min (finish), according to the manufacturer's instructions (16) or by a modified technique (18). The elutriation medium consisted of phosphate-buffered saline (PBS) supplemented with 1 to 2 g percent bovine serum albumin and with or without 7.5 percent ACD at 330 to 350 mOsm at pH 7.2. Granulocyte separations by CCE were carried out on (a) cells from citrated or defibrinated whole blood (10 to 30 ml) (15,16), (b) leukocyte buffy coat concentrates (60 to 120 ml whole blood) (18), or (c) cells from dextran-isopaque 1 g sedimentation (10 to 30 ml whole blood) (15).

If required (e.g., platelet or anticoagulant removal), leukocytes were washed twice, as described previously (15). Leukocytes were suspended in autologous citrated plasma, or 4:4:2 medium (4 parts MEM, 4 parts HBSS minus Ca^{2+} and Mg^{2+}, and 2 parts autologous serum or citrated plasma) (15).

Preservation

Liquid and freeze preservation of granulocytes were carried out by methods described previously (14,15). These methods allowed the comparison of isolation and preservation systems that either eliminated or included anticoagulant exposure and/or centrifugal washing. In this manner, granulocytes isolated by the methods described could be stored in any other medium to determine the effects of additives or supplements on preservation independent of the initial isolation medium.

Electron Microscopy

Granulocytes were fixed in suspension by the dropwise addition of an equal volume of 4.5 percent distilled glutaraldehyde in 0.1 M cacodylate-buffered 7.5 percent sucrose over a 2- to 3-min period at room temperature with gentle agitation. This material was processed further for transmission electron microscopy, as previously described (15).

Myeloperoxidase ultracytochemistry in dog granulocytes was determined by the method of Bainton and Farquhar (2). Granulocytes were suspension-fixed in an equal volume of 3 percent glutaraldehyde with 0.05 cacodylate-buffered 2 percent sucrose at 4°C for 10 min before further processing, as described elsewhere (2). The mean number of azurophilic granules was determined by counting the number of peroxidase-positive granules in uniform, thin secretions (600 to 700 Å) of at least 50 cells from three different preparations.

RESULTS
Post-isolation Characteristics

Granulocytes isolated by either gravity sedimentation or elutriation are similar and normal in ultrastructural appearance prior to preservation (Figures 28-1, 28-2, and 28-3). These representative electron micrographs show the well-segmented nuclei enclosed by a flattened nuclear envelope, with peripherally marginated, electron-dense heterochromatin and the electron-lucent euchromatin located medially but extending to the nuclear pores. The cytoplasm contains mitochondria, Golgi apparatus, glycogen, microtubules, and the characteristic azurophilic (primary) and specific (secondary) granules. The structure and distribution of these granules are best illustrated after myeloperoxidase (MPO) staining (Figures 28-4, 28-5, and 28-6). Granulocytes from either the defibrination (Figure 28-4) or the ACD-anticoagulation procedure (Figure 28-5) exhibited similar granule distribution ($\bar{x}$ no. MPO + granules/thin section = 35) and degree of degranulation (loss of MPO reaction product) Table 28-1). These cells also exhibit the cytoplasmic mottling that occurs after these methods of isolation and preparation for ultracytochemistry. Control cell preparations (without H_2O_2 substrate or diaminobenzidene) also exhibited mottling. The granulocytes isolated by either of the elutriation procedures showed an even distribution of the electron-dense, myeloperoxidase-positive granules (means number of granules/thin section = 34), with no detectable degranulation and no mottling in the cytoplasm (Table 28-1 and Figure 28-6).

Liquid Preservation (4°C)

In order to visualize the ultrastructural changes during liquid preservation and study the storage medium requirements, granulocytes isolated by the described methods were stored in autologous serum, citrated autologous plasma, or 4:4:2 medium (with autologous serum or citrated plasma). Only with citrated autologous plasma or 4:4:2 medium (2 parts citrated autologous plasma) was granulocyte ultrastructure (relative to prestorage controls) maintained over the 7-day liquid storage period. Autologous serum, MEM alone (or with 20 percent autologous serum), or 4:4:2 medium (with 2 parts autologous serum) did not preserve the morphology as well. Granulocytes collected from whole blood after defibrination (no anticoagulation or centrifugal washing exposure) and then stored with at least 20 percent citrated autologous plasma (4:4:2 medium) showed the same degree of azurophilic (MPO+) degranulation (28 MPO+ granules/thin section) (Table 28-1 and Figure 28-7) as did the ACD anticoagulation method (with ACD and cen-

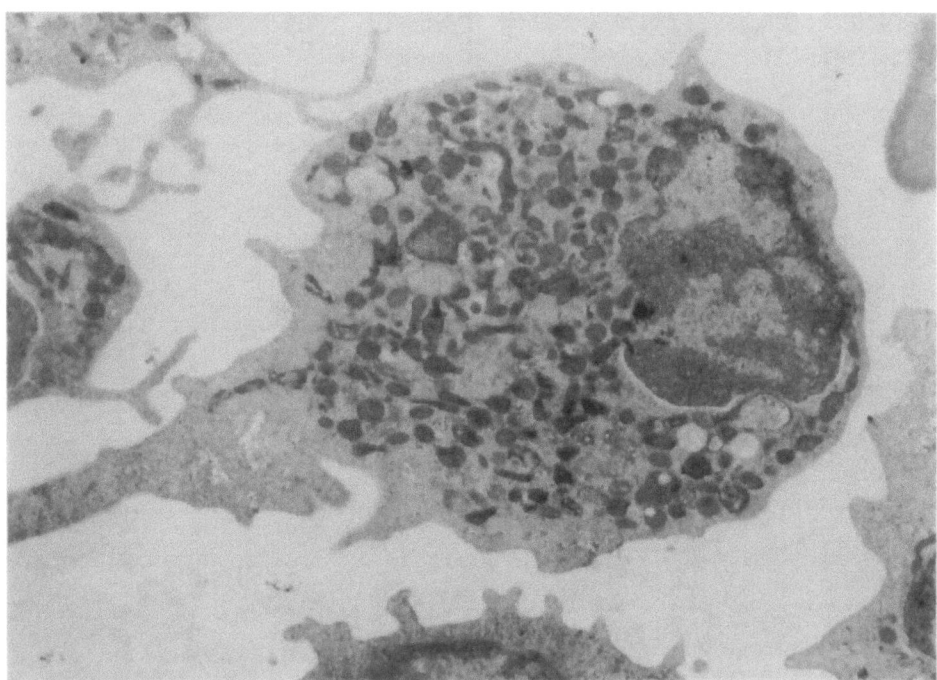

FIGURE 28-1. An electron micrograph of a dog granulocyte isolated form defibrinated whole blood by 1*g* sedimentation on a dextran-isopaque column to remove red blood cells. This cell is utilized as a typical control.

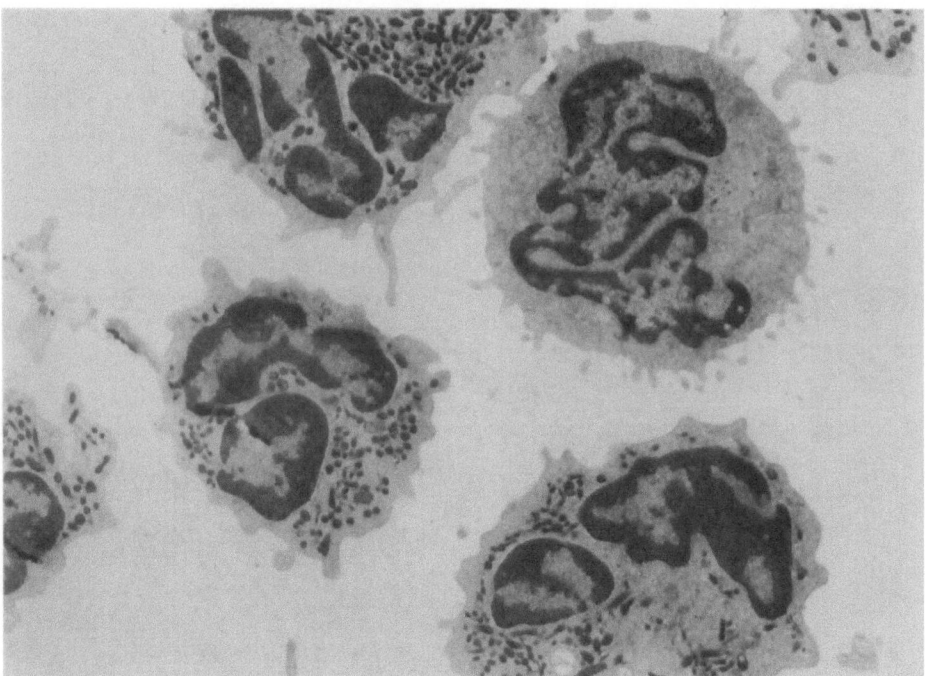

FIGURE 28-2. An electron micrograph of dog granulocytes and a monocyte after isolation from whole blood with 7.5 percent ACD and 1*g* sedimentation on a dextran-hypaque column to remove red blood cells. Control cells.

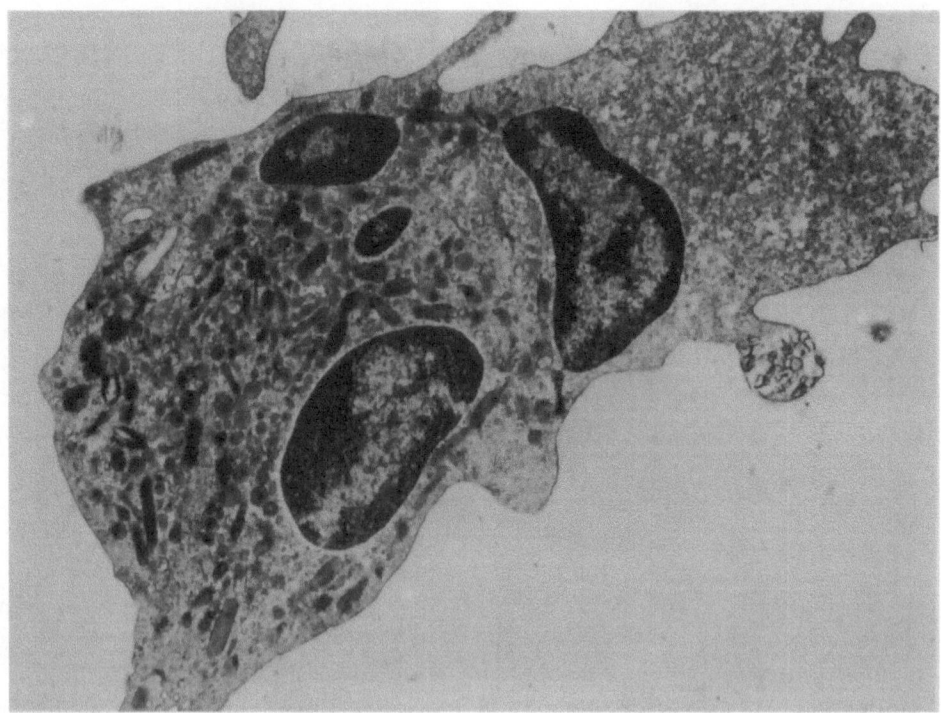

FIGURE 28-3. An electron micrograph of a motile dog granulocyte after isolation from whole blood with 7.5 percent ACD and counterflow elutriation. Some membranous material appears to be fixed in the process of being extruded. Control cell.

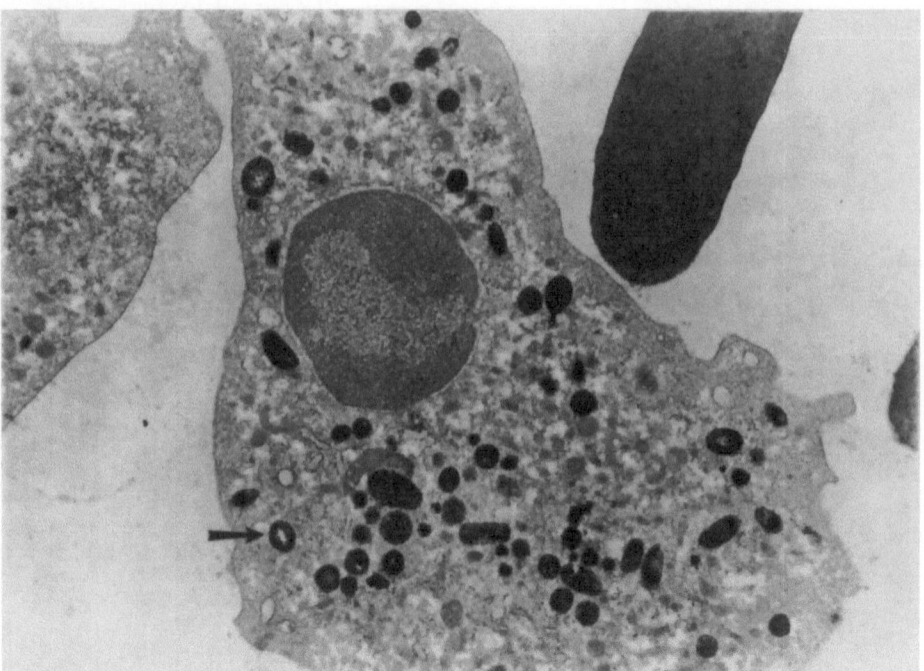

FIGURE 28-4. This dog granulocyte was cytochemically stained for myeloperoxidase (MPO, azurophil marker enzyme) after isolation by defibrination and 1g sedimentation of whole blood. The azurophils are the electron-dense granules (MPO+); specific granules are left unstained and electron-lucent. A loss of the reaction product can be seen in some granules (arrow).

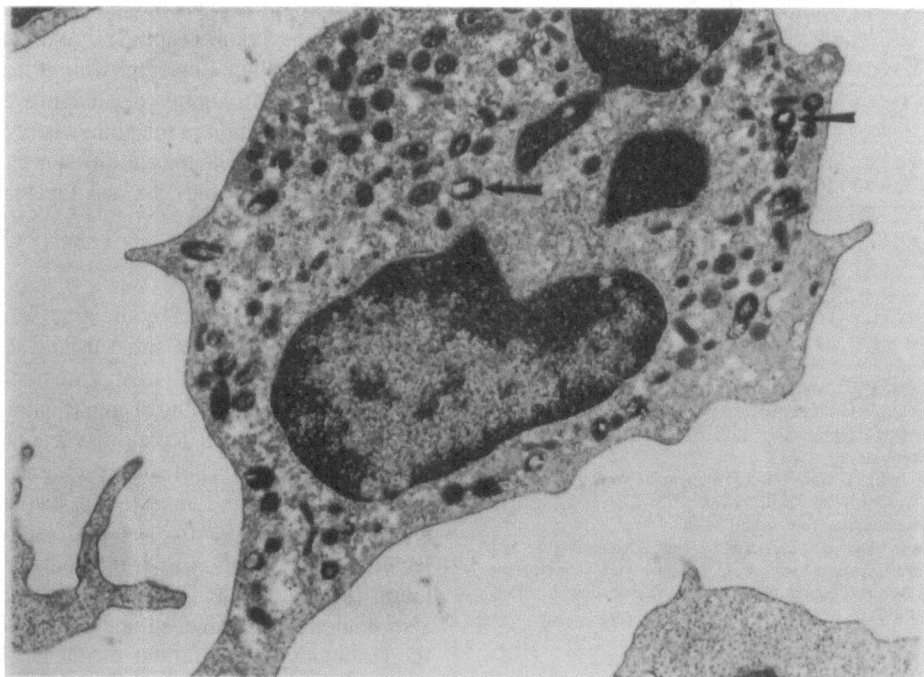

FIGURE 28-5. An electron micrograph of a dog granulocyte with azurophils cytochemically stained for myeloperoxidase after isolation from whole blood with 7.5 percent ACD and 1g sedimentation of red blood cells on dextran-isopaque. A partial loss of the reaction product from azurophilic granules can be seen (arrows).

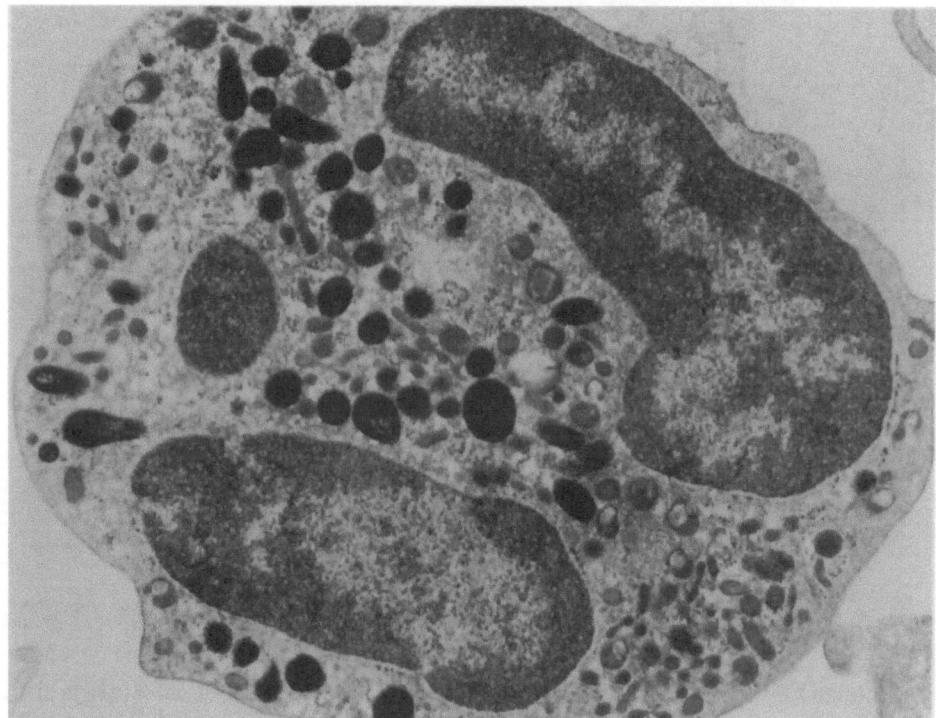

FIGURE 28-6. This electron micrograph shows a granulcyte isolated with ACD and counterflow centrifugal elutriation. Such granulocytes consistently possess large numbers of electron-dense MPO+ granules, with little loss in the reaction product.

TABLE 28-1 The Number of Myeloperoxidase-positive (MPO+) Granules in Uniform Thin Sections of Dog Granulocytes according to the Method of Isolation and Storage Medium

ISOLATION METHOD	STORAGE METHOD		
	Day 0 mean (range)	Day 7[a] mean (range)	Day 7[b] mean (range)
1[c]	35 (12–60)	28 (10–58)	6 (0–15)
2[d]	36 (8–55)	30 (10–59)	10 (2–18)
3[e]	34 (10–57)	34 (8–58)	17 (5–30)

[a]Storage media consisted of 4 parts MEM, 4 parts HBSS minus Ca^{2+} and Mg^{2+}, and 2 parts autologous citrated plasma.

[b]Storage medium consisted of 4 parts MEM, 4 parts HBSS minus Ca^{2+} and Mg^{2+}, and 2 parts autologous serum.

[c]Dog whole blood was collected by jugular venipuncture and anticoagulated by defibrination. After 1 g sedimentation, isolated leukocytes were stored at 4°C.

[d]Dog whole blood was collected by jugular venipuncture and coagulation was prevented with ACD. After 1 g sedimentation, the isolated leukocytes were stored at 4°C.

[e]Dog whole blood was collected by jugular venipuncture and coagulation was prevented with ACD. Granulcoytes were isolated by either of two procedures of elutriation described in Methods and stored at 4°C.

trifugal washing exposure, 30 MPO+ granules/thin section) (Table 28-1 and Figure 28-8). However, granulocytes collected from ACD-anticoagulated whole blood, elutriated, and stored in 4:4:2 (with 20 percent autologous serum) for up to 7 days showed marked MPO+ degranulation (17 MPO+ granules/thin section) (Table 28-1 and Figure 28-9). Storage in nutritional medium with at least 20 percent citrated autologous plasma prevented this marked degranulation to the point where there were no discernible morphological differences with prestorage controls (Table 28-1 and Figures 28-4, 28-5, and 28-6).

Freeze Preservation

Granulocytes isolated by the described methods survive the freeze-thaw and washing (DMSO removal) process and retain their ultrastructural characteristics to a remarkable degree (Figure 28-10 and 28-11). However, the degree of vacuolation is increased (Figure 28-11) in some preparations of elutriated granulocytes. Ingested particulate matter (cell debris) is frequently seen in granulocytes collected by defibrination as a method of anticoagulation (Figure 28-10). Unusual granule structure is also common after freeze-thaw and washing in all the granulocyte preparations examined (Figure 28-10). In those surviving granulocytes that show more extensive damage, both nuclear and cytoplasmic changes are observed (Figures 28-11 and 28-12). These granulocytes are swollen, exhibit chromatin-

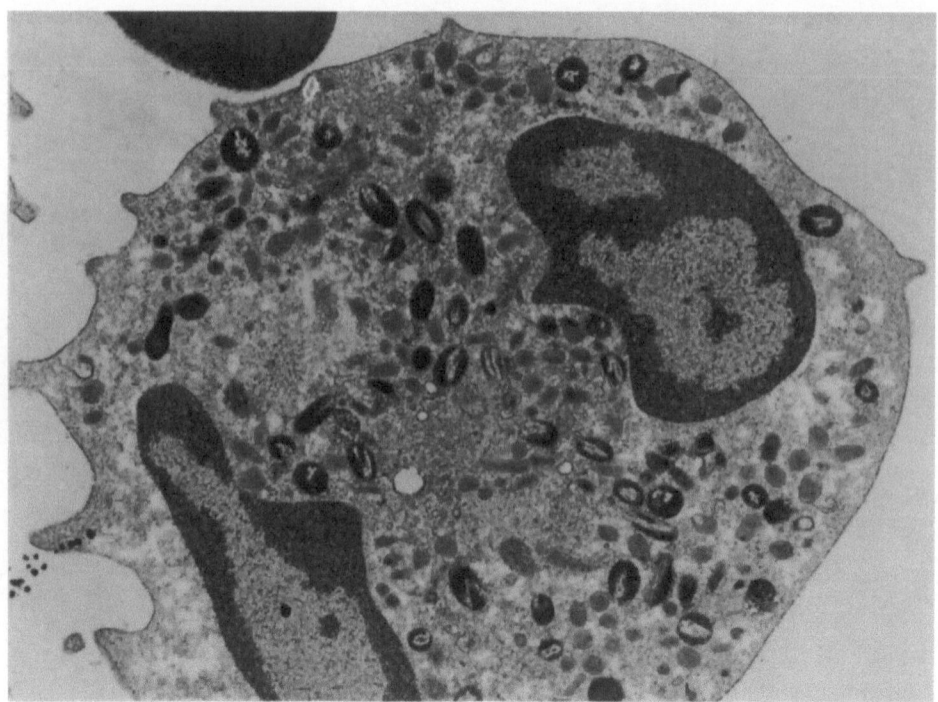

FIGURE 28-7. A granulocyte that still retains its normal morphology after 7 days' storage at 4°C in 4:4:2 medium (2 parts citrated autologous serum with 7.5 percent ACD added) is shown in this electron micrograph. This cell was isolated from defibrinated whole blood that was sedimented at 1g over a dextran-isopaque gradient. Substantial amounts of the azurophil granule (MPO+) content have been lost.

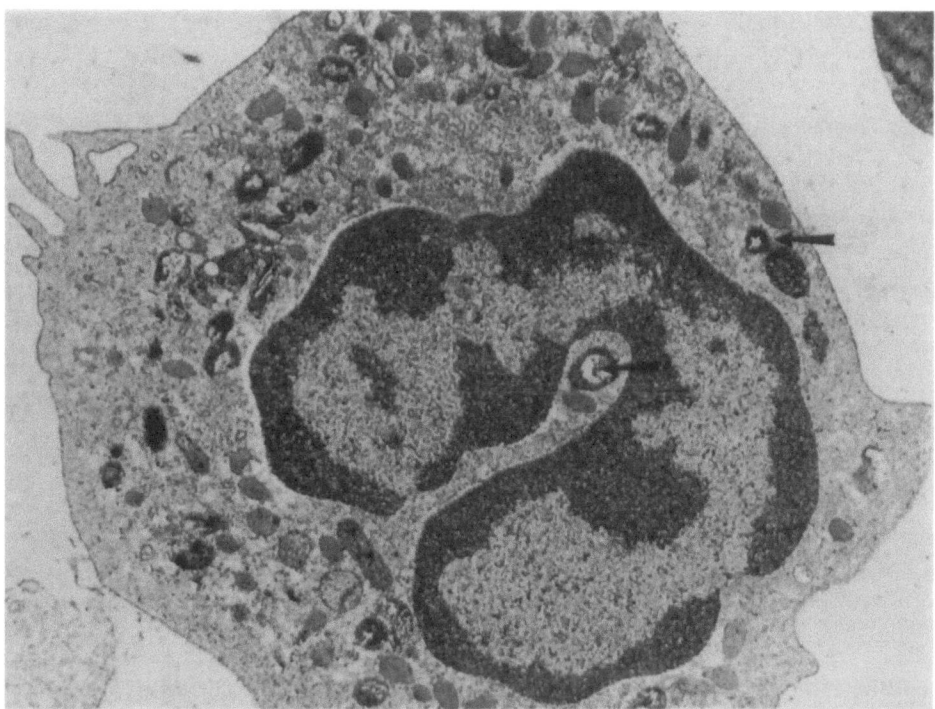

FIGURE 28-8. This electron micrograph shows a granulocyte that was isolated from whole blood with ACD, sedimented at 1g over a dextran-isopaque column to remove red cells, and stored 7 days at 4°C in 4:4:2 medium (2 parts citrated autologous plasma). The granulocyte exhibits normal ultrastructure except for a loss of the MPO+ reaction product from the azurophils and an unusual azurophil structure (arrows).

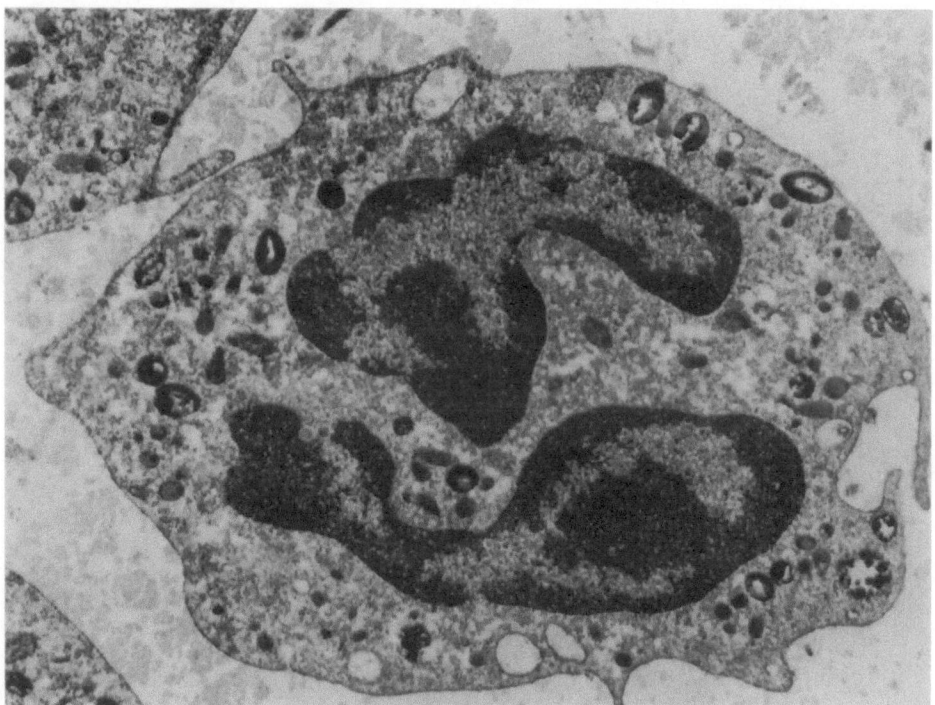

FIGURE 28-9. The granulocyte in this electron micrograph is from buffy coat or whole blood stored in 4:4:2 medium (2 parts autologous serum) after counterflow elutriation. It shows substantial azurophil degranulation (loss of MPO+ product) after being stored for 7 days at 4°C.

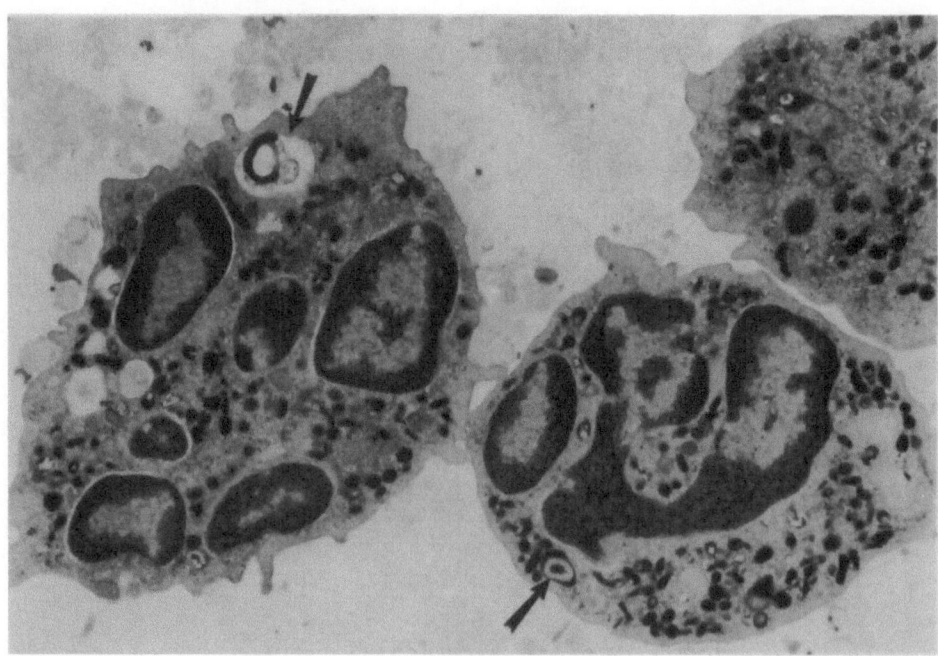

FIGURE 28-10. Dog granulocytes isolated from defibrinated whole blood by 1*g* sedimentation on dextran-isopaque column and then frozen, thawed, and washed. These cells exhibit ultrastructural characteristics in this electron micrograph similar to prefreezing controls. Ingested debris and unusual granule formations are frequently seen (arrows).

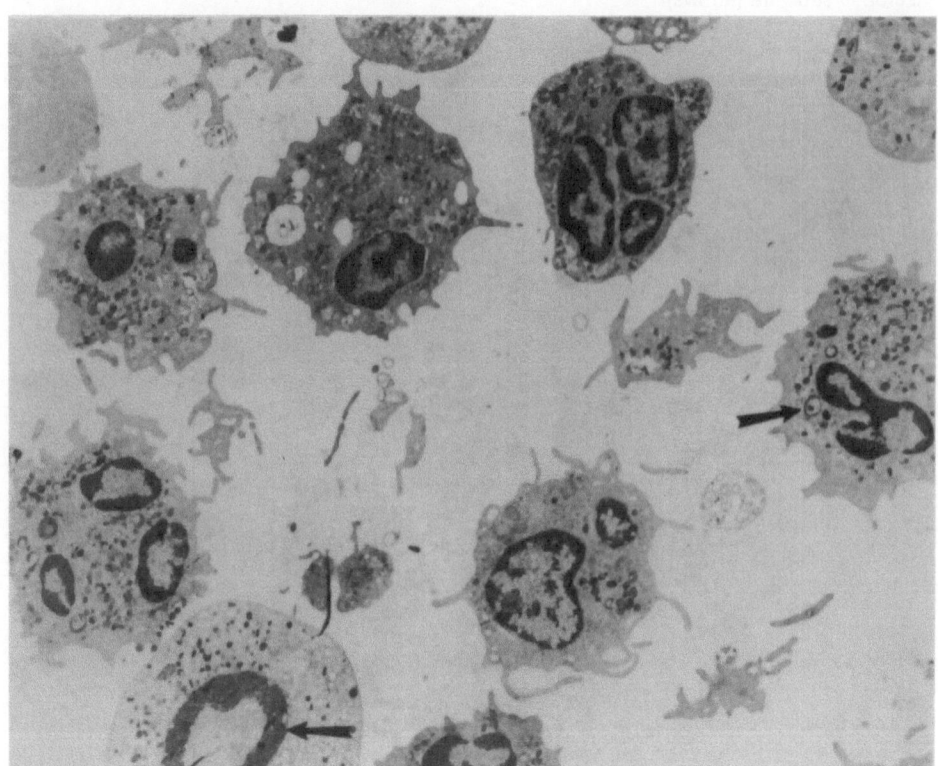

FIGURE 28-11. Elutriated granulocytes that are frozen, thawed, and washed, exhibit a morphology similar to base-line controls as demonstrated in this electron micrograph. Cytoplasmic vacuolization, abnormal granule structure, and chromatinorrhexis can be seen in some cells (arrows).

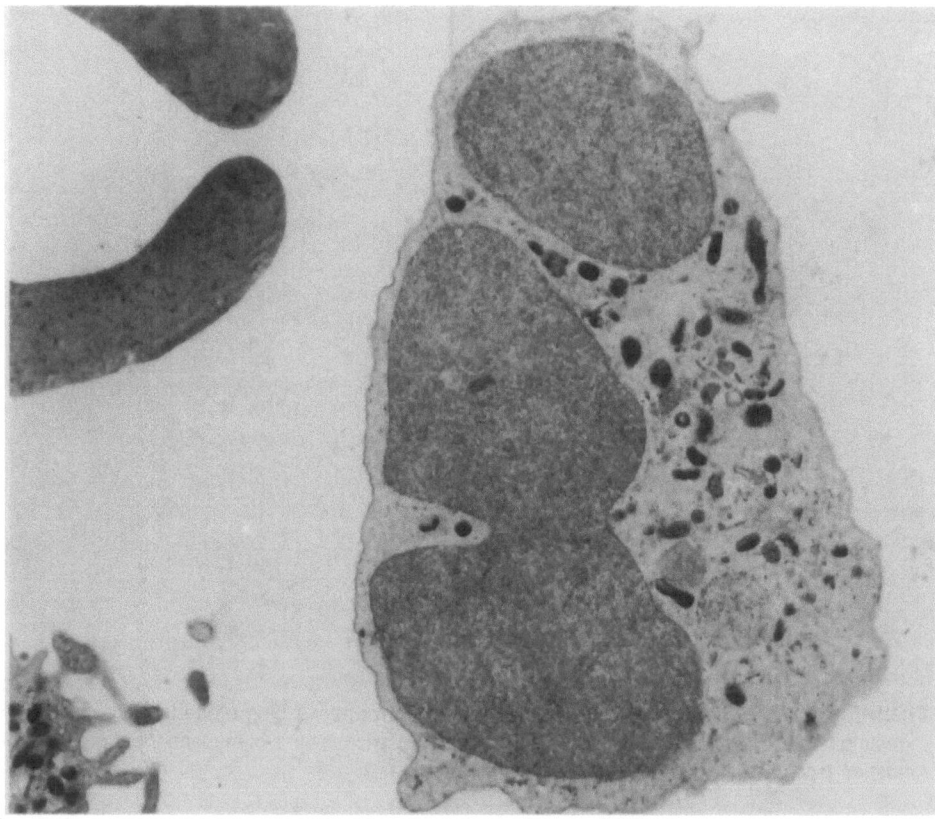

FIGURE 28-12. Freeze-thaw injury in 10 to 15 percent surviving dog granulocytes, as demonstrated in this electron micrograph, produces chromatinolysis and increased electron density in the granules, which masked the myeloperoxide reaction used in these experiments. The controls exhibited similar granule electron density.

orrhexis (Figure 28-11) progressing to chromatinolysis (Figure 28-12), electron-lucent cytoplasm, and increasingly electron-dense granules. This increase in the electron density of the granules made it difficult to determine MPO+ granules, since control preparation (without H_2O_2 substrate) was similar in appearance when prepared by the methods described above. The granulocytes were still phagocytic after freeze preservation, as illustrated by Figure 28-13.

DISCUSSION

Liquid preservation of human granulocytes was first demonstrated by Tullis (33), who provided critical information on temperature, divalent cations, and the need for multiple assays to determine granulocyte function. More recently, Skeel et al. (31) and Price and Dale (29) demonstrated that granulocytes stored at 4°C were more functional *in vitro* and *in vivo* than cells stored at 22 to 25°C in the media used in their respective studies. Crowley and

Valeri (7) found that liquid preservation (4°C) in ACD-plasma increased recovery and function of human granulocytes. McCullough and colleagues (24–26) have studied granulocyte storage in whole blood under blood bank conditions. They found that, for short periods, exposure to ACD and CPD anticoagulants did not greatly affect chemotactic, phagocytic, or bactericidal functions.

In this study we have utilized this information on storage methodology and have shown that the ultrastructural morphology of dog granulocytes after liquid and freeze preservation is also influenced by the method of isolation and storage medium. French et al. (14,15) previously demonstrated that cellular function and cell survival of dog granulocytes were also affected by the method of isolation and preservation as well as by the storage medium. During liquid storage (4°C), granulocyte loss was greater when dog cells were stored in autologous serum or in 4:4:2 medium (with 2 parts autologous serum) (up to 62 percent after 7 days) as compared to human cells similarly treated (7,29,31,33). Granulocyte loss in the leukocyte sus-

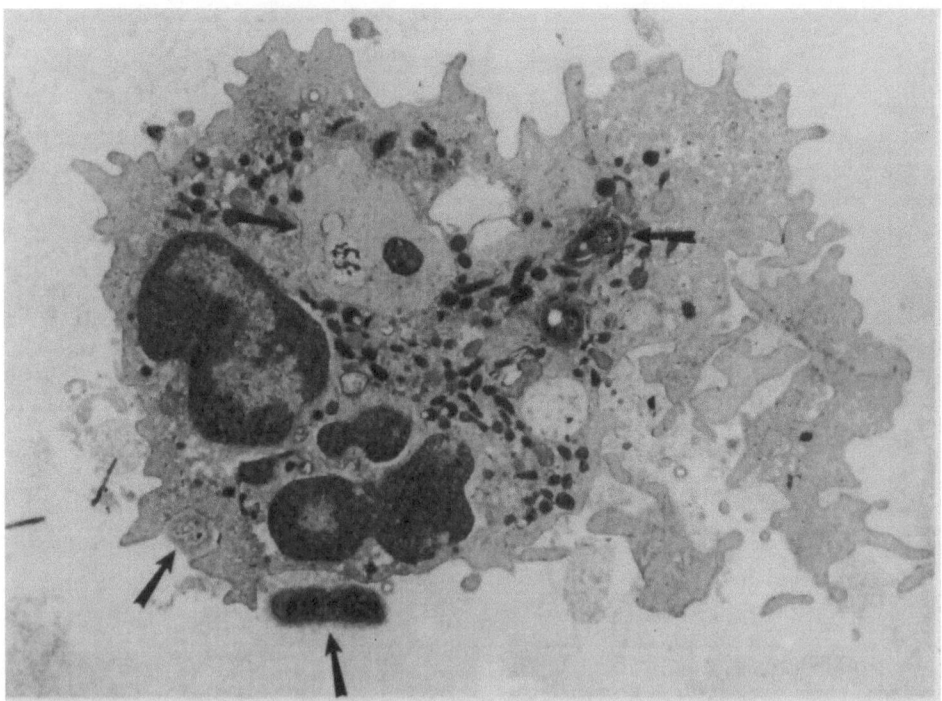

FIGURE 28-13. Granulocytes that survive the freeze-thaw and washing procedure are capable of ingesting bacteria, as seen in this electron micrograph. Various states of ingestion, phagolysosome formation (see arrows), and degranulation can be seen.

pension was never greater than 20 percent after 7 days' storage in autologous citrated plasma or 4:4:2 medium (with 2 parts autologous citrated plasma), regardless of the isolation method employed. Both the O_2 consumption associated with phagocytosis and the bactericidal capacity were also highest in granulocytes isolated and stored by this method (14).

It is important to note that in this investigation using liquid preservation, we attempted to compare isolation systems that either eliminated or included anticoagulant exposure and washing by centrifugation as sources of trauma (15,24,26,33). Liquid-preserved granulocytes without at least 20 percent citrated autologous plasma showed a shortened *in vitro* survival, more advanced pyknosis, perinuclear swelling, vacuolation, and degranulation as compared to cells stored in autologous citrated plasma or 4:4:2 medium (with 2 parts autologous citrated plasma) regardless of the method of isolation employed. The myeloperoxidase-H_2O_2 system is a potent microbicidal system (20), and loss of a significant number of these granules would seriously compromise granulocyte antimicrobial function.

The cryopreservation of granulocytes has been reported and discussed by several investigators (3–6,9,19,27), and their results have been mixed. A

variety of methods for freezing, storing (temperature and length of time), and assaying function were employed in these studies, making it difficult to compare their success. More recently, Crowley et al. (8) demonstrated reduction in motility and phagocytosis in frozen human granulocytes. Malinin (23) studied morphological injury and concluded that human granulocytes cannot withstand freezing in the presence of DMSO. However, Lionnetti et al. (21,22) demonstrated freeze preservation of human granulocytes with increased functional characteristics after thawing and DMSO removal.

Dog granulocytes isolated by the described methods and then preserved by freezing methods similar to those described in the literature cited above after 4- or 24-hr liquid preservation (4:4:2 medium, with 2 parts citrated autologous plasma) do not differ ultrastructurally from those frozen immediately after isolation (under 2 hr). Freeze-preserved granulocytes showed a similar magnitude of degranulation, abnormal granule structure, and cytoplasmic vacuolation. A few granulocytes per thin section also showed nuclear changes, such as chromatinorrhexis and chromatinolysis. These nuclear changes are visualized more frequently in freeze-preserved human cells (8,23). Freeze-preserved dog granulocytes exhibited phagocytic ability and overall excellent ultrastructural characteristics

(32,34). However, it was previously demonstrated that cell functions were affected differentially (15) according to the method of cell isolation.

The alteration of granule structure in granulocytes surviving freeze-thaw and washing may be due to intracellular ice formation and concomitant solute effects (30) on the azurophilic granules, which are lysosomal in nature (2,28).

In conclusion, dog granulocytes seem to be more resistant to liquid- and freeze-preservation methods than human granulocytes preserved under similar conditions. Understanding the differences between dog and human granulocytes may be important in developing successful human granulocyte preservation methods. It is also important to develop an animal model with maximum experimental control for *in vivo* determinations of the efficacy of preserved granulocytes against infection.

SUMMARY

Liquid- and freeze-preserved dog granulocytes were studied by ultrastructural analysis. Liquid preservation for short-term holding before transfusion or after freeze preservation for long-term holding are necessary to increase the *in vitro* half-life of granulocytes in order to store them for use in neutropenic patients. Granulocytes were isolated from whole blood by either defibrination or anticoagulation with ACD and sedimented or elutriated. They were resuspended in autologous serum, ACD-plasma, or 40 percent MEM, 40 percent HBSS, and 20 percent autologous serum or plasma (4:4:2 medium). These methods allowed the comparison of isolation systems, which either eliminated or included ACD exposure and washing by centrifugation. Granulocytes were liquid-preserved or freeze-preserved for 0 to 7 days after liquid holding for up to 24 hr, as described previously. They were suspension-fixed in cacodylate-buffered glutaraldehyde and then prepared for ultracytochemical determination of myeloperoxidase (MPO) and for transmission electron microscopy. Granulocytes isolated by sedimentation or elutriation are normal in ultrastructural appearance prior to preservation. Granulocytes isolated by all methods described showed few changes for up to 24 hr, and cells frozen after 4 to 24 hr liquid holding do not significantly differ from those frozen immediately after isolation. Cells isolated by all the methods tried and stored with or without autologous serum (at least 20 percent) showed greater cytoplasmic vacuolation, advanced pyknosis, and chromatin changes and degranulation of MPO+ granules than did cells stored with autologous citrated plasma or serum (at least 20 percent). Freeze-preserved cells

showed a magnitude of degranulation and granule changes and vacuolation similar to the liquid-held granulocytes. In conclusion, dog granulocytes differ in morphology according to the method of isolation and storage medium, but unlike human cells, they survive to a remarkable degree. An understanding of the differences between human and dog granulocytes should provide valuable insight into developing successful preservation methods for human cells.

ACKNOWLEDGMENT

The authors wish to acknowledge the technical expertise of Joe L. Parker in the preparation of the cells for electron microscopy.

REFERENCES

1. Alavi, J.B., Root, R.K., Djerassi, I., Evans, A.E., Gluckman, S.J., MacGregor, R.R., Guerry, D., Schreiber, A.D., Shaw, J.M., Koch, P., and Cooper, R.A. A randomized clinical trial of granulocyte transfusions for infection in acute leukemia. *New Eng. J. Med., 296*:706, 1977.
2. Bainton, D.F., and Farquhar, M.G. Differences in enzyme content of azurophil and specific granules of polymorphonuclear leukocytes. II. Cytochemistry and electron microscopy of bone marrow cells. *J. Cell Biol., 39*:299, 1968.
3. Bannatyne, R.M., and Umamaheswaran, B. Bactericidal function of cyropreserved neutrophils. *Cryobiology, 10*:338, 1973.
4. Bouroncle, B.A., Aschenbrand, J.F., and Todd, R.F. Comparative study of the effectiveness of dimethyl sulfoxide and polyvinylpyrrolidone in the preservation of human leukemic blood cells at −80°C. *Cryobiology, 6*:409, 1970.
5. Cavins, J.A., Djerassi, I., Roy, A.J., and Klein, E. Preservation of viable human granulocytes at low temperatures in dimethyl sulfoxide. *Cryobiology, 2*:129, 1965.
6. Cavins, J.A., Djerassi, I., Aghai, E., and Roy, A.J. Current methods for the cyropreservation of human leukocytes (granulocytes). *Cryobiology, 5*:60, 1968.
7. Crowley, J.P., and Valeri, C.R. Recovery and function of granulocytes stored in plasma at 4°C for one week. *Transfusion, 14*:574, 1974.
8. Crowley, J.P., Rene, A., and Valeri, C.R. The recovery, structure and function of human blood leukocytes after freeze-preservation. *Cryobiology, 11*:395, 1974.
9. Davies, J.D. The preservation of white cells. *Cryobiology, 5*:70, 1968.
10. Dale, D.C., Reynolds, H.Y., Pennington, J.E. et al. Granulocyte transfusion therapy of experimental *Pseudomonas* pneumonia. *J. Clin. Invest., 54*:664, 1974.

11. Debelak, K.M., Epstein, R.B., and Andersen, B.R. Granulocyte transfusions in leukopenic dogs: *In vivo* and *in vitro* function of granulocytes obtained by continuous-flow filtration leukaphoresis. *Blood, 43*:757, 1974.

12. Epstein, R.B., Clift, R.A., and Thomas, E.D. The effect of leukocyte transfusions on experimental bacteremia in the dog. *Blood, 34*:782, 1969.

13. Epstein, R.B., Waxman, F.J., Bennett, B.T., and Andersen, B.R. *Pseudomonas* septicemia in neutropenic dogs. I. Treatment with granulocyte transfusions. *Transfusion, 14*:51, 1974.

14. French, J.E., Flor, W.J., and Grissom, M.P. Liquid preservation of dog granulocytes at 4°C in various storage media. *Exp. Hematol., 4*(Suppl.):129, 1976.

15. French, J.E., Flor, W.J., Grissom, M.P., Parker, J.L., Sajko, G., and Ewald, W.G. Recovery, structure, and function of dog granulocytes after freeze-preservation with dimethyl sulfoxide. *Cryobiology, 14*:1, 1977.

16. Grissom, M.P., French, J.E., and Ewald, W.G. Use of centrifugal elutriation in blood cell separations. *Exp. Hematol., 4*(Suppl.):11, 1976.

17. Herzig, R.H., Herzig, G.P., Graw, R.G., Jr., Bull, M.I., and Ray, K.R. Successful granulocyte transfusion therapy for gram-negative septicemia: A prospectively randomized controlled study. *New Eng. J. Med., 296*:701, 1977.

18. Jemionek, J.F., Contreras, T.J., French, J.E., and Hartwig, V. Improved technique for increased granulocyte recovery from canine whole blood samples by counterflow centrifugation-elutriation. I. *In vitro* analysis. *Exp. Hematol.* (in press).

19. Kessel, D., Blakely, S., Cavins, J.A. et al. Stability of human leukocyte processes to freeze-thawing. *Cryobiology, 4*:209, 1968.

20. Klebanoff, S.J. Antimicrobial mechanisms in neutrophilic polymorphonuclear leukocytes. *Sem. Hematol., 12*:117, 1975.

21. Lionetti, F.J., Hunt, S.M., Gore, J.M., and Curby, W.A. Cryopreservation of human granulocytes. *Cryobiology, 12*:181, 1975.

22. Lionetti, F.J., Hunt, S.M., Lin, P.S., Kurtz, S.R., and Valeri, C.R. Preservation of human granulocytes. II. Characteristics of granulocytes obtained by counterflow centrifugation. *Transfusion, 17*:465, 1977.

23. Malinin, T.I. Injury of human polymorphonuclear granulocytes frozen in the presence of cyroprotective agents. *Cryobiology, 9*:123, 1972.

24. McCullough, J., Yunis, E.J., Benson, S.J., and Quie, P.G. Effect of blood-bank storage on leukocyte function. *Lancet, ii*:1333, 1969.

25. McCullough, J., Carter, S.J., and Quie, P.G. Effects of anticoagulants and storage on granulocyte function in bank blood. *Blood, 43*:207, 1974.

26. McCullough, J., Weiblen, B.J., and Quie, P.G. Chemotactic activity of human granulocytes preserved in various anticoagulants. *J. Lab. Clin. Med., 84*:902, 1974.

27. Pegg, P.J. The preservation of leucocytes for cytogenetic and cytochemical studies. *Brit. J. Hematol., 11*:586, 1965.

28. Persidsky, M.D., and Ellett, M.H. Lysosomes and cell cryoinjury. *Cryobiology, 8*:345, 1971.

29. Price, T.H., and Dale, D.C. Neutrophil preservation: the effect of short term storage on *in vivo* kinetics. *J. Clin. Invest., 59*:475, 1977.

30. Rapatz, G., and Luyet, B. Electron microscope study of slowly frozen suspensions of human leucocytes. *Biodynamica, 11*:69, 1971.

31. Skeel, R.T., Yankee, R.A., Spivak, W.A., Novikovs, L., and Henderson, E.S. Leukocyte preservation. I. Phagocytic stimulation of the hexose monophosphate shunt as a measure of cell viability. *J. Lab. Clin. Med., 73*:327, 1969.

32. Spicer, S.S., and Hardin, J.H. Ultrastructure, cytochemistry, and function of neutrophil leukocyte granules. *Lab. Invest., 20*:488, 1969.

33. Tullis, J.L. Preservation of leukocytes. *Blood, 8*:563, 1953.

34. Zucker-Franklin, D. Electron microscopic studies of human granulocytes: Structural variations related to function. *Sem. Hematol., 5*:109, 1968.

29

Physical Separation and Kinetics of Colony-forming Cells in Diffusion Chambers *in Vivo* (CFU-d) and Colony-forming Cells in Agar *in Vitro* (CFU-c)

Niels Jacobsen,
Hal E. Broxmeyer, and
M. A. S. Moore

INTRODUCTION

Normal human bone marrow contains cells capable of forming hemopoietic colonies in fibrin clot diffusion chambers implanted into the peritoneum (i.p.) of sublethally irradiated mice (2,5,9,10). Such colony-forming units in diffusion chambers (CFU-d) give rise to neutrophilic, eosinophilic, megakaryocytic, and fibroblast-like colonies, which are usually scored after 14 days of culture (9). The formation of neutrophilic colonies is stimulated when the host mice are irradiated (450 to 600 R) prior to chamber implantation, and the chambers are reimplanted into other irradiated mice after 7 days of i.p. culture (9). This enhancement is due to an effect on precursor cells, which initiate DNA synthesis in cultures in irradiated mice but do not synthesize DNA in non-irradiated mice; they are apparently influenced by diffusible, species nonspecific stimulatory or inhibitory host factors (6).

Quantitatively, neutrophilic colonies predominate, and after 14 days of the i.p. culture, the majority of the cells are at the myelocyte stage of maturation. Very rarely, mixed neutrophilic-eosinophilic or granulocytic-megakaryocytic colonies are found. Usually such colonies are approximately 5 percent of the total and may result from several CFU-d, each committed to a particular pathway of differentiation. Erythropoietic colonies do not occur under the experimental conditions employed (9).

This chapter describes the relationship between CFU-d and CFU-c, which form granulocyte-macrophage colonies in agar culture *in vitro* in response to colony-stimulating activity (CSA) derived from peripheral blood leukocyte feeder layers (12). The CFU-d and CFU-c can be separated by velocity sedimentation, primarily according to size (11). Evidence will be presented that suggests that the CFU-d may be identical to a population of cells that give rise to CFU-c in suspension culture (4,13).

MATERIALS AND METHODS
Cell Separation
Normal human bone marrow cells were obtained form healthy volunteers by aspiration from the posterior superior iliac spine. Buffy coat cells were separated by a density "cut" procedure (1) in phosphate-buffered bovine serum albumin solution into high and low density cells. The low density fraction (under 1.070 g/cm³), which contained all the colony-forming units (CFU), was separated by velocity sedimentation as described by Miller and Phillips (11). Cells were loaded onto a gradient of 0.4 to 2 percent bovine serum albumin in phosphate-buffered saline and allowed to sediment at 1 g and 4°C

for 4 to 5 hrs. Thirty-milliliter fractions of cells with different sedimentation velocities were collected and assayed for precursor cells. Unseparated low density cells were assayed in parallel with the separated cell fractions. The recovery of CFU after velocity separation was on an average 70 percent, equal to the recovery of total nucleated cells.

Assays for Precursor Cells

Figure 29-1 illustrates the assays employed to measure precursor cell populations. The medium used for the present series of experiments was modified McCoy's 5A medium (12) supplemented with 10 to 20 percent fetal calf serum.

CFU-c Cultures were performed in 35-mm Petri dishes using the technique described by Robinson and Pike (12). Cells were suspended in 0.3 percent agar medium, and 1-ml aliquots were plated onto feeder layers containing normal blood leukocytes (10^6 cells in 1 ml of 0.5 percent agar medium) as an exogenous source of CSA. Cultures were incubated at 37°C in a humidified atmosphere of 7.5 percent CO_2 in air and scored for clusters (3 to 50 cells) and colonies (over 50 cells) after 7 and 14 days. An average of four plates were scored per point.

CFU-d Cells were suspended in medium containing 0.5 percent fibrinogen and 150 μl aliquots were inoculated into diffusion chambers, made from Millipore GS filters, with a pore size of 0.22 μm. Thrombin was added to induce fibrin clot formation within the chambers, which were sealed and implanted i.p. into 600 R ^{137}Cs-irradiated CD₁ female Swiss mice. Two chambers were implanted into each mouse. After 7 days, the chambers were removed, their exteriors were cleaned, and they were reimplanted into other irradiated mice. Cultures were harvested after a total of 14 days, using a previously described modified cytocentrifuge technique (5), stained with May-Grünwald Giemsa stain, and scored for neutrophilic and eosinophilic colonies (over 30 cells) and megakaryocytic aggregates (at least 2 cells). An average of 12 chambers were scored per point (5,7,9,10).

Cells that give rise to CFU-c in suspension culture Cells were suspended in medium supplemented with 15 percent conditioned medium obtained from cultures of low density adherent blood leukocytes (less than 1.070 g/cm³, 3 to 5 × 10⁵ cells/ml) after 3 days of incubation at 37°C *in vitro*. The cell suspensions were cultured in Falcon plastic tubes at 37°C in a humidified 7.5 percent CO_2

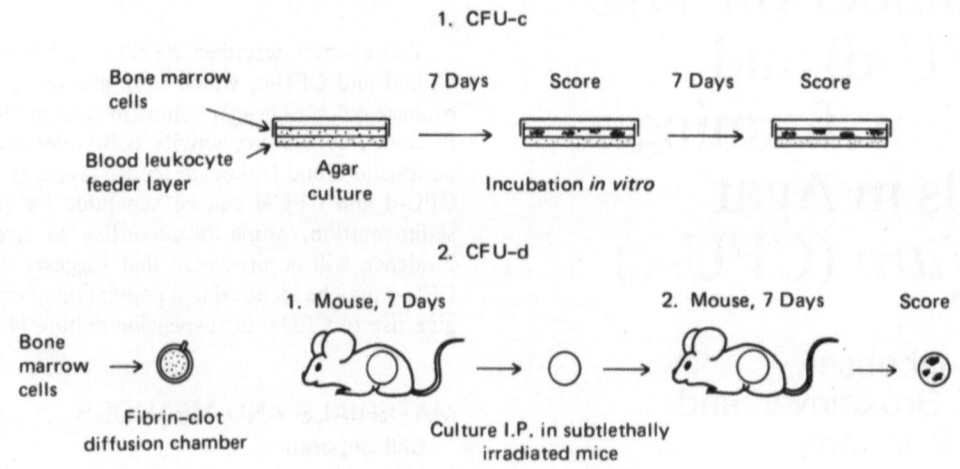

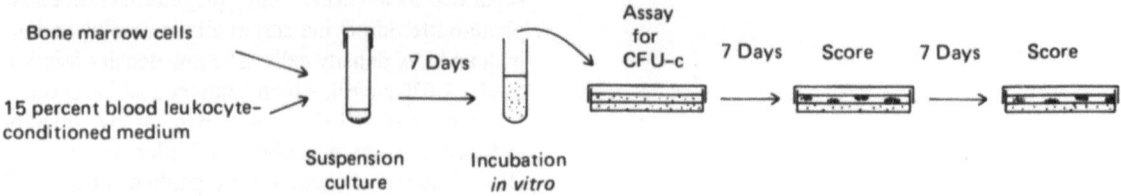

FIGURE 29-1. Diagrammatic presentation of assays for granulopoietic precursor cells. For details, see the Material and Method section.

atmosphere for 7 days. They were subsequently washed in fresh medium and plated in agar cultures on feeder layers as described above.

CFU kinetics in diffusion chamber pre-cultures Cells were inoculated into fibrin clot diffusion chambers or into initially fibrin-free chambers and cultured in the peritoneal cavity of 600 R irradiated mice or in non-irradiated mice. At predetermined intervals, the chambers were removed, the cells were harvested under sterile conditions, as previously described (6), and they were washed three times in fresh medium. The number of CFU-c and CFU-d recovered from these pre-cultures was assessed by plating aliquots in agar cultures or inoculating cells into fibrin clot diffusion chambers as described above.

RESULTS

Normal low density bone marrow cells were separated by velocity sedimentation, and the number of precursor cells in each fraction was assessed by standard tests using culture in diffusion chambers, in agar, and in suspension. The velocity sedimentation profiles obtained are shown in Figure 29-2. Neutrophilic CFU-d sedimented as a homogeneous population of cells, with a peak sedimentation rate of 5.2 to 5.4 mm/hr. A very similar profile was obtained with eosinophilic colonies. Agar colonies were derived from cells that sedimented more rapidly than the CFU-d. By counting *in vitro* cultures on day 7 and on day 14, two different CFU-c populations could be identified. Cells that gave rise to colonies on day 14 sedimented with a peak at approximately 6.2 mm/hr, whereas the colonies scored on day 7 of incubation were derived from very rapidly sedimenting cells, which formed a profile with peak numbers at 7.2 mm/hr.

Clusters scored on day 7 in agar cultures could be partially segregated from the colonies (Figure 29-2). Compared with the colonies, the clusters were more heterogeneously distributed over the fractions, suggesting that more than one cell type could give rise to clusters on day 7. Analysis of the profiles have suggested that clusters formed by rapidly sedimenting cell fractions were derived from the same cell compartment that also gave rise to colonies formed by day 7. Clusters formed by more slowly sedimenting cells, however, could be formed by day 14 CFU-c, and at least some of these clusters continued to proliferate in culture and were counted as colonies on day 14 of incubation (7).

The majority of day-7 agar colonies contained neutrophilic granulocytes. Day-14 colonies contained neutrophils and/or macrophages. More

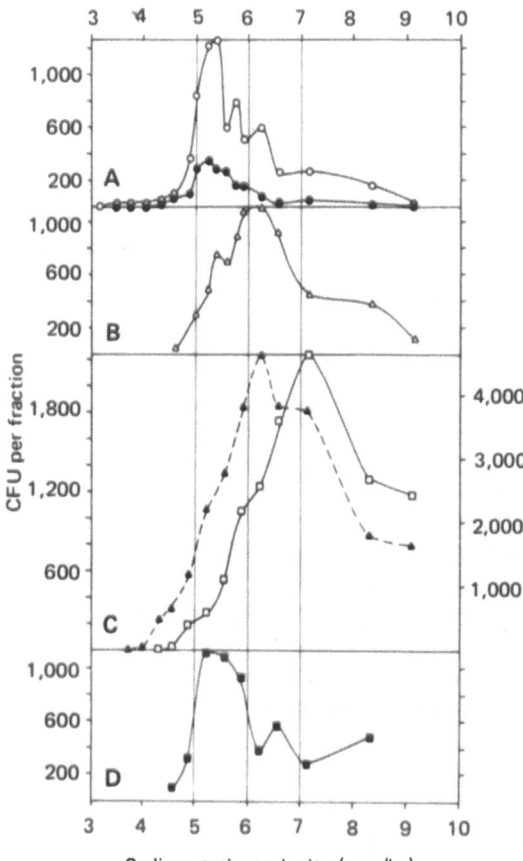

FIGURE 29-2. Velocity sedimentation profiles of granulopoietic precursor cell. Low density cells (less than 1.070 g/cm³) of normal human marrow were separated by velocity sedimentation, and the cell fractions were cultured to assess granulpoietic precursor cells. (A) Cells (CFU-d) were cultured in fibrin clot diffusion chambers implanted in 600 R-irradiated mice for a total of 14 days, and the cultures were then scored for neutrophilic (o—o) and eosinophilic (●—●) colonies. (B) and (C) Other cells (CFU-c) were plated in agar cultures stimulated by leukocyte feeder layers. The profiles show colonies scored after day 14 (△—△) and clusters (▲—▲) and colonies (□—□) counted on day 7 respectively. Finally, cell fractions were placed in suspension cultures containing light density, adherent blood cell-conditioned medium. (D) After 7 days, the suspension culture cells (CFU-c) were washed and plated in agar; these cultures (■—■) were scored for colonies after a further 7 days of culture. The figure shows the mean number of precursor cells per fraction obtained in three independent experiments.

rarely, eosinophilic colonies (10%) were found on day 14.

The last profile shown in Figure 29-2 depicts a population of cells capable of generating CFU-c in suspension culture. Velocity sedimentation fractions were incubated in suspension culture *in vitro* for 7 days in the presence of adherent blood leukocyte-conditioned medium. The cells were subsequently washed and plated in agar to determine the number of CFU-c in each culture. There was a marked decrease in fractions originally containing peak numbers of CFU-c. Instead, an absolute increase in cultures of slowly sedimenting cell fractions (5.2 mm/hr) was observed, suggesting the existence of a population of cells that could be distinguished from CFU-c by their sedimenting velocity and that gave rise to CFU-c in suspension culture. The profile seen in Figure 29-2 was obtained by counting colonies 7 days after plating the cell suspensions in agar cultures. When the agar cultures were counted 14 days after plating, essentially the same profile resulted, indicating that day-7 and day-14 CFU-c in suspension culture were derived from cells with an identical sedimentation velocity. As seen in Figure 29-2, these precursors of CFU-c had the same sedimentation velocity as granulocytic CFU-d.

In order to elucidate further the relationship between CFU-d and CFU-c, velocity sedimentation fractions were placed in fibrin clot diffusion chambers instead of suspension cultures and cultured in irradiated mice for 7 days. The cells were then harvested and plated in agar to measure the number of CFU-c produced by each fraction in the chambers. As in suspension culture, the maximum increase of day-7 and day-14 CFU-c was observed in the fractions that initially contained peak numbers of CFU-d (8). Similar results were obtained with fibrin-free diffusion chamber CFU-c cultured in irradiated or non-irradiated mice.

Figure 29-3 shows the results of a more detailed study of the kinetics of colony-forming cells in diffusion chambers. Normal low density bone marrow cells were separated by velocity sedimentation and the cells that sedimented between 4.7 and 5.6 mm/hr were pooled. In this way, a cell sample enriched with CFU-d but with low numbers of CFU-c was obtained. The cells were inoculated into diffusion chambers, which were implanted into non-irradiated mice. At predetermined time intervals, the cells were harvested from the culture chambers and plated in agar to assess the number of CFU-c or in fibrin clot diffusion chambers for further culture in irradiated mice to assess the CFU-d. Cells capable of cluster or colony formation in agar disappeared from the diffusion chambers during the first day of culture. This decline was followed by reappearance of the cells in the cultures. Cells that formed clusters by day 7 *in vitro* were detected on day 2 and increased further to a maximum on day 4. At the same time, an increase of day-14 CFU-c was observed, followed approximately 3 days later by the appearance of cells that formed colonies on day 7

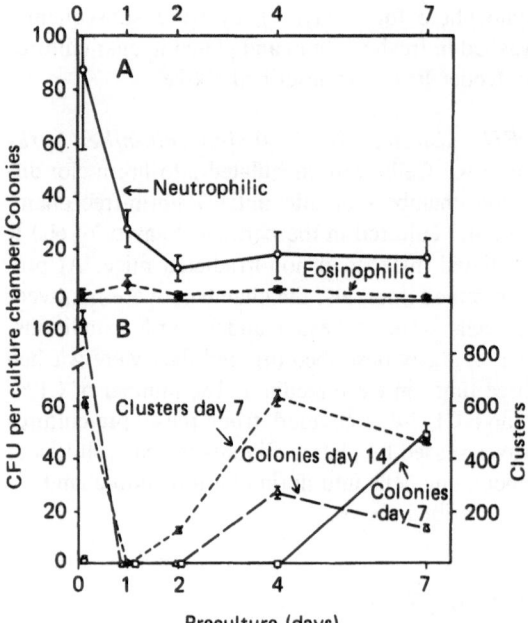

FIGURE 29-3. Kinetics of granulopoietic precursor cells in diffusion chambers. Low density normal marrow cells were separated by velocity sedimentation (4.7 to 5.6 mm/hr), and the cells sedimenting between 4.7 and 5.6 mm/hr were pooled, and 31.5×10^4 cells were inoculated into fibrin-free diffusion chamber cultures, which were implanted into non-irradiated mice. At predetermined time intervals, the cultures were harvested. The numbers of CFU-d (A) and CFU-c (B), measured by transferring the cells to fibrin clot diffusion chambers, and to agar cultures, respectively. Key: (A) Neutrophilic cells (o—o); eosinophilic cells (•—•). (B) Clusters, day 7 (▲---▲); colonies, day 7 (□—□); colonies, day 14 (△---△).

in agar (day-7 CFU-c). The described sequence of reappearance of *in vitro* colony- and cluster-forming cells was highly reproducible, and remarkably similar when the cells were cultured in suspension *in vitro* or in diffusion chambers in irradiated rather than non-irradiated mice. The CFU-c recovered from culture chambers formed colonies of the same type as those formed by cells in freshly aspirated bone marrow.

The number of neutrophilic CFU-d that could be recovered from diffusion chambers decreased initially. After 2 days, no significant change in their number was observed. A similar decrease in CFU-d was observed when velocity fractions with peak numbers of CFU-c (rapidly sedimenting cells) were cultured in diffusion chambers.

When the cells were cultured in chambers in non-irradiated mice, a difference in the kinetics of eosinophilic and neutrophilic CFU-d was apparent. In contrast to neutrophilic CFU-d, the former cell type increased slightly in number during the initial culture period.

To elucidate the relationship between granulocytic CFU-d and cells that formed megakaryocytic aggregates (at least 2 cells) in diffusion chambers, the velocity sedimentation profiles of the cells were compared (Figure 29-4). Megakaryocytic aggregates were displaced toward more slowly sedimenting cells compared with granulocytic CFU-d. However, there was evidence that the formation of megakaryocytic aggregates was not only dependent of the number of megakaryocytic precursors, but was also influenced by other cells, which were heterogeneously distributed in the fractions. This was demonstrated by mixing velocity fractions from various parts of the profile and by the co-culture of the cell mixtures in fibrin clot diffusion chambers. The capital letters in Figure 29-4 show the fractions combined in such co-cultures. The number of megakaryocytic aggregates observed in mixed diffusion chamber cultures was compared with the expected number, which was calculated by adding the number of aggregates formed in cultures of each of the two fractions separately (Table 29-1). There was a significant enhancement of megakaryocytic aggregate formation when fraction A was mixed with fraction B or C. This suggests that there were stimulatory cells present in either fraction A or in fractions B and C and that the profile of megakaryocytic aggregates seen in Figure 29-4 did not represent the true distribution of megakaryocytic CFU-d in the fractions.

DISCUSSION

The results presented in Figure 29-2 demonstrated that CFU-d could be separated from the vast majority of CFU-c and that two different CFU-c populations were distinguishable by means of velocity sedimentation. Experiments reported elsewhere (7), in which various fractions were mixed and co-cultured in diffusion chambers and in agar, have excluded the possibility that the observed profiles of CFU-c and granulocytic CFU-d resulted from interactions with other cells that stimulated or in-

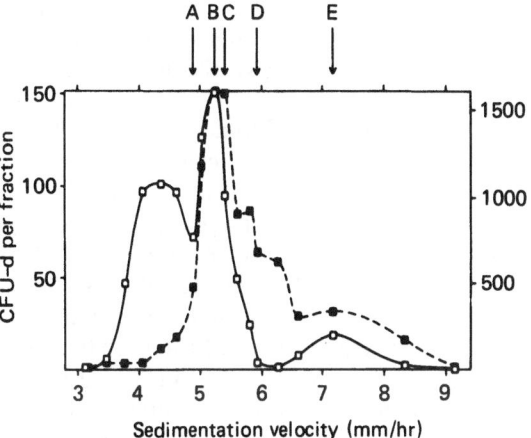

FIGURE 29-4. Profiles of megakaryocytic (□—□) and granulocytic (■—■) CFU-d. Low density cells separated by velocity sedimentation and cultured for 14 days in fibrin clot diffusion chambers in irradiated mice. (A)-(D) denote the fractions that were mixed and co-cultured in diffusion chambers to identify interacting cells that affected megakaryocytic aggregate formation in culture (compare Table 29-1).

hibited colony formation in some of the fractions. Rather, the profiles shown in Figure 29-2 reflected true physical differences between the colony-forming cells. Cell cycle studies, using tritiated thymidine ([³H]TdR) with a high specific activity as an agent to measure the number of colony-forming cells in DNA synthesis (S phase), have shown that the three profiles represent independent cell populations rather than different components of the mitotic cycle of the same cell type (7). Since sedimentation velocity depends primarily upon cell size (11), the data suggest that CFU-d, day-14 CFU-c, and day-7 CFU-c cells increase in size in the sequence mentioned.

Sutherland et al. (13) have provided evidence for the existence in mouse bone marrow of a population of cells capable of giving rise to CFU-c in suspension culture. This cell population, which was later demonstrated in human marrow (4), was shown to sediment more slowly than CFU-c by ve-

TABLE 29-1 Co-culture of Velocity Sedimentation Fractions in Diffusion Chambers

FRACTION COMBINATIONS[a]	MEGAKARYOCYTIC AGGREGATES[b]		SIGNIFICANCE
	Observed	Expected	
A + B	14.4 ± 2.0	5.9 ± 0.9	$p < 0.001$
A + C	20.0 ± 2.7	4.9 ± 0.8	$p < 0.001$
B + D	4.0 ± 1.0	3.5 ± 0.7	
C + D	1.7 ± 0.8	2.5 ± 0.6	
B + E	3.8 ± 1.0	3.4 ± 0.7	

[a]Letters refer to the fractions indicated in Figure 29-4.
[b]Number of aggregates observed in fraction mixtures compared with the number expected, as calculated by addition of the results obtained with each separate fraction (± 1 SEM).

locity sedimentation. As seen in Figure 29-2, peak numbers of CFU-d were found in the same fractions as the cells that gave rise to the maximum increase of CFU-c. It is therefore possible that granulocytic CFU-d are identical with the cells that form CFU-c in suspension culture. Further evidence that CFU-d were precursors of CFU-c was obtained by the culture of velocity sedimentation fractions of normal marrow in diffusion chambers and the assessment of the number of colony-forming cells produced in the cultures by each fraction (8). There was a decrease of CFU-d in all fractions, irrespective of the number of CFU-d and CFU-c initially present. After 7 days of diffusion chamber culture, however, CFU-c were found only in the cultures that had initially contained CFU-d, and the profile of CFU-c recovered after the culture of velocity fractions coincided with the original granulocytic CFU-d profile.

Additional information about the kinetics of colony-forming cells in diffusion chambers or in suspension culture has been obtained by assessment of the proportion of CFU sensitive to pulse incubation with high specific activity [³H]TdR, an agent specifically toxic to colony-forming cells in DNA synthesis. When cells were harvested after culture in diffusion chambers in irradiated mice or in suspension culture and exposed to [³H]TdR before they were plated in agar culture a 65 percent reduction of CFU-c was observed, indicating that a high proportion of the cells were in S phase. However, no effect of [³H]TdR was observed on CFU-c harvested from diffusion chambers in nonirradiated mice, suggesting that the cells were in the resting state under these culture conditions (8). Thus, the reappearance of CFU-c in diffusion chambers, illustrated in Figure 29-3, occurred without detectable cell proliferation, which strongly supports the suggestion that the cells were recruited from a more immature cell compartment (3).

A highly reproducible sequence of regeneration of day-14 and day-7 CFU-c was observed. Day-14 CFU-c reappeared early in the cultures, reaching maximum numbers on day 4. Approximately 3 days later, a parallel increase of day-7 CFU-c was observed. The same kinetic pattern of day-14 and day-7 CFU-c was found in diffusion chambers implanted into irradiated mice and in suspension cultures (8). Therefore, it is possible that there exists a parent-progeny relationship between day-14 and day-7 CFU-c.

As seen in Figure 29-3, the increase of day-7 cluster-forming cells occurred simultaneously with that of day-14 CFU-c, and the two curves were geometrically similar. This is consistent with the contention that some of the day-7 cluster-forming

cells were identical with day-14 CFU-c, which was also supported by the velocity separation results (Figure 29-2).

The decrease of neutrophilic CFU-d (Figure 29-3) was observed during the first 2 days of diffusion chamber culture. However, increasing numbers of in vitro CSA-responding cells (CFU-c) were not observed until days 1 to 2, and the maximum increase occurred between day 2 and day 4 during a time interval when CFU-d numbers reached plateau levels. This is not inconsistent with the hypothesis that CFU-d give rise to CFU-c, since there may be undetected cell populations that are intermediate in the CFU-d–day-14 CFU-c maturation sequence. In the mouse, the increase in CFU-c in suspension culture was maximal in velocity sedimentation fractions, with peak numbers of multipotent stem cells, identified by the spleen colony assay (13). The observation that CFU-d and cells that gave rise to CFU-c had similar velocity profiles may therefore suggest a close relationship between granulocytic CFU-d and the as yet unidentified multipotent stem cells in man.

SUMMARY

Normal human bone marrow contains cells that form hemopoietic colonies in fibrin clot diffusion chambers implanted into the peritoneum (i.p.) of sublethally irradiated mice. Such colony-forming units in diffusion chambers (CFU-d) can be separated by velocity sedimentation from CFU-c. The latter gives rise to granulocyte-macrophage colonies in agar cultures in vitro in response to the colony-stimulating activity (CSA) of the peripheral blood leukocytes. The CFU-D, which form granulocytic colonies, sediment more slowly than the CFU-c. Heterogeneity exists within the CFU-c compartment, the cells of which can be classified as rapidly sedimenting cells that form colonies on day 7 in agar culture (day-7 CFU-c) and cells that form colonies on day 14 and have intermediate sedimentation velocities (day-14 CFU-c).

Megakaryocytic aggregates in diffusion chambers are found in fractions of cells that sediment more slowly than granulocytic CFU-d. However, there is evidence that the velocity profile of megakaryocytic CFU-d is influenced by interacting cells that affect megakaryocyte aggregate formation.

The CFU-c harvested from in vitro suspension cultures of bone marrow fractions are derived from cells that have the same sedimentation velocity as neutrophilic and eosinophilic CFU-d. Cultures of slowly sedimenting cells in diffusion chambers in mice generate increasing numbers of day-14

CFU-c after 2 days and day-7 CFU-c after approximately 3 days. In contrast, neutrophilic CFU-d decrease during culture. These results support the hypothesis that granulocytic CFU-d gives rise to day-14 CFU-c and day-7 CFU-c in suspension culture *in vitro* and in diffusion chambers.

ACKNOWLEDGMENTS

This work was supported by NCI Grants CA-17353, CA-17085, and The Gar Reichman Foundation.

We would like to thank Mrs. Sabariah Schrader and Miss Valery Magolis for their excellent technical assistance.

REFERENCES

1. Broxmeyer, H.E., Baker, F.L., and Galbraith, P.R. *In vitro* regulation of granulopoiesis in human leukemia: Application of an assay for colony inhibiting cells. *Blood, 47*:389, 1976.

2. Elmgreen, J., Jacobsen, N., and Knudtzon, S. The effect of syngeneic peripheral blood cells on the formation of colonies by normal human bone marrow cells in diffusion chambers in mice. *Scand. J. Haematol., 17*:379, 1976.

3. Hoelzer, D., Harriss, E.B., Slade, M., and Kurrle, E. Growth of *in vitro* colony-forming cells from normal human peripheral blood leukocytes cultured in diffusion chambers. *J. Cell. Physiol., 89*:89, 1976.

4. Iscove, N.N., Messner, H., Till, J.E., and McCulloch, E.A. Human marrow cells forming colonies in culture: Analysis by velocity sedimentation and suspension culture. *Ser. Haematol., 2*:37, 1972.

5. Jacobsen, N. Chamber centrifugation: A harvesting technique for estimation of the growth of human haemopoietic cells in diffusion chambers. *Br. J. Haematol., 29*:171, 1975.

6. Jacobsen, N. Proliferation of diffusion chamber colony-forming units (CFUD) in cultures of normal human bone marrow in diffusion chambers in mice. *Blood, 49*:415, 1977.

7. Jacobsen, N., Broxmeyer, H.E., Grossbard, E., and Moore, M.A.S. Diversity of human granulopoietic precursor cells: Separation of cells that form colonies in diffusion chambers (CFU-d) from populations of colony-forming cells *in vitro* (CFU-c) by velocity sedimentation. Submitted for publication

8. Jacobsen, N., Broxmeyer, H.E., Grossbard, E., and Moore, M.A.S. In manuscript

9. Jacobsen, N., and Fauerholdt, L. Human granulocytopoietic colonies in diffusion chambers in mice: Growth of colonies and the effect of host irradiation. *Scand. J. Haematol., 16*:101, 1976.

10. Jacobsen, N., and Fauerholdt, L. Quantitative aspects of an *in vivo* diffusion chamber assay for normal human hemopoietic colony-forming units. *Exp. Haematol., 5*:171, 1977.

11. Miller, R.G., and Phillips, R.A. Separation of cells by velocity sedimentation. *J. Cell. Physiol., 73*:191, 1969.

12. Robinson, W.A., and Pike, B.L. Colony growth of human bone marrow cells *in vitro*. In F. Stohlman, ed., *Hemopoietic Cellular Proliferation*. New York and London: Grune and Stratton, 1970, p. 249.

13. Sutherland, D.J.A., Till, J.E., and McCulloch, E.A. Short-term cultures of mouse marrow cells separated by velocity sedimentation. *Cell Tissue Kinet, 4*:479, 1971.

Index